MW01254061

THE PSYCHOLOGY OF JUSTICE AND LEGITIMACY:

The Ontario Symposium
Volume 11

ONTARIO SYMPOSIUM ON PERSONALITY AND SOCIAL PSYCHOLOGY

E. T. Higgins, C. P. Herman, M. P. Zanna, Eds.
Social Cognition: The Ontario Symposium, Volume 1

M. P. Zanna, E. T. Higgins, C. P. Herman, Eds.
Consistency in Social Behavior: The Ontario Symposium, Volume 2

C. P. Herman, M. P. Zanna, E. T. Higgins, Eds.
Physical Appearance, Stigma, and Social Behavior: The Ontario Symposium, Volume 3

J. M. Olson, C. P. Herman, M. P. Zanna, Eds.
Relative Deprivation and Social Comparison: The Ontario Symposium, Volume 4

M. P. Zanna, J. M. Olson, C. P. Herman, Eds.
Social Influence: The Ontario Symposium, Volume 5

J. M. Olson, M. P. Zanna, Eds.
Self-Inference Processes: The Ontario Symposium, Volume 6

M. P. Zanna, J. M. Olson, Eds.
The Psychology of Prejudice: The Ontario Symposium, Volume 7

C. Seligman, J. M. Olson, M. P. Zanna, Eds.
The Psychology of Values: The Ontario Symposium, Volume 8

S. J. Spencer, S. Fein, M. P. Zanna, J. M. Olson, Eds.
Motivated Social Perception: The Ontario Symposium, Volume 9

R. M. Sorrentino, D. Cohen, J. M. Olson, M. P. Zanna, Eds.
Culture and Social Behavior: The Ontario Symposium, Volume 10

D. R. Bobocel, A. C. Kay, M. P. Zanna, J. M. Olson, Eds.
The Psychology of Justice and Legitimacy: The Ontario Symposium, Volume 11

THE PSYCHOLOGY OF JUSTICE AND LEGITIMACY:

The Ontario Symposium Volume 11

edited by
D. Ramona Bobocel
Aaron C. Kay
Mark P. Zanna
James M. Olson

New York London

Psychology Press
Taylor & Francis Group
270 Madison Avenue
New York, NY 10016

Psychology Press
Taylor & Francis Group
27 Church Road
Hove, East Sussex BN3 2FA

Psychology Press is an imprint of Taylor & Francis Group, an Informa business

Printed in the United States of America on acid-free paper
10 9 8 7 6 5 4 3 2

International Standard Book Number: 978-1-84872-878-3 (Hardback)

Library of Congress Cataloging-in-Publication Data

The psychology of justice and legitimacy / editors: Ramona Bobocel, Aaron C. Kay.
p. cm. -- (Ontario Symposium on Personality and Social Psychology)
Papers presented at the Eleventh Ontario Symposium on Personality and Social Psychology, held at the University of Waterloo, Ontario, Canada, August 16 to 18, 2007.
Includes bibliographical references and index.
ISBN 978-1-84872-878-3 (hardcover : alk. paper)
1. Justice--Psychological aspects. 2. Self. 3. Justice. I. Bobocel, Ramona. II. Kay, Aaron C. III. Ontario Symposium on Personality and Social Psychology (11th : 2007 : University of Waterloo)

BF789.J8P75 2010
155.9'2--dc22 2009031898

Visit the Taylor & Francis Web site at
http://www.taylorandfrancis.com

and the Psychology Press Web site at
http://www.psypress.com

Contents

Preface

The Eleventh Ontario Symposium on Personality and Social Psychology was held at the University of Waterloo, Ontario, Canada, August 16 to 18, 2007. Thirteen internationally renowned scholars presented their recent theoretical and empirical work on the topic of the symposium, which was "The Psychology of Justice and Legitimacy." The presentations were intellectually stimulating, and the atmosphere was most convivial. In addition to the invited presenters, approximately 100 audience members (20 faculty and 80 graduate or postdoctoral students) were in attendance. We were extremely pleased that audience members included students and psychology faculty from Canada, the United States, Europe, and Japan. Since presenting their papers, the invited speakers have had the opportunity to incorporate feedback received during the conference, and subsequently from the editors. The current volume thus comprises the updated versions of the papers presented at the conference.

Over the past decade, social psychological interest in the topics of legitimacy and social justice has blossomed considerably, no doubt in response to the international turmoil, violence, and increasing ideological polarization that has occurred since the beginning of this decade. These events have led to a natural increase in interest in the psychological underpinnings of people's reactions to injustice and illegitimacy, including the behavioral and psychological consequences of the motivation to view individual outcomes and governmental systems as just and legitimate. This interest has resulted in an exponential increase in journal articles and conference presentations devoted to understanding the psychological bases of people's (a) concerns with justice and reactions to instances of injustice (labeled the psychological study of *social justice*) and (b) ideological justifications of illegitimacy and the varying conceptions of what constitutes a legitimate social system (labeled the psychological study of *legitimacy*). Although both of these hot topics in social psychology are clearly related at conceptual and theoretical levels, these two rich literatures are rarely integrated.

Working from the assumption that people prefer fair and just social outcomes, the empirical study of social justice has traditionally sought to address four primary theoretical questions: (a) What types of outcomes, distributions, or procedures are (and are not) considered fair? (b) How do people react to observed or experienced acts of unfairness, and why? (c) When and why do people justify or rationalize injustice? (d) What underlying processes motivate people's preference for fairness? Historically, these questions are addressed in reference to specific instances of injustice among individuals.

Researchers interested in processes of legitimacy direct their attention to similar questions. Whereas social justice researchers, however, have tended to focus on how people make sense of particular instances of injustice, legitimacy researchers have tended to focus primarily on people's reactions to unfair systems of intergroup relations, including unequal status and power hierarchies, imbalances in the distribution of societal resources, and the social systems governing these societal deficiencies. Despite this difference in focus on injustice at the system level versus instance level, research and theory on legitimacy have much in common with the social justice tradition. In particular, legitimacy too is based on the notion that people are motivated to prefer justice over injustice, and it seeks to uncover the mechanisms through which people confront injustice, including their tendency to rationalize injustice and their motivations for doing so.

Such theoretical and empirical overlap indicates that social justice and legitimacy research can inform each other on important conceptual and practical issues. Traditionally, developments in each domain have occurred in relative isolation of one another. The current volume of the Ontario Symposium brings together the work of leading researchers in fields of social justice and legitimacy to facilitate the cross-pollination and integration of these related fields. This volume contains 13 chapters that address both broad theoretical issues and cutting-edge empirical advances.

In the book, we begin with chapters from the social justice perspective and move to those adopting the legitimacy perspective. Consistent with our belief that these fields are conceptually and empirically interrelated, however, this dichotomy is often blurred. In Chapter 1, Linda Skitka, Nicholas Aramovich, Brad Lytle, and Edward Sargis integrate past theory and research on justice reasoning into a more general theoretical whole. Skitka and her coauthors review three major theoretical perspectives on justice reasoning and integrate these approaches into a contingency theory of justice, arguing that how people define fairness will depend on their current motivations, namely, whether they are interested in maximizing material gain, gauging their standing within valued social groups, or acting in accordance with core beliefs about moral right and wrong. Their approach is important because it highlights the idea that a person's definition of fairness, and fairness judgments, is flexible and will differ as a function of the person's current perceptual frame of reference.

Similar to the preceding argument that justice reasoning will vary as a function of the perceiver's frame of reference, in Chapter 2, Ramona Bobocel and Agnes Zdaniuk suggest that people will respond differently to those who treat them unfairly depending on the aspect of their self-identity that is most accessible in memory. Drawing on research demonstrating that the experience of injustice can threaten people's feelings of self-worth, the authors suggest that responses to injustice can be motivated by a desire to restore a positive self-view and, accordingly, that responses to injustice should vary systematically depending on the source of one's feelings of positive self-regard. The authors review their research that supports this reasoning, thus highlighting the role of the victim's self-identity, and of the motivation to maintain a positive self-view, in understanding reactions to unfair treatment.

Whereas the preceding chapter focuses on how people react to the threat to self-worth that may arise from the experience of injustice, the next two chapters—by Mitchell Callan and John Ellard and by Carolyn Hafer and Leanne Gosse—address the general question of how people's desire for justice influences their thoughts and actions when confronted with instances of injustice. In Chapter 3, Callan and Ellard review their research on strategies that people use in daily life to cope with the threat to their fundamental need to believe in a just world. For example, in one series of experiments, they demonstrate that people's desire for justice for *others* can influence their causal explanations for others' undeserved bad experiences and their memory of such experiences. In a second series of experiments, they demonstrate that people's concerns for *own* justice can increase their desire to gamble and their desire for consumer goods. The authors' research thus elucidates several novel ways in which the need to preserve one's fundamental belief in a just world can manifest itself in everyday experience.

Hafer and Gosse tackle a different question about just-world motivation in Chapter 4. They build on existing theory and research by focusing on the novel question of how to predict when, and for whom, individuals might prefer certain just-world preservation strategies to other strategies. Through their review, the authors elucidate a number of situational and individual difference predictors of preservation strategy. Given that previous just-world research has not explicitly examined the issue of predicting which strategies people might prefer to preserve their just-world beliefs in the face of threat, Hafer and Gosse's chapter provides an important road map for future theory and research.

In Chapter 5, Dale Miller, Daniel Effron, and Sonya Zak have us consider another factor that influences how people respond to injustice. The authors address why people often do not act on the moral outrage they may experience when confronted with injustice, arguing that people must also feel entitled to act on their outrage, a feeling they label *psychological standing*. They provide a convincing argument for the concept of psychological standing as an important construct that can serve as both a barrier and a contributor to social protest behavior. Miller and colleagues' analysis is important in furthering our understanding of the conditions under which people are more or less likely to move from the experience of moral outrage when confronted with injustice to engaging in behaviors to remedy the injustice.

Of course, for moral outrage to ensue, people first must acknowledge that an injustice has occurred. Whereas considerable research has investigated the conditions that make it more or less likely for people to recognize injustices happening to themselves, less attention has been given to understanding what dictates when people will perceive the treatment of others as just versus unjust. In Chapter 6, James Olson, Irene Cheung, Paul Conway, and Carolyn Hafer present a nuanced analysis of the conditions that lead one to be more or less concerned with injustice occurring elsewhere, that is, not to the self. In particular, Olson and colleagues challenge previous evidence documenting a "scope of justice" and instead argue that it is only under incredibly extreme circumstances that people will consider rules of fairness irrelevant or inapplicable to others; past evidence of exclusion from one's scope of justice, Olson, Cheung, Conway, and Hafer suggest, may be better interpreted as specific applications of justice rules, such as deservingness. The authors reflect on the ambiguities and confounds of past research in this domain and suggest new directions for those interested in this important topic.

The next two chapters integrate classic justice theorizing with recent developments in the legitimacy literature to explain why and how social inequalities are often maintained. In Chapter 7, Danielle Gaucher, Aaron Kay, and Kristin Laurin discuss how societal injustice is perpetuated by people's need to justify the status quo. The authors first review evidence for the existence of a system justification motive; they then review research—much of which comes from their own laboratory—on both the motivational antecedents and the consequences of the system justification motive. Their research suggests that although system justification has an adaptive function, it also can lead to several insidious social and interpersonal consequences. The authors' research thus explains how it is that people who strongly value fairness and justice can, paradoxically, contribute to the perpetuation of societal injustice.

In Chapter 8, John Jost, Ido Liviatan, Jojanneke van der Toorn, Alison Ledgerwood, Anesu Mandisodza, and Brian Nosek offer evidence in support of the claim made by system justification theory that tendencies to legitimize the status quo—that is, tendencies to view social systems as more fair, legitimate, and desirable than they may be in reality—are motivational and not due to purely cognitive factors or social pressure. After first reviewing reasons for why people would hold such a motivation, including both existential and epistemic needs, the authors discuss six lines of evidence to support this motivational account. Understanding the motivational properties of system justification, the authors also suggest, are crucial to identifying when people will and will not seek social change.

The next two chapters deal with the intersection of morality, justice, and legitimacy. Kees van den Bos adopts a self-regulatory perspective to explain human behavior in contexts of morality and justice in Chapter 9. Borrowing from biological models on human homeostasis, he suggests that events that disrupt the balanced state of the individual, such as uncertainty or mortality salience, will elicit behaviors aimed at restoring this balance. This process, Van den Bos argues, can explain many of the reactions that follow from exposure to injustice or other world-view-threatening information. In addition, whereas Van den Bos proposes that it is generally under conditions of homeostasis that people will behave most rationally, calmly, and fairly, this balanced state can also hold negative repercussions, such as the disinclination to intervene in moral dilemmas.

In Chapter 10, John Darley and Dena Gromet consider the motivations that underlie ordinary people's decisions to punish an offender who has violated the law by intentionally inflicting harm on another person. This topic is important, as the authors argue, because the legitimacy of a society's legal codes depends on the extent to which the legal codes agree with citizens' moral intuitions about punishment. Darley and Gromet review research revealing that people's decisions are not carefully reasoned but instead driven by automatic intuitions of just deserts or, in other words, by a retributive justice impulse. They also review their recent studies demonstrating that people are able to consider justice more broadly than punishing the victim, for example, by showing concern for restorative justice, but only when people have the ability and motivation to consider such additional responses.

Tom Tyler offers us a broad analysis of the relation between perceptions of legitimacy, values, and rule following in Chapter 11. According to Tyler, the traditional modes of encouraging individuals to follow the rules and laws of society are failing. Tyler suggests that rather than implementing sanctions and punishments for those people who do not comply, implementing rules and regulations that individuals see as legitimate and congruent with their personal values will most effectively regulate behavior, insofar as they will tap into people's intrinsic motivations to abide by the rules of their group, organization, or society. Drawing on classic Lewinian theory and his and his collaborators' research, Tyler presents a compelling case for a value-based model of social regulation.

At the same time, it is also possible that the rules imposed by governmental institutions may themselves be illegitimate. In Chapter 12, Stephen Wright and Donald Taylor make such an argument in the context of educational policy in Arctic Canada. Wright and Taylor present a fascinating set of field evidence, collected in Arctic villages, to buttress their position that dominant-only education (in this case, educating Inuit children in English rather than their native tongue) is detrimental to the development and psychological well-being of students. Given these negative repercussions, and the lack of any measurable benefits, the authors suggest such policies are unfair, unjust, and unwarranted and should be considered a case of discrimination. The implications for wider issues of cultural preservation, and "linguicide," are also discussed.

What happens when a long-term social injustice finally ends? In the final chapter (Chapter 13), Karina Schumann and Michael Ross consider the question of how successors of historical injustice respond to the original injustice; in particular they focus on the political apology. Schumann and Ross point out that despite the popular perception that government apologies for historical injustices will promote reconciliation and forgiveness, social psychological theory suggests that government apologies may not necessarily have such effects. The authors identify the antecedents of political apologies and their nature, as well as conduct a content analysis of 14 actual government apologies. They also consider reactions to political apologies, from the perspective of both the previously victimized group as well as the nonvictimized majority.

We believe these chapters illustrate both the diversity and the richness of research in the fields of social justice and legitimacy. Our hope is that—in uniting these two domains—this volume will stimulate new directions in theory and research that seek to explain how and why people make sense of injustice at all levels of analysis. Ten previous Ontario Symposia on Personality and Social Psychology have been held. The series is designed to bring together scholars from Canada, the United States, and around the world who work in the same substantive area in an effort to integrate research findings and identify common concerns. Participation of Canadian and international faculty and graduate students has been gratifying. We hope that the symposia have contributed to, and will continue to stimulate, the growth of personality and social psychology in Ontario and Canada.

The first Ontario Symposium, held at the University of Western Ontario in August 1978, dealt with social cognition (see Higgins, Herman, & Zanna, 1981); the second, held at the University of Waterloo in October 1979, had the theme of variability and consistency in social behavior (see Zanna, Higgins, & Herman, 1982); the third, held at the University of Toronto in May 1981, addressed the social psychology of physical appearance (see Herman, Zanna, & Higgins, 1986); the fourth, held at the University of Western Ontario in October 1983, was concerned with relative deprivation and social comparison processes (see Olson, Herman, & Zanna, 1986); the fifth, held at the University of Waterloo in August 1984, dealt with social influence processes (see Zanna, Olson, & Herman, 1987); the sixth, held at the University of Western Ontario in June 1988, focused on self-inference processes (see Olson & Zanna, 1990); the seventh, held at the University of Waterloo in June 1991, examined the topic of prejudice (see Zanna & Olson, 1994); the eighth, held at the University of Western Ontario in August 1993, examined the psychology of values (see Seligman, Olson, & Zanna, 1996); the ninth, held at the University of Waterloo, May 2000, dealt with motivated social perception (see Spencer, Fein, Zanna, & Olson, 2003); and the tenth, held at the University of Western Ontario in June 2002, focused on culture and social behavior (see Sorrentino, Cohen, Olson, & Zanna, 2005).

As in previous years, the Social Sciences and Humanities Research Council of Canada, whose continuing support has been the backbone of the series, provided primary financial support for the Eleventh Ontario Symposium. We are also deeply indebted to the Department of Psychology and the Faculty of Arts at the University of Waterloo for their financial and administrative support. We also thank graduate students Jillian Banfield (social psychology) and Katrina Goreham (industrial-organization psychology) at the University of Waterloo for aiding us in conducting the conference. Finally, we thank Paul Dukes and the editorial team at Psychology Press for their support and editorial guidance.

D. Ramona Bobocel
Aaron C. Kay
Mark P. Zanna
James M. Olson

References

Herman, C. P., Zanna, M. P., & Higgins, E. T. (Eds.). (1986). *Physical appearance, stigma, and social behavior: The Ontario Symposium* (Vol. 3). Hillsdale, NJ: Lawrence Erlbaum.

Higgins, E. T., Herman, C. P., & Zanna, M. P. (Eds.). (1981). *Social cognition: The Ontario Symposium* (Vol. 1). Hillsdale, NJ: Lawrence Erlbaum.

Olson, J. M., Herman, C. P., & Zanna, M. P. (Eds.). (1986). *Relative deprivation and social comparison: The Ontario Symposium* (Vol. 4). Hillsdale, NJ: Lawrence Erlbaum.

Olson, J. M., & Zanna, M. P. (Eds.). (1990). *Self-inference processes: The Ontario Symposium* (Vol. 6). Hillsdale, NJ: Lawrence Erlbaum.

Seligman, C. V., Olson, J. M., & Zanna, M. P. (Eds.). (1996). *The psychology of values: The Ontario Symposium* (Vol. 8). Mahwah, NJ: Lawrence Erlbaum.

Sorrentino, R. M., Cohen, D., Olson, J. M., & Zanna, M. P. (Eds.). (2005). *Culture and social behavior: The Ontario Symposium* (Vol. 10). Mahwah, NJ: Lawrence Erlbaum.

Spencer, S. J., Fein, S., Zanna, M. P., & Olson, J. M. (Eds.). (2003). *Motivated social perception: The Ontario Symposium* (Vol. 9). Mahwah, NJ: Lawrence Erlbaum.

Zanna, M. P., Higgins, E. T., & Herman, C. P. (Eds.). (1982). *Consistency in social behavior: The Ontario Symposium* (Vol. 2). Hillsdale, NJ: Lawrence Erlbaum.

Zanna, M. P., & Olson, J. M. (Eds.). (1994). *The psychology of prejudice: The Ontario Symposium* (Vol. 7). Hillsdale, NJ: Lawrence Erlbaum.

Zanna, M. P., Olson, J. M., & Herman, C. P. (Eds.). (1987). *Social influence: The Ontario Symposium* (Vol. 5). Hillsdale, NJ: Lawrence Erlbaum.

CHAPTER 1

Knitting Together an Elephant

An Integrative Approach to Understanding the Psychology of Justice Reasoning

LINDA J. SKITKA, NICHOLAS P. ARAMOVICH, BRAD L. LYTLE, and EDWARD G. SARGIS

University of Illinois at Chicago

Abstract: Why do people care about justice? How do people reason about what is fair or unfair? To answer these questions, justice researchers have developed theories of justice reasoning based on their assumptions about people's needs, desires, and motivations. For example, theories of social exchange assume people are rationally self-interested and will evaluate fairness through the lens of maximizing rewards. Alternatively, theories of procedural fairness assume people fundamentally need to belong to groups and will focus on the fairness of procedures as an indication of their worth to the group. Moral theories of justice reasoning assume people have fundamental beliefs about right and wrong and that people evaluate fairness in accordance with these beliefs. This chapter reviews these three theoretical perspectives and integrates them into a contingency theory of justice. The contingency theory of justice posits that how people define fairness depends on the current perspective of the perceiver (material, social, or moral perspective). Specifically, the authors propose that the perspective and motivations of the perceiver impact the factors people use to decide whether something is fair or unfair. The contingency theory of justice can account for the complexity and flexibility of people's justice reasoning and how justice judgments vary both between and within persons over time. In addition, the theory suggests that an important area of future research inquiry is exploring how people cope with differences in their fairness judgments and how they resolve conflicts and arrive at a consensus that everyone can agree is fair.

Keywords: justice, fairness, morality, *moralis*, contingency theory

Theories that have attempted to explain why people care about fairness and the factors that people use to decide whether fairness has been achieved have a rich and vibrant history dating back to the earliest philosophers. Moreover, over time, modern social psychological inquiry into questions of how people decide whether something is fair or unfair, and the consequences of these judgments, has already cycled through a number of different major shifts in theoretical focus. One goal of the current chapter is to provide a brief historical overview of psychological justice theory and research focusing on three major metaphors that have guided various shifts in research focus and attention: (a) *homo economicus,* that is, a metaphor of human motivation focused on what people "get" out of social relationships; (b) *homo socialis,* that is, a metaphor of human motivation focused less on material goals and outcomes and instead on people's need for status, for standing, and to belong; and (c) a relatively new metaphor for thinking about fairness, *homo moralis,* that is, that people are sometimes motivated to enforce or live up to their core conceptions of moral right and wrong (cf. Skitka, Bauman, & Mullen, 2008). To a considerable degree, justice theorists present these different metaphors for human motivation as competing, rather than complementary, accounts for what people most care about and therefore as competing conceptions of the factors people weigh most heavily when deciding whether something is fair or unfair.

A second and more ambitious goal of this chapter is to generate a more general model of justice reasoning that integrates these different perspectives into a greater theoretical whole. The working premise of our contingency model of justice is that people are both flexible and complex and that human psychology is not driven by single motives or frames of reference. Sometimes people will be concerned about maximizing material gain, other times they will be more concerned about their social status and standing in the group, and yet other times they will be concerned about neither of these things and will be motivated by living up to or defending personal conceptions of the moral good. How they define what is fair or unfair, and the factors that will weigh most heavily in their fairness judgments, will vary as a function of which perceptual frame of reference they currently see as most relevant to the situation at hand. Before going into further details about the contingency theory of justice we propose here, we first provide some historical context and background for it.

Metaphors Guiding Justice Research: A Brief Historical Overview

Research in the psychological and social sciences is often guided by initial assumptions, or guiding metaphors, about human nature (Lakatos, 1978). Guiding metaphors strongly influence what is to be observed and scrutinized, what questions are considered interesting and important, how these questions are to be structured, as well as how the results of scientific investigations are interpreted (Kuhn, 1962). As briefly mentioned earlier, one can argue that various programs of justice research have tended to be guided by different assumptions about the key motives that drive human behavior and therefore that shape how people think about fairness and why they care about it. We briefly review three of these programs of theory and research next.

Homo Economicus: Justice as Social Exchange

The metaphor of *homo economicus*, that is, the idea that people are rationally self-interested utility maximizers, represented a hard-core assumption of classic social exchange theories, as well as early theories of distributive justice (e.g., Adams, 1965; Blau, 1964; Homans, 1961; Walster, Walster, Berscheid, & Austin, 1978). These theories assume that people approach life as a series of negotiated exchanges and that human relationships and interactions are best understood by applying subjective cost-benefit analyses and comparisons of alternatives. These theories posit that issues of equity and justice arise whenever two or more persons exchange valued resources, whether these resources consist of goods, services, money, or even love and affection. Although based on an assumption that people are rationally self-interested, these theories also propose that properly socialized persons learn that to maximize rewards in the long run, they need to understand and adhere to norms of fairness in their relationships with others (e.g., Walster et al., 1978). Groups maximize their collective gain by evolving accepted systems for fairly apportioning the costs and benefits of social cooperation among members. Therefore, these theories propose that (a) groups evolve norms about fair exchange, (b) groups generally reward members who treat others according to these norms and punish those (i.e., incur greater costs to) who do not, and (c) participating in unfair exchange causes psychological distress that in turns motivates attempts to restore fairness (Walster et al., 1978). Not surprisingly, justice theories that use economic exchange as a guiding theoretical metaphor have primarily inspired studies that examine people's reactions to what they get out of a given encounter or relationship and how perceptions of either under- or overbenefit lead people to change either their costs or their benefits to restore a psychological sense of balance or fairness.

Considerable research has been consistent with the notion that people do tend to track relative costs and benefits, and these cost-benefit calculations influence perceptions of fairness and a host of fairness-related behavior and reactions. For example, people (a) attend to and care about how much they contribute to and get out of their social relationships (Konow, 1996; Walster et al., 1978), including what they contribute to and get out of their closest and most intimate relationships, such as dating and marriage (Rusbult, 1983; Sprecher & Schwartz, 1994); (b) take into account deservingness criteria (e.g., relative contributions) when deciding how to fairly allocate resources to others (for reviews, see Brewer & Kramer, 1985; Cook, 1975); (c) incur costs to punish someone who violates standards of fair allocation behavior (e.g., Fehr & Fischbacher, 2004; Fehr & Schmidt, 1999); and (d) perceive getting both more and less than others do for similar effort to be unfair, and they will change their levels of contribution to a relationship if they feel either under- or overbenefited (see Walster et al., 1978, for a review). In summary, a vast amount of research has been consistent with the notion that people care about the fairness of distributions of costs and benefits, and people use notions of economic exchange to understand the fairness of their social relationships.

Homo Socialis: Needs for Status, Standing, and Belonging

A shift in the metaphor that guided justice theorizing and research in social psychology occurred during the early 1980s with the introduction of the group value model of procedural fairness. The guiding metaphor of this program of research was the notion that people more often seek to satisfy relational motives, such as needs to feel valued, respected, and included in important social groups, than pursue material self-interest; that is, *homo economicus* yielded to *homo socialis*. Therefore, research began to focus more on leaders' or authorities' behaviors and decision making and how these affected recipients' reasoning about fairness, and it focused less on outcome distributions and the factors that determined them (e.g., Lind & Tyler, 1988; Tyler & Lind, 1992). By using the "need to belong" as a lens for examining what mattered in the psychology of justice, researchers broadened their understanding of how and why people make fairness judgments. For example, researchers identified the pervasive influence of procedural treatment (such as variations in opportunities for voice, being treated with dignity and respect, and freedom from bias) on perceptions of fairness, working from the assumption that procedural treatment provides more relevant information for judging one's relative standing than do material outcomes.

Considerable research has been consistent with the *homo socialis* prediction that procedural treatment and people's concern with needs to feel valued and respected as group members influence perceptions of fairness and a host of fairness-related behavior and reactions. For example, people

(a) spontaneously mention issues about treatment and lack of respect more than they do specific outcomes when they are asked to recall specific instances of injustice or unfairness (Lupfer, Weeks, Doan, & Houston, 2000; Mikula, Petri, & Tanzer, 1990); (b) use procedural treatment and not just decision outcomes when evaluating the fairness of authorities and institutions (Lind & Tyler, 1988; McFarlin & Sweeney, 1992); (c) become more committed to organizations when they believe they are treated well, even if they receive nonpreferred outcomes (Greenberg, 1990; Skarlicki & Folger, 1997; Tyler, 1989); and (d) identify more strongly with procedurally fair groups and authorities, which in turn relates to a host of other consequences, such as cooperating with the rules and going the extra mile to serve the groups' interests (see Tyler & Blader, 2003, for a review). In summary, a large amount of research has been consistent with the notion that people's fairness reasoning is shaped by more than the material or concrete outcomes they receive from a social exchange. People also care about how decisions are made and the degree to which decision makers signal that people are valued and respected members of the group. Researchers may not have discovered the importance of procedures and treatment to justice judgments had they remained solely committed to understanding justice from the perspective of only economic exchange.

Homo Moralis: Moral Authenticity and Conceptions of the Good

Now another shift seems to be underway. A number of theorists and researchers have turned their attention to the role that moral concerns play in people's justice reasoning and behavior (e.g., Brosnan & de Waal, 2003; Cropanzano, Byrne, Bobocel, & Rupp, 2001; de Waal, 1996; Fehr & Fischbacher, 2004; Folger, 2001; Folger, Cropanzano, & Goldman, 2005; Skitka & Houston, 2001; Skitka et al., in press). Theorists across a host of different disciplinary traditions seem to be converging on the insight that managing the particular challenges of group living (e.g., aggression, competition, cooperation, deception, and the undermining role of self-interest) led to the adaptation, through natural selection, of humans to care about morality independent of their self-interest and belongingness needs. People who learned to manage the balance between competition and cooperation, develop conceptions of moral right and wrong, and punish those who broke contracts or other justice arrangements had a clear adaptive advantage over those who failed to develop traits that allowed them to manage these challenges (see Robinson, Kurzban, & Jones, 2007, for a detailed review).

The connections between morality and justice are also clear in other theories and schools of thought as well. A working definition of justice and what it means to people could reasonably start with morality, righteousness, virtues, and ethics rather than with self-interest, belongingness, or other nonmoral motivations (Skitka & Bauman, 2008). For example, Plato's conception of individual justice was distinctively moral. Plato considered actions to be just if they sustained or were consonant with ethics and morality rather than baser motives, such as appetites (e.g., lust, greed; Jowett, 1999). In addition to having strong roots in classical philosophy, the connection between conceptions of justice and morality has been a consistent theme in moral development theory and research. For example, Kohlberg's theory of moral development (e.g., Kohlberg, 1973) posits that justice is an essential feature of moral reasoning and that "justice operations" are the processes people use to resolve disputes between conflicting moral claims. From this developmental perspective, people progress toward moral maturity as they become more competent and sophisticated in their approach to justice operations. In short, an alternative guiding metaphor in justice research and closely related areas is *homo moralis*, that is, the notion that people have an intrinsic propensity for caring about and acting on conceptions of morality.

Although *homo moralis* is a newer area of empirical inquiry than research inspired by the *homo economicus* or *socialis* metaphors, the notion that people's sense of morality or immorality affects how they reason about fairness has received some empirical support. For example, when people's outcome preferences are experienced as strong moral convictions, rather than as equally strong but nonmoral preferences, people's perceptions of outcome fairness and decision acceptance are shaped more by whether outcomes are consistent with perceivers' moral priorities than by whether authorities are perceived as acting in procedurally proper or improper ways (e.g., Skitka, 2002; Skitka & Mullen, 2002) or whether authorities' behavior has been experimentally manipulated to be high or low in decision and treatment quality (e.g., Bauman & Skitka, 2007). People also use whether decision outcomes are consistent with their personal moral standards to judge the legitimacy and fairness of authorities (see Skitka and Bauman, 2008, for a review). For example, Skitka et al. (in press) found that people's predecision moral convictions about the appropriateness of physician-assisted suicide were stronger predictors of their subsequent perceptions of the outcome fairness and decision acceptance of a U.S. Supreme Court decision than were their predecision perceptions of the legitimacy, the procedural fairness, or their trust of the Supreme Court. Moreover, the degree to which the Court ruled consistently or inconsistently with people's moral convictions also predicted how legitimate, procedurally fair, and trustworthy the Court was perceived as being after the decision.

The guiding metaphors of *homo economicus*, *socialis*, and *moralis* have each been useful for theory and hypothesis generation, and each has provided a frame for thinking about how and why people might care about justice. In turn, each has led to testable hypotheses. Moreover, each perspective has received considerable support. That said, each metaphor in itself is limited in scope and range. Justice research may therefore be like the poet John Godfrey Saxe's blind men of Indostan, who each examined a different part of an elephant and therefore came to different conclusions about the fundamental character of "elephantness" ("it is very like a tree," "a wall," or "a snake"). The tree, wall, and snake descriptions are each accurate but limited descriptions of an elephant (its leg, body, and trunk, respectively). Similarly, justice researchers guided by the *homo economicus, socialis,* and *moralis* metaphors have found considerable evidence consistent with their predictions and claims, but these metaphors may provide a limited view of the total "justice elephant." We turn next to an attempt to (a) "knit together" these different conceptions of the justice elephant into a conceptual whole, (b) provide some background for why this integration makes particular sense in terms of what we know more broadly about social psychology and social cognition, and (c) explore some of the implications of taking a more integrative, contingency-based approach to understanding the psychology of justice.

Knitting Together the Justice Elephant: Introducing a Contingency Theory of Justice

We accept the validity of the empirical research conducted within each of these different programs of research. People do approach justice from the perspective of economic exchange and worry about what they get. People also approach justice from the perspective of a need to belong, to feel valued, and to care about treatment and procedures in addition to the fairness of material outcomes. People sometimes think, however, they know the right or wrong outcome, and this certainty can lead them to focus on whether the right outcome is achieved, to the relative neglect of how that outcome is achieved. What is missing is a clear overarching theoretical framework that is consistent with each of these seemingly inconsistent conclusions about the critical foundations of the psychology of justice. A theoretical perspective that could account for these discrepancies and knit together a coherent theoretical "whole elephant" is a contingent one: How people define what is fair or unfair and the factors that weigh most strongly in predicting fairness judgments and behavior depend on the current perspective of the perceiver. The factors people weigh most heavily in a given fairness judgment will importantly depend on *which* set of motivational concerns currently is most important or cognitively accessible to them in a given situation (cf. Deutsch, 1985; Lerner, 1977).

Shifting From Competing Metaphors to Contingencies

We argue that people's justice reasoning is likely to be contingently shaped by at least three major categories of motivational concerns, that is, the *economicus*, *socialis*, and *moralis* motivational concerns already described. According to our contingency theory of justice (which is an updated version of the accessible identity model, or AIM, proposed by Skitka, 2003),[1] various justice prototypes, schemata, or norms are likely to be stored in memory in close association with people's *economicus*, *socialis*, and *moralis* goals and needs. Therefore, we posit that conceptions of justice are partially dependent on other goals simultaneously pursued and currently activated in people's working memory. How people think about fairness will therefore contingently depend on whether perceivers are currently working from an *economicus*, *socialis*, or *moralis* perspective. Moreover, people's normative expectations about what constitutes justice in any given situation will depend—that is, be contingent on—whether they see the situation as relevant to their own and others' *economicus*, *socialis*, or *moralis* needs and goals.

The Economicus Contingency

Material goals and concerns refer to people's desire to satisfy their basic needs, such as for food, shelter, clothing, and so on, as well as their desire to accumulate possessions, property, and wealth as valuable ends in themselves rather than in the service of needs for social status (e.g., Belk, 1988; James, 1890). Materialistic goals and concerns are among the most normatively "self-interested" or "selfish" of human motivations. That said, considerable research has indicated that pure self-interest and a Hobbesian war of all against all is avoided by people's acceptance of the need for fairness in economic and material exchange (e.g., Lane, 1986; Walster et al., 1978).

An *economicus* perspective is most likely to be activated in situations when (a) people's basic material needs are perceived as not being met or are under threat (Maslow, 1993); (b) the social situation presents a real or perceived potential for material loss or gain; (c) the relational context is defined in market pricing terms, for example, people are involved in negotiating prices, wages, or the exchange of fungible resources (Fiske, 1991); (d) the goal of the social system is to maximize productivity (Deutsch, 1985; Lerner, 1977); and (e) the other goals or concerns are not particularly salient. Proportionality of entitlements to contributions, or equity rules, most frequently defines fairness in market contexts (e.g., Fiske, 1991; Lane, 1986; Rainwater, 1974; Walster et al., 1978) and therefore presumably when people's material concerns are especially salient and accessible in memory as well.

Consistent with some of the ideas just outlined, some research has suggested that distributive justice concerns (i.e., whether outcomes are allocated in fair ways) emerge as stronger predictors of overall perceptions of organizational justice in cultures higher in materialism, whereas perceptions of interactional justice (i.e., interpersonal treatment, usually operationalized in terms of how authorities treat subordinates) are stronger predictors of overall perceptions of organizational justice in cultures lower in materialism (Kim & Leung, 2007). Also consistent with the notion that a *homo economicus* perspective is more likely to be dominant when basic material needs are not being met is the finding that higher levels of materialism are associated with lower levels of per capita gross national income (Abramson & Inglehart, 1995).

The relative salience of *economicus* goals and concerns will also shape which conceptions of procedural justice are likely to be salient or emphasized in a given situation. For example, Thibaut and Walker (1975) argued that one reason why people might care about procedural fairness is that it provides them with a real or perceived sense of process control over outcomes, something that serves people's material interests in the long run. Therefore, when material needs and goals are especially salient, people may define fairness more in terms of variables related with process control (e.g., voice, consistency, lack of bias) than variables that reflect relational concerns, such as whether people are treated with dignity and respect.

In summary, the contingency theory of justice makes a number of predictions about when material goals and concerns are especially likely to dominate people's justice reasoning, as well as some specific predictions about the kinds of justice considerations that are likely to be most salient when people are focused on their own or others' material needs and goals. It is important to point out that the contingency model does not predict that people will be concerned solely about outcomes when material concerns loom larger in people's minds than either social or moral concerns. Rather, the model predicts that when material motives are more highly accessible to people than either their social or their moral motivations, people will be more likely to define fairness in terms of equitable outcomes *and* in terms of procedures that maximize rather than minimize process control.

The Socialis Contingency

In addition to having materialistic needs and goals, people also have strong needs for connection and to belong (Baumeister & Leary, 1995; Leary, Tambor, Terdal, & Downs, 1995) that are typically expressed through their desire to find and maintain relationships with others (Maslow, 1993). People's relational motives prompt concerns about their being valued, respected, and included in important social groups (de Cremer & Blader, 2006).

A *socialis* perspective is most likely to be activated in situations when (a) people's basic material needs are already satisfied (Maslow, 1993); (b) the needs to belong and be included are not being met and therefore are especially strong (De Cremer & Blader, 2006); (c) the social situation poses a real or perceived potential for loss or gain in inclusion, social status, or respect; and/or (d) the situation primes the need for, or goal of, maintaining group harmony and minimizing conflict (e.g., Deutsch, 1985).

Existing research has suggested that when perceivers view a situation from a more *socialis* than *economicus* perspective, they are more likely to define distributive justice according to norms focused on need or equality than norms focused on contributions or inputs (Deutsch, 1975, 1985). For example, people who are primed with solidarity and group harmony goals (Deutsch, 1985) or who are chronically higher in communal or interpersonal orientation (Major & Adams, 1983; Watts, Messé, & Vallacher, 1982) are more likely to allocate material rewards equally than equitably and to rate equal allocations as more fair than equitable ones. Other research has indicated that conceptions of distributive justice also vary as a function of the social role of the perceiver and the degree to which roles prime relational concerns. For example, when one's social role as a parent is more highly activated than other roles or concerns, one is more likely to perceive allocations based on need as more fair than those based on equity or equality (Drake & Lawrence, 2000; Prentice & Crosby, 1987).

There is also evidence that indicates that the relative salience of *socialis* needs and concerns affects how people define procedural fairness. Vast amounts of research now support the notion that people's justice reasoning is more strongly influenced by interactional treatment (e.g., the degree to which involved parties are treated with dignity and respect) when their relational needs are especially salient or high rather than low. For example, people's fairness reasoning is influenced more strongly by variations in interactional treatment when (a) social identity needs are particularly strong (Brockner, Tyler, & Cooper-Schneider, 1992, Study 1; Huo, Smith, Tyler, & Lind, 1996; Platow & von Knippenberg, 2001; Wenzel, 2000), (b) perceivers are of low rather than high status (Chen, Brockner, & Greenberg, 2003), (c) status concerns are primed (van Prooijen, van den Bos, & Wilke, 2002), and (d) people are high rather than low in interdependent self-construal and interdependent self-construal is primed (Brockner, Chen, Mannix, Leung, & Skarlicki, 2000; Holmvall & Bobocel, 2008).

It should be noted that groups—like individuals—can also have material, social, and moral goals and needs. For example, one group might be concerned about building a stronger financial endowment, another group might be concerned about building its status or brand recognition, and a third group might be concerned about protecting the environment. Therefore, taking a group perspective does not mean that people will necessarily be more motivated by *socialis* than *economicus* or *moralis* considerations. Taking a group perspective rather than an individual perspective has the potential to prime any one of these needs and goals, depending on the primary concerns or goals of the group (cf. Clayton & Opotow, 2003;

Skitka, 2003). Which justice norms or values are most likely to be activated when people take a more group perspective will depend on the dominant goal orientation of the group and whether the group is focused on *economicus*, *socialis*, or *moralis* needs and goals.

The Moralis Contingency

Economicus and *socialis* motivations are primarily defined in terms of people's wants or desires. People want to accumulate material goods and wealth; similarly, people want to belong and to have status and standing in important groups. In contrast to a motivational focus on wants and desires, moral motivations are instead based more on feelings of "ought" and "should" and desires to express and defend conceptions of basic right and wrong (Reed & Aquino, 2003; Steele, 1988, 1999).

There is some evidence that people are most likely to be motivated by moral concerns when (a) their basic material and social needs are at least minimally satisfied (Maslow, 1993); (b) moral intuitions or emotions are aroused (e.g., disgust, moral outrage, guilt, shame; Haidt, 2003); (c) they see examples of undeserved harm, especially when the harm is intentionally inflicted (Pittman & Darley, 2003); (d) there is a real or perceived threat to people's conception of the moral order (e.g., Tetlock, 2002); (e) their conception of themselves as morally authentic or virtuous is questioned or undermined (Steele, 1988, 1999; Zhong & Liljenquist, 2006); (f) the social context primes concerns about the greater good or people's conceptions of virtue (e.g., Aquino, Reed, Thau, & Freeman, 2007); or (g) they are reminded of their mortality (e.g., Jonas, Schimel, Greenberg, & Pyszczynski, 2002).

When people take a moral perspective to defining whether justice has been done, they are more likely to reason from a belief that duties and rights follow from the greater moral purpose that rules, procedures, and authority dictate rather than from the rules, procedures, and authorities themselves (Kolhberg, 1973; Rest, Narvaez, Bebeau, & Thoma, 1999). Morally based justice reasoning is not by definition antiestablishment or antiauthority, it just is not de facto dependent on establishment, convention, rules, or authorities. Instead, a *moralis* perspective focuses people on the way they believe things "ought" to be or "should" be done. Therefore, when people take a *moralis* perspective, their fairness reasoning is less likely to be based on clearly defined normative rules or norms and more likely to be based on people's gut intuitions of right or wrong (Haidt, 2001; Haidt, Koller, & Dias, 1993) and aroused moral emotions (e.g., Mullen & Skitka, 2006). Morality is also often defined in terms of "postconventional" reasoning (Kohlberg, 1973; Rest et al., 1999), meaning that people's moral judgments are more likely to be somewhat idiosyncratic rather than rooted in strong and shared normative conventions. For example, not everyone will have their moral sensibilities aroused in the same context or in response to the same issues (for more detail, see Lester, 2000; Skitka, Bauman, & Sargis, 2005).

Research that has tested the implications of the *homo moralis* guiding metaphor for why people might care about fairness has revealed that when people's outcome preferences are held with strong rather than weak moral conviction, their judgments of the fairness of both outcomes and procedures tend to be shaped more by whether preferred outcomes are achieved than by whether they are achieved by proper or improper procedures. For example, when people had a moral mandate about defendant guilt, punishments doled out by either due process or vigilantism were seen as equally fair so long as they achieved the "correct" outcome (the guilty were punished, the innocent were not; Skitka & Houston, 2001). Similarly, whether authorities made the morally correct decision in a controversial custody case was a stronger predictor of postresolutions perceptions of outcome fairness, decision acceptance, and procedural fairness than were predecision perceptions of procedural fairness (Skitka & Mullen, 2002).

The empirical discovery that the "objective" fairness of procedures does little to offset perceptions of unfairness when outcomes fail to match perceivers' a priori conceptions of basic right and wrong provides valuable insight into the intractability of any number of public policy debates (e.g., abortion, capital punishment, gay marriage). Maximally fair procedures and the imprimatur of legitimate authorities do little to offset people's sense of injustice when these procedures and authorities yield outcomes that people perceive as fundamentally immoral and wrong (see Skitka et al., 2008, for a review).

In addition to being related to judgments of distributive and procedural fairness, there are also reasons to believe that moral considerations may play an especially important role in judgments of retributive justice. Specifically, Pittman and Darley (2003) argued that the magnitude of punishment assigned to transgressions depends on the magnitude of harm and the degree to which it is intended. Accidental harm arouses little moral emotion and leads people to primarily focus on compensatory justice (perhaps because accidental harms are associated more with *economicus* concerns about restitution), whereas intentional harms are associated with increased levels of moral outrage, especially as they increase in severity and therefore represent more severe violations of people's conceptions of moral order. Increased moral outrage, in turn, predicts more punitive reactions and sentencing goals. Mullen and Skitka (2006) found—consistent with the notion that emotion may be driving people's justice reasoning when they are morally engaged—people's anger about immoral outcomes was a better predictor of people's justice reasoning than a host of alternative explanations, such as the degree to which people engaged in post hoc reappraisals of whether procedures were fair (e.g., motivated reasoning) or the degree to which they identified with the parties involved. Although the *moralis* program of justice research is still relatively new, existing research has been consistent with the notion that fairness reasoning is influenced by the relative degree to which people have a moral stake in a given decision.

Complementary Perspectives

In addition to integrating several different theoretical traditions, the contingency model of justice is also consistent with other major theories that propose that people's reasoning is likely to vary as a function of their motivational perspective. We turn to a brief review of these complementary theoretical perspectives in an attempt to further establish the reasonableness of taking a more contingent rather than competing motivational approach to understanding how people think about justice. These other theories do not focus their predictions on people's justice reasoning per se but provide some independent validation that the *economicus*, *socialis*, and *moralis* contingencies account for important differences in how people approach social judgments and interactions.

At least three other programs of theory and research are consistent with the contingency model's emphasis on the idea that people are fluid and flexible in how they approach and perceive their social worlds and, more specifically, that the *economicus*, *socialis*, and *moralis* perspectives provide useful heuristics for organizing hypotheses about how people's goals and needs can influence how they make sense of their social worlds. We turn next to briefly review each of these perspectives: (a) self-schema and identity theories, (b) domain theory, and (c) moral schema theory.

Self-Schema and Identity Theories

Self-schema, categorization, and regulation theorists assume that self-definition is a dynamic and basic categorization process that has important implications for virtually all human thoughts, feelings, and behavior through its activation of personal strivings or goals (Emmons, 1986). Research on the self has revealed that (a) people have multiple levels of self or identity and therefore multiple layers of identity-relevant goals, (b) not all aspects of identity (and related goals) can be equally accessible at any given time, (c) the relative accessibility of a given identity (and therefore various goals) in the working self-concept or working memory is influenced by the perceiver's past experience and present expectations in combination with cues from the social context, and (d) a shift in identity focus similarly shifts the accessibility of associated expectations, motives, values, knowledge, and goals (for relevant reviews and research, see Baumeister, 1999; Brewer, 1991; Carver & Scheier, 1998; Markus & Kunda, 1986; McGuire, McGuire, & Cheever, 1986; Showers, 2002; Turner, 1999).

Moreover, there tends to be a great deal of similarity in the structure or major categories of the self-concept across individuals (Bugental & Zelen, 1950; Rentsch & Heffner, 1994). For example, James (1890) proposed that the self was best classified into three overlapping but distinguishable categories: the material, social, and "spiritual" aspects of self.[2] James posited that people's sense of material self consists of the body and its adornment and their home and hearth, acquisitions, and accumulated wealth. People define and sustain their material self by endeavoring to acquire and maintain things such as property, goods, and wealth (James, 1890). In contrast, the social self is defined in terms of the groups people belong to, their social role in those groups, and the reflected appraisal or standing that they have vis-à-vis other group members. People have as many different social selves as there are distinct groups of persons about whose opinion they care. The goals and strivings of the social self are met by the roles that people have, seek out, and see as important and through their ability to live up to the demands associated with those roles (e.g., Leary et al., 1995; Turner, 1985). James's (1890) spiritual identity refers to "the most enduring and intimate part of the self, that which we most verily seem to be … it is what we think of our ability to argue and discriminate, of our moral sensibility and conscience, of our indomitable will" (p. 315). People's sense of personal or spiritual identity is shaped to a considerable degree by people's need to live up to internalized notions of moral "ought" and "should" and a desire to live up to both public and private conceptions of moral authenticity (Bandura, 1986; Higgins, 1987; Steele, 1988, 1999).

The overlap between James's (1890) categories of the self and the different motivational frameworks and guiding metaphors that have driven justice research is quite clear. There are relatively transparent similarities between the descriptions of *homo economicus*, *socialis*, and *moralis* and James's conception of material, social, and personal or spiritual aspects of self or identity. Given that one self-system or schema and its attendant goals and strivings tend to dominate people's "working self-concept" at any given point in time (see Markus & Wurf, 1987, for a review), it seems reasonable to propose that how people think about and define fairness may therefore be importantly shaped or dependent on which self-schema currently dominates a perceiver's working self-concept (Skitka, 2003).

In summary, one way to generate contingent predictions that integrate theory and research from the *homo economicus*, *socialis*, and *moralis* perspectives is to posit a fundamental organizing role of the self in justice reasoning (Skitka, 2003). How people define and think about fairness is likely to be stored in close connections to different aspects of the self and self-related goal systems in memory and therefore is differentially likely to be activated or accessible as a function of which aspect of the self currently dominates the perceivers' working self-concept. Although there is considerable research showing close connections between self-awareness and the activation of various identity-relevant concerns about justice, it is not yet clear whether the self or activation of identity-relevant concerns is necessary, rather than simply sufficient, to activate justice reasoning or different kinds of justice reasoning. Specifically, activating material, social, or moral identities may shift people's conceptions or definitions of justice (a sufficient cause); so too, however, might priming material, social, and moral constructs unrelated to anything about the perceivers' identity (in which case, identity priming is not a necessary cause of shifting the contingencies people use to judge fairness).

Domain and schema theories of moral development have independently posited the importance of three very similar contingencies to the ones we have been discussing here, without positing a key role for the self or identity. We turn to a review of these theories next.

Domain and Moral Schema Theories

Domain Theory Theory and research in moral development arrive at very similar and contingent categories to describe how people make moral judgments, of which they would consider justice to be one part. Domain theorists argue that there are three core systems or domains of social judgment: (a) personal prerogative or matters of taste or preference, (b) social convention, and (c) morality (for reviews, see Nucci, 1996; Turiel, 2002). Preferences are by definition subjective and in the eye of the beholder. Preferences represent instances when people favor one outcome more than another, but there is no expectation that others would or should feel the same way. Others' preferences about the same object are not seen as either right or wrong; they are simply different. For example, it is acceptable for you to prefer apples even though I prefer oranges because people are entitled to have different tastes about fruit. Similarly, some people's positions about abortion may also be based in a sense of preference or self-interest rather than a sense of normative convention or morality. For example, someone may support legalized abortion because she prefers having a backstop birth control option to not having one. Another person's view on abortion may be based more on a sense of normative convention than on either preference or morality. For example, this person may oppose legalized abortion because it is illegal in his state or because church authorities oppose the practice, but this person may feel no personal preference or moral connection to the issue. If abortion was legalized tomorrow or the church was to reverse its position, individuals whose position is based on normative convention and authority would likely reverse their position on the issue as well. Someone who sees abortion as a moral issue, however, is likely to see abortion as universally right or wrong and will persist in believing it is right or wrong even if this position is inconsistent with normative conventions, local custom, and rule of law. The relative authority independence of a position is what best distinguishes whether a position is viewed as moral or conventional.[3]

Young children's and adults' abilities to make distinctions across these domains replicate across a wide array of nationalities and religious groups (for reviews, see Nucci, 2001; Smetana, 1993; Tisak, 1995). Also intriguing is the finding that adult psychopaths and children who exhibit psychopathic tendencies do not make distinctions between the conventional and moral domains (Blair, 1995, 1997).[4]

The distinctions that domain theorists make between social judgments that involve preferences, normative conventions, and morality map reasonably well onto some of the findings and distinctions made by theory and research that have taken *homo economicus, socialis,* and *moralis* perspectives. Preferences are not always the fodder of economic exchange, but they tend to be more self-oriented and materialistic in orientation than judgments related to normative convention or morality. Matters of normative convention, by definition, reflect a primary focus on community standards, authorities, rules, and so on and map well onto the concerns of *homo socialis*. Finally, one of the key distinctions between morality and normative convention in domain theory is the extent to which the latter tends to be independent of concerns about authority or conformity to group norms, something very consistent with current justice research and theorizing about how morality affects justice reasoning.

Moral Schema Theory Another neo-Kohlbergian approach to understanding moral development suggests that judgmental domains might be better understood as cognitive schemata (Rest et al., 1999). Rest et al.'s (1999) moral schema theory posits that people use three kinds of schemata to make sociomoral judgments, specifically, the personal interest schema, the norm maintenance schema, and the postconventional schema. The personal interest schema develops in early childhood, the norm maintenance schema develops during adolescence, and the postconventional schema develops in late adolescence and adulthood. Once the schemata are formed, people can use each of them to guide their judgment and behavior and theoretically can move fluidly between them as a function of how well features of situations and social relationships map onto, and therefore prime the activation of, one or another core schema.

When people apply the personal interest schema, they tend to focus on either their own self-interests or their personal stake in a situation or to justify the behavior of others in terms of their perceptions of others' personal interests. The norm maintenance schema focuses on the need for norms that address more than the personal preferences of those involved in a given situation and places a heavy focus on (a) the needs of cooperative social systems and the group, (b) a belief that living up to these norms and standards will pay off in the long run, and (c) a strong duty orientation, whereby one should obey and respect authorities and authority hierarchies.

Finally, the postconventional schema primes a sense of moral obligation based on the notions that laws, roles, codes, and contracts are all relatively arbitrary social arrangements that facilitate cooperation but that there is a variety of ways these coordination rules could be constructed to achieve the same ends. Just because the existing rules exist does not mean that people think these arrangements are right when a situation primes a more postconventional schema. When activated, a postconventional schema leads people more toward an orientation that duties and rights follow from the greater moral purpose behind conventions, not from the conventions themselves. In summary, postconventional thinking focuses people on ideals, conceptions of the ultimate moral good or imperative, something people presume (not always correctly so) that reasonable others will also share and understand or could easily be persuaded to share or understand (Rest et al., 1999).

In summary, there are clear similarities between the *homo economicus*, *socialis*, and *moralis* metaphors that have separately guided different periods and programs of research on adult conceptions of justice. The metaphors that have been used in justice research show a remarkable degree of overlap with the categories of social judgment that guide contemporary theories of moral development, despite each program developing quite independently of the other. Taken together with theory and research on the self, contemporary theories of moral development provide further support for the notion that it is theoretically useful to focus on contingent predictions rather than competing conceptions of how people think about justice. Rather than remaining wedded to competing single-motive accounts for why people care about fairness and therefore the factors most likely to influence perceptions of whether social encounters are fair or unfair, there is a host of reasons to believe that people's perceptions of fairness depend to some degree on the goals or concerns that are currently most important to them.

Implications and Directions for Future Research

Hastorf and Cantril (1954) conducted a study that is often cited as providing the case for the value of social psychology. Specifically, they studied Dartmouth and Princeton students' perceptions of an actual football game played in 1951 between the Dartmouth Indians and the Princeton Tigers. It was a particularly rough game with many penalties. The Princeton quarterback had to leave the game with a broken nose and a concussion in the second quarter of the game; the Dartmouth quarterback's leg was broken in a backfield tackle in the third quarter. One week after the game, Hastorf and Cantril surveyed Princeton and Dartmouth students who saw the game, as well as a sample of students who viewed a film of it.

Despite seeing the same game, participants viewed it very differently. The Princeton students saw the Dartmouth team make more than twice as many rule infractions as the Dartmouth students saw, whereas the Dartmouth students saw a reverse pattern of infractions. Sixty-nine percent of Princeton students described the game as "rough" and "dirty," whereas a majority of Dartmouth students felt that even though the Dartmouth team played rough, the play was generally "clean" and "fair." These results indicated that people actively constructed different realities as a function of their perspective; as Hastorf and Cantril (1954) put it, "There is no such 'thing' as a 'game' existing 'out there' in its own right which people merely 'observe.' The game 'exists' for a person and is experienced by him only insofar as certain happenings have significances in terms of his purpose" (p. 133).

Hastorf and Cantril's (1954) findings, and other similar findings, led psychologists and other scholars to the realization that people do not react to each other's actions in a stimulus-response pattern without the mediating influence of interpretation. People interpret their social worlds, and therefore human interaction is mediated through people's *understanding* and interpretation of what social interactions mean (Blumer, 1969). Similarly, a guiding premise of the contingency theory of justice is that there is not an objective reality or set of circumstances that is fair or unfair. People do not always interpret social interactions such as football games, performance evaluations, negotiations of the price of a car or home, or their intimate relationships from similar perspectives. Instead, people actively construct their perceptions of fairness and unfairness, and these active constructions are influenced by different fairness norms and the various goals, needs, expectations, and histories people carry with them into their social interactions.

In addition to regrounding justice theory in classic conceptions of symbolic interactionism, the contingency theory of justice can account for the mundane reality that people often disagree about whether a given situation was handled fairly or unfairly. Until people arrive at some consensus about the nature of the judgment to be made and what goals they wish to achieve in a given a context, it is not surprising that people approach the same situation with very different conceptions of fairness. The notion that people are likely to approach the same situation from different perspectives, which in turn shapes the fairness norms or considerations they apply to it, suggests that future research should extend beyond the study of how individuals in isolation make fairness judgments. Further research should begin to explore how people socially negotiate and arrive at a consensus about how to make decisions fairly or whether fairness has been achieved in specific circumstances. A truism of early theories of distributive justice is that people do not make justice judgments in social vacuums. Instead, justice judgments are inherently *social* judgments and require social comparison information (e.g., Walster et al., 1978). The social aspect of deciding whether something is fair or unfair may go beyond relying on passive comparisons to see if others received outcomes proportional to inputs or similar treatment or opportunities for voice. Instead, people's natural fairness reasoning may rely more on active gathering of social comparison information. That is, people will attempt to seek information about how others interpret a given situation and whether they too see it as fair or unfair. Moreover, it will be interesting to explore whether influence attempts that frame issues more in *economicus*, *socialis*, or *moralis* terms are differentially likely to influence people to concede to others' views that a given situation is fair or unfair.

Few if any contemporary studies of fairness have focused on the importance of social comparison and consensus in how people think about fairness. Because the contingency model of justice predicts a certain amount of fluidity and flexibility in how people are likely to think about fairness and focuses more explicitly on socially constructed fairness judgments that depend on perspective, the model suggests that social comparisons and consensus seeking may be especially important areas for future research in the psychology of justice.

Lastly, we should note that the contingency model of justice reasoning we are positing here has considerable overlap with the AIM proposed by Skitka (2003). Skitka, however, emphasized a primary role for material, social, and personal or moral identities as a way to organize predictions about when *economicus*, *socialis*, and *moralis* concerns were likely to be activated and particularly relevant as guides to perceptions of fairness (see also the section on identity reviewed in the current chapter). The contingency model proposed here is more parsimonious and to some degree more flexible because it leaves the role of identity involvement open for empirical test rather than using identity as a central organizing theme for when people's *economicus*, *socialis*, and *moralis* perspectives are most likely to be salient and therefore applied to specific contexts. The contingency model suggests that situational priming of a specific identity or identity-relevant concerns—or individual differences in the chronic accessibility of different identities or schemata—may be sufficient to activate different fairness norms or rules, but identity involvement per se may or may not be a necessary condition for leading people to be concerned about fairness.

Authors' Note

Preparation of this chapter was facilitated by grant support from the National Science Foundation to the first author (NSF-0518084, NSF-0530380). Correspondence about this chapter should be directed to Linda J. Skitka, University of Illinois at Chicago, Department of Psychology lskitka@uic.edu.

Notes

1. One major difference between the AIM and the contingency model proposed here is the relative emphasis on identity as a foundation for how justice contingencies are stored in memory and therefore the degree to which identity activation is a necessary, rather than a sufficient, cause of activation of different contingencies of needs or goals. The AIM proposed that these conceptions were stored in close connections with the material, social, and moral senses of identity. The contingency model acknowledges that the self may be one of any number of possible ways these needs and goals may be organized in memory, but it considers the necessary versus sufficient role of identity as a more open and testable question rather than a core foundation of the theory. We discuss these differences in some detail later in the chapter.
2. It should be noted that James (1890) was explicit that the spiritual self was not to be confused with religiosity. The term *spiritual* was meant to represent more inner-directed and autonomous senses of self than either the extrinsically focused material self or the socially constructed and focused social self, and therefore it has sometimes been referred to by others as people's sense of "personal" or "moral" self (e.g., Skitka, 2003).
3. There is considerable research that has used the "moral–conventional" test of authority independence to differentiate conventional and moral-based thinking. Preferences and matters of personal taste, however, are also likely to be authority independent as well. People, however, are unlikely to justify preferences in terms of "it's wrong" (which could be a rule-based or conventional definition of wrong or a moral one). The authority independence test allows for a distinction between conventional and moral kinds of right and wrong.
4. To our knowledge, researchers have not tested whether psychopaths make distinctions between preferences and other domains.

References

Abramson, P. R., & Inglehart, R. (1995). *Value change in global perspective.* Ann Arbor: University of Michigan Press.

Adams, J. S. (1965). Inequity in social exchange. In L. Berkowitz (Ed.), *Advances in experimental social psychology* (Vol. 2, pp. 267–299). New York: Academic Press.

Aquino, K., Reed, A., II, Thau, S., & Freeman, D. (2007). A grotesque and dark beauty: How moral identity and mechanisms of moral engagement influence cognitive and emotional reactions to war. *Journal of Experimental and Social Psychology*, *43*, 385–392.

Bandura, A. (1986). *Social foundations of thought and action: A social cognitive view.* Englewood Cliffs, NJ: Prentice Hall.

Bauman, C. W., & Skitka, L. J. (2007). *Fair but wrong: Procedural and moral influences on fairness judgments and group rejection.* Unpublished manuscript.

Baumeister, R. (1999). The nature and structure of the self: An overview. In R. Baumeister (Ed.), *The self in social psychology* (pp. 1–20). Philadelphia: Psychology Press.

Baumeister, R. F., & Leary, M. R. (1995). The need to belong: Desire for interpersonal attachments as a fundamental human motive. *Psychological Bulletin, 117,* 497–529.

Belk, R. W. (1988). Possessions and the extended self. *Journal of Consumer Research, 15,* 139–168.

Blair, R. (1995). A cognitive developmental approach to morality: Investigating the psychopath. *Cognition, 57,* 1–29.

Blair, R. (1997). Moral reasoning and the child with psychopathic tendencies. *Personality and Individual Differences, 26,* 731–739.

Blau, P. (1964). *Exchange and power in social life.* New York: Wiley.

Blumer, H. (1969). *Symbolic interactionism: Perspective and method.* Berkeley: University of California Press.

Brewer, M. B. (1991). The social self: On being the same and different at the same time. *Personality and Social Psychology Bulletin, 17,* 475–482.

Brewer, M. B., & Kramer, R. M. (1985). The psychology of intergroup attitudes and behavior. *Annual Review of Psychology, 36,* 219–243.

Brockner, J., Chen, Y., Mannix, E. A., Leung, K., & Skarlicki, D. P. (2000). Culture and procedural fairness: When the effects of what you do depend on how you do it. *Administrative Science Quarterly, 45,* 138–159.

Brockner, J., Tyler, T. R., & Cooper-Schneider, R. (1992). The influence of prior commitment to an institution on reactions to perceived unfairness: The higher they are, the harder they fall. *Administrative Quarterly, 37,* 241–261.

Brosnan, S. F., & de Waal, F. B. M. (2003). Monkeys reject equal pay. *Nature, 425,* 297–299.

Bugental, J. F. T., & Zelen, S. L. (1950). Investigations into the “self-concept”: I. The W-A-Y technique. *Journal of Personality, 18,* 483–498.

Carver, C. S., & Scheier, M. F. (1998). *On the self-regulation of behavior.* New York: Cambridge University Press.

Chen, Y., Brockner, J., & Greenberg, J. (2003). When is it a “pleasure to do business with you”? The effects of status, outcome favorability, and procedural fairness. *Organizational Behavior and Human Decision Processes, 92,* 1–21.

Clayton, S., & Opotow, S. (2003). Justice and identity: Changing perspectives on what is fair. *Personality and Social Psychology Review, 7,* 298–310.

Cook, K. S. (1975). Expectations, evaluations, and equity. *American Sociological Review, 40,* 372–388.

Cropanzano, R., Byrne, Z. S., Bobocel, D. R., & Rupp, D. E. (2001). Moral virtues, fairness heuristics, social entities, and other denizens of organizational justice. *Journal of Vocational Behavior, 58,* 164–209.

De Cremer, D., & Blader, S. L. (2006). Why do people care about procedural fairness? The importance of belongingness in responding and attending to procedures. *European Journal of Social Psychology, 36,* 211–228.

Deutsch, M. (1975). Equity, equality, and need: What determines which value will be used as the basis of distributive justice. *Journal of Social Issues, 31*, 137–149.

Deutsch, M. (1985). *Distributive justice: A social psychological perspective.* New Haven, CT: Yale University Press.

De Waal, F. (1996). *Good natured: The origins of right and wrong in humans and other animals.* Cambridge, MA: Harvard University Press.

Drake, D. G., & Lawrence, J. A. (2000). Equality and distributions of inheritance in families. *Social Justice Research, 13*, 271–290.

Emmons, R. A. (1986). Personal strivings: An approach to personality and subjective well-being. *Journal of Personality and Social Psychology, 51*, 1058–1068.

Fehr, E., & Fischbacher, U. (2004). Third-party punishment and social norms. *Evolution and Human Behavior, 25*, 63–87.

Fehr, E., & Schmidt, K. M. (1999). A theory of fairness, competition, and cooperation. *The Quarterly Journal of Economic Behavior, 114*, 817–868.

Fiske, A. P. (1991). *Structures of social life: The four elementary forms of human relations.* New York: Free Press.

Folger, R. (2001). Fairness as deonance. In S. W. Gilliland, D. D. Steiner, & D. P. Skarlicki (Eds.), *Research in social issues in management* (Vol. 1, pp. 3–33). Greenwich, CT: Information Age.

Folger, R., Cropanzano, R., & Goldman, B. (2005). What is the relationship between justice and morality? In J. Greenberg & J. Colquitt (Eds.), *Handbook of organizational justice research* (pp. 218–248). Mahwah, NJ: Lawrence Erlbaum.

Greenberg, J. (1990). Employee theft as a reaction to underpayment inequity. *Journal of Applied Psychology, 75*, 561–568.

Haidt, J. (2001). The emotional dog and its rational tail: A social intuitionist approach to moral judgment. *Psychological Review, 108*, 814–834.

Haidt, J. (2003). The moral emotions. In R. J. Davidson, K. R. Scherer, & H. H. Goldsmith (Eds.), *Handbook of affective sciences* (pp. 852–870). New York: Oxford University Press.

Haidt, J., Koller, S. H., & Dias, M. G. (1993). Affect, culture, and morality, or is it wrong to eat your dog? *Journal of Personality and Social Psychology, 65*, 613–628.

Hastorf, A., & Cantril, H. (1954). They saw a game: A case study. *Journal of Abnormal and Social Psychology, 49*, 129–134.

Higgins, E. T. (1987). Self-discrepancy: A theory relating self and affect. *Psychological Review, 94*, 319–340.

Holmvall, C. M., & Bobocel, D. R. (2008). What fair procedures say about me: Self-construals and reactions to procedural fairness. *Organizational Behavior and Human Decision Processes, 105*, 147–168.

Homans, G. C. (1961). *Social behavior: Its elementary forms.* New York: Harcourt Brace.

Huo, Y. J., Smith, H. J., Tyler, T. R., & Lind, E. (1996). Superordinate identification, subgroup identification, and justice concerns: Is separatism the problem; is assimilation the answer? *Psychological Science, 7*, 40–45.

James, W. (1890). *The principles of psychology.* New York: Holt.

Jonas, E., Schimel, J., Greenberg, J., & Pyszczynski, T. (2002). The Scrooge effect: Evidence that mortality salience increases prosocial attitudes and behavior. *Personality and Social Psychology Bulletin, 28*, 1342–1353.

Jowett, B. (1999). *Plato: The republic.* New York: Barnes and Noble.

Kim, T. Y., & Leung, K. (2007). Forming and reacting to overall fairness: A cross-cultural comparison. *Organizational Behavior and Human Decision Processes, 104*, 83–95.

Kohlberg, L. W. (1973). The claim to moral adequacy of a highest stage of moral development. *Journal of Philosophy, 70*, 630–646.

Konow, J. (1996). A positive theory of economic fairness. *Journal of Economic Behavior and Organization, 31*, 13–35.

Kuhn, T. S. (1962). *The structure of scientific revolutions.* Chicago: University of Chicago Press.

Lakatos, I. (1978). *The methodology of scientific research programmes: Philosophical papers* (Vol. 1). Cambridge, UK: Cambridge University Press.

Lane, R. E. (1986). Market justice, political justice. *American Political Science Review, 80*, 383–402.

Leary, M. R., Tambor, E. S., Terdal, S. K., & Downs, D. L. (1995). Self-esteem as an interpersonal monitor: The sociometer hypothesis. *Journal of Personality and Social Psychology, 68*, 518–530.

Lerner, M. J. (1977). The justice motive: Some hypotheses as to its origins and forms. *Journal of Personality, 45*, 1–52.

Lester, D. (2000). Is there a consistent pro-life attitude? *Advances in Psychology, 1*, 1–9.

Lind, E. A., & Tyler, T. R. (1988). *The social psychology of procedural fairness.* New York: Plenum.

Lupfer, M. B., Weeks, K. P., Doan, K. A., & Houston, D. A. (2000). Folk conceptions of fairness and unfairness. *European Journal of Social Psychology, 30*, 405–428.

Major, B., & Adams, J. B. (1983). Role of gender, interpersonal orientation, and self-presentation in distributive-justice behavior. *Journal of Personality and Social Psychology, 45*, 598–608.

Markus, H., & Kunda, Z. (1986). Stability and malleability of the self-concept. *Journal of Personality and Social Psychology, 51*, 858–866.

Markus, H., & Wurf, E. (1987). The dynamic self-concept: A social psychological perspective. *Annual Review of Psychology, 38*, 299–337.

Maslow, A. H. (1993). *The farther reaches of human nature.* New York: Penguin Books.

McFarlin, D. B., & Sweeney, P. D. (1992). Distributive and procedural justice as predictors of satisfaction with personal and organizational outcomes. *Academy of Management Journal, 35*, 626–637.

McGuire, W. J., McGuire, C. V., & Cheever, J. (1986). The self in society: Effects of social contexts on the sense of self. *British Journal of Social Psychology, 25*, 259–270.

Mikula, G., Petri, B., & Tanzer, N. (1990). What people regard as unjust. *European Journal of Social Psychology, 22*, 133–149.

Mullen, E., & Skitka, L. J. (2006). Exploring the psychological underpinnings of the moral mandate effect: Motivated reasoning, identification, or affect? *Journal of Personality and Social Psychology, 90*, 629–643.

Nucci, L. P. (2001). *Education in the moral domain.* New York: Cambridge University Press.

Nucci, L. P. (1996). Morality and the personal sphere of actions. In E. S. Reed, E. Turiel, & T. Brown (Eds.), *Values and knowledge* (pp. 41–60). Mahwah, NJ: Erlbaum.

Pittman, T. S., & Darley, J. M. (2003). The psychology of compensatory and retributive justice. *Personality and Social Psychology Review, 7*, 324–336.

Platow, M. J., & von Knippenberg, D. A. (2001). A social identity analysis of leadership endorsement: The effects of leader in-group prototypicality and distributive intergroup fairness. *Personality and Social Psychology Bulletin, 27*, 1508–1519.

Prentice, D. A., & Crosby, F. (1987). The importance of context for assessing deservingness. In J. C. Masters & W. P. Smith (Eds.), *Social comparison, social justice, and relative deprivation: Theoretical, empirical, and policy perspectives* (pp. 165–182). Hillsdale, NJ: Lawrence Erlbaum.

Rainwater, L. (1974). *What money buys: Inequality and the social meaning of income.* New York: Basic Books.

Reed, A., II, & Aquino, K. (2003). Moral identity and the expanding circle of moral regard toward out-groups. *Journal of Personality and Social Psychology, 84*, 1270–1286.

Rentsch, J. R., & Heffner, T. S. (1994). Assessing self-concept: Analysis of Gordon's coding scheme using "Who Am I?" responses. *Journal of Social Behavior and Personality, 9*, 283–300.

Rest, J. R., Narvaez, D., Bebeau, M. J., & Thoma, S. J. (1999). *Postconventional moral thinking: A neo-Kohlbergian approach.* Mahwah, NJ: Lawrence Erlbaum.

Robinson, P. H., Kurzban, R., & Jones, O. D. (2007). The origins of shared intuitions of justice. *Vanderbilt Law Review, 60*, 1634–1688.

Rusbult, C. E. (1983). A longitudinal test of the investment model: The development (and deterioration) of satisfaction and commitment in homosexual involvements. *Journal of Personality and Social Psychology, 45*, 101–117.

Showers, C. J. (2002). Integration and compartmentalization: A model of self-structure and self-change. In D. Cervone & W. Mischel (Eds.), *Advances in personality science* (pp. 271–291). New York: Guilford.

Skarlicki, D. P., & Folger, R. (1997). Retaliation in the workplace: The roles of distributive, procedural, and interactional justice. *Journal Applied Psychology, 82*, 434–443.

Skitka, L. J. (2002). Do the means always justify the ends or do the ends sometimes justify the means? A value protection model of justice reasoning. *Personality and Social Psychology Bulletin, 28*, 588–597.

Skitka, L. J. (2003). Of different minds: An accessible identity model of justice reasoning. *Personality and Social Psychology Review, 7*, 286–297.

Skitka, L. J. (2006). *Legislating morality: How deep is the U.S. Supreme Court's reservoir of good will?* Paper presented at the meeting of the International Society for Justice Research, Berlin, Germany.

Skitka, L. J., & Bauman, C. W. (2008). Is morality always an organizational good? A review of morality in the context of organizational justice theory and research. In S. W. Gilliland, D. D. Steiner, & D. P. Skarlicki (Eds.), *Justice, morality, and social responsibility: Research in social issues in management* (Vol. 6). Greenwich, CT: Information Age.

Skitka, L. J., Bauman, C. W., & Mullen, E. (2008). Morality and justice: An expanded theoretical perspective and review. In K. A. Hedgvedt & J. Clay-Warner (Eds.), *Advances in group processes* (Vol. 25, pp. 1–27). Bingley, UK: Emerald Group.

Skitka, L. J., Bauman, C. W., & Sargis, E. G. (2005). Moral conviction: Another contributor to attitude strength or something more? *Journal of Personality and Social Psychology, 88*, 895–917.

Skitka, L. J., & Houston, D. (2001). When due process is of no consequence: Moral mandates and presumed defendant guilt or innocence. *Social Justice Research, 14*, 305–326.

Skitka, L. J., & Mullen, E. (2002). Understanding judgments of fairness in a real-world political context: A test of the value protection model of justice reasoning. *Personality and Social Psychology Bulletin, 28*, 1419–1429.

Skitka, L. J., Bauman, C. W., & Lytle, B. L. (in press). The limits of legitimacy: Moral and religious convictions as constraints on deference to authority. *Journal of Personality and Social Psychology.*

Smetana, J. (1993). Understanding of social rules. In M. Bennett (Eds.), *The development of social cognition: The child as psychologist.* New York: Guilford.

Sprecher, S., & Schwartz, P. (1994). Equity and balance in the exchange of contributions in close relationships. In M. J. Lerner & G. Mikula (Eds.), *Entitlement and the affectional bond: Justice in close relationships* (pp. 11–43). New York: Plenum.

Steele, C. (1988). The psychology of self-affirmation: Sustaining the integrity of the self. In L. Berkowitz (Ed.), *Advances in experimental social psychology* (Vol. 21, pp. 261–302). New York: Academic Press.

Steele, C. M. (1999). The psychology of self-affirmation: Sustaining the integrity of the self. In R. F. Baumeister (Ed.), *The self in social psychology: Key readings in social psychology* (pp. 372–390). Philadelphia: Psychology Press/Taylor & Francis.

Tetlock, P. E. (2002). Social functionalist frameworks for judgment and choice: Intuitive politicians, theologians, and prosecutors. *Psychological Review, 109*, 451–471.

Thibaut, J., & Walker, J. (1975). *Procedural justice: A psychological analysis*. Hillsdale, NJ: Lawrence Erlbaum.

Tisak, M. (1995). Domains of social reasoning and beyond. In R. Vasta (Ed.), *Annals of child development* (Vol. 11). London: Jessica Kingsley.

Turiel, E. (2002). *The culture of morality: Social development, context, and conflict.* Cambridge, UK: Cambridge University Press.

Turner, J. C. (1985). Social categorization and the self-concept: A social cognitive theory of group behavior. In E. J. Lawler (Ed.), *Advances in group processes: Theory and research* (Vol. 2, pp. 77–122). Greenwich, CT: JAI Press.

Turner, J. C. (1999). Some current issues in research on social identity and self-categorization theories. In N. Ellemers, R. Spears, & B. Doosje (Eds.), *Social identity* (pp. 6–34). Malden, MA: Blackwell.

Tyler, T. R. (1989). The psychology of procedural justice: A test of the group value model. *Journal of Personality and Social Psychology, 57*, 850–863.

Tyler, T. R., & Blader, S. L. (2003). The group-engagement model: Procedural justice, social identity, and cooperative behavior. *Personality and Social Psychology Review, 7*, 349–361.

Tyler, T. R., & Lind, E. A. (1992). A relational model of authority in groups. In M. Zanna (Ed.), *Advances in experimental social psychology* (Vol. 4, pp. 595–629). Boston: McGraw-Hill.

Van Prooijen, J. W., van den Bos, K., & Wilke, H. A. M. (2002). Procedural justice and status: Status salience as antecedent of procedural fairness effects. *Journal of Personality and Social Psychology, 83*, 1353–1361.

Walster, E., Walster, G. W., Berscheid, E., & Austin, W. (1978). *Equity: Theory and research*. Boston: Allyn and Bacon.

Watts, B. L., Messé, L. A., & Vallacher, R. R. (1982). Toward understanding sex differences in pay allocation: Agency, communion, and reward distribution behavior. *Sex Roles, 8*, 1175–1187.

Wenzel, M. (2000). Justice and identity: The significance of inclusion for perceptions of entitlement and the justice motive. *Personality and Social Psychology Bulletin, 26*, 157–176.

Zhong, C. B., & Liljenquist, K. (2006). Washing away your sins: Threatened morality and physical cleansing. *Science, 8*, 1451–1452.

CHAPTER 2

Injustice and Identity

How We Respond to Unjust Treatment Depends on How We Perceive Ourselves

D. RAMONA BOBOCEL
University of Waterloo

AGNES ZDANIUK
University of Guelph

Abstract: People vary greatly in how they respond toward transgressors—from forgiving to seeking revenge. What determines how victims respond to those who have treated them unfairly? In this chapter, the authors argue that victims' reactions to their transgressors are guided by how victims define the self. In the first part of the chapter, the authors discuss the theoretical rationale underlying the central premise. They then summarize supportive evidence from a number of their studies that demonstrates that people's reactions toward transgressors differ systematically, depending on the strength of people's *independent* and *interdependent* self-construals. Taken together, these findings are consistent with the idea that reactions to unjust treatment can be motivated by the victim's desire to restore positive self-regard following threat. In the final section, the authors discuss the theoretical implications of the research and the implications for future work on the study of justice.

Almost all people can recall a time when they felt unfairly treated. People differ, however, in how they react toward their transgressors. Perhaps the most destructive response is to retaliate or seek revenge. Revenge is defined as an "action taken in response to a perceived harm or wrongdoing by another person that is intended to inflict harm, damage, discomfort or injury to the party judged responsible" (Aquino, Tripp, & Bies, 2001, p. 53; see also Vidmar, 2001). Revenge is associated with negative consequences not only for the offender but also for the victim, as well as for their relationship. For example, seeking revenge has been associated with lower life satisfaction and well-being (e.g., McCullough, 2001) and with greater difficulty maintaining interpersonal relationships (Rose & Asher, 1999).

In contrast, one of the most constructive responses is to forgive. Although there is some disagreement on the definition of forgiveness, McCullough, Pargament, and Thorensen (2000) observed one critical assumption: "When people forgive, their responses toward (or, in other words, what they think of, feel about, want to do to, or actually do to) people who have offended or injured them become more positive and less negative" (p. 9). Research has identified various benefits of forgiveness for the victim, the offender, and their relationship. For example, forgiveness is associated with greater mental and psychological well-being (see Toussaint & Webb, 2005, for a recent review), greater physical health (Witvliet, Ludwig, & Van der Laan, 2001), higher life satisfaction (e.g., McCullough, 2001), and lower depression (e.g., Brown, 2003; Krause & Ellison, 2003). Forgiveness has been positively associated with a willingness to accommodate the offender and to sacrifice on the offender's behalf (Karremans & Van Lange, 2004). Research has even shown that forgiveness can lead to a generalized prosocial orientation that extends beyond the victim–offender relationship (Karremans, Van Lange, & Holland, 2005).

What determines how victims respond toward those who have treated them unfairly, in particular whether victims seek revenge or forgive? In this chapter, we review one line of our recent research that focuses on the role of the victim's self-identity in moderating reactions toward a transgressor. We argue that whether people respond in a more interpersonally destructive manner, such as by retaliating or seeking revenge, or whether they respond more constructively, such as by forgiving, is influenced by how they define the self. In the sections that follow, we briefly discuss the theoretical rationale for our central premise, then we review the primary findings of several studies that we have conducted to examine our ideas, and finally we discuss the theoretical implications of our research and the implications for future justice research.

Part I: The Experience of Injustice and Its Relation to Self-Worth and Self-Identity

Historically, researchers have distinguished between the concepts of distributive injustice (i.e., perceived unfairness in the distribution of resources) and procedural injustice (i.e., perceived unfairness in the procedures and processes used to make decisions). In more recent years, research has demonstrated that people can also perceive unfairness in how they are treated interpersonally, such as when decision-making authorities or other interaction partners are insincere or disrespectful or violate interpersonal codes of conduct. These latter ideas are a core component of relational models of procedural justice (e.g., Lind, 2001; Lind & Tyler, 1988; Tyler & Lind, 1992) and are also embodied in the concept of interactional injustice (Bies, 2005; Bies & Moag, 1986). As argued by Miller (2001), disrespectful treatment can be perceived as unjust for two reasons: It can deprive people of something that they believe they are entitled to (i.e., respectful treatment), and it can create a social imbalance by subjecting people to something they do not deserve.

Several theories have been advanced to explain the psychological mechanisms underlying people's reactions to perceived unfairness in the distribution of outcomes, the processes and procedures used to make decisions, or interpersonal treatment. Although important differences among the theories exist, many share the assumption that the experience of unfairness threatens identity-relevant concerns. For example, according to equity theory, inequity in social exchanges threatens one's sense of deservingness, value, and self-esteem (Adams, 1965; Walster, Walster, & Berscheid, 1978). The group-value, relational, and group engagement models of procedural justice (e.g., Lind & Tyler, 1988; Tyler & Blader, 2003; Tyler & Lind, 1992) argue that people care about procedural fairness because it communicates information about the quality of their relationships with others, thus shaping their social, or group-based, identity, which ultimately has implications for their self-esteem. Likewise, according to fairness heuristic theory (Lind, 2001; Van den Bos, Lind, & Wilke, 2001), perceived unfairness is related to feelings of exploitation, rejection, and loss of identity. Consistent with these ideas, a number of studies have demonstrated that the experience of injustice is associated with diminished self-worth (e.g., Brockner et al., 2003; De Cremer, Van Knippenberg, Van Knippenberg, Mullenders, & Stinglhamber, 2005; Koper, Knippenberg, Bouhuijs, Vermunt, & Wilke, 1993; Schafer, 1988; Smith & Tyler, 1997; Smith, Tyler, Huo, Ortiz, & Lind, 1998; Tyler, Degoey, & Smith, 1996).

Thus, according to several theoretical approaches to the study of justice, and supported by the available empirical data, the experience of injustice can be conceptualized as a threat to self. Drawing on the idea that people are motivated to restore positive self-views following threat (e.g., Tesser, 2000), we reasoned that responses to injustice should differ systematically as a function of how victims *define the self*.

According to the literature on the self (e.g., Markus & Kitayama, 1991; Singelis, 1994; Triandis, 2001), people within any culture vary flexibly in the extent to which they define the self in terms of their *differences from others* and in terms of their *interconnections to others*. For individuals with a strong *independent* self-construal, self-identity is based on one's unique abilities or attributes and on the importance of distinguishing oneself from others. When conflict arises, these individuals use dominating conflict styles and seek to maximize their own interests over the needs of others. Most important for the present purposes, feelings of positive self-regard are derived from social dominance, achievement, and seeing oneself as different from—and better than—others.

In contrast, individuals with a strong *interdependent* self-construal emphasize their relationships and connections to others. They give group goals priority over individual goals, and their behaviors are determined by group norms, duties, and obligations. When conflict arises, they use cooperating conflict styles. Moreover, feelings of positive self-regard are derived from "belonging, fitting in, occupying one's proper place, promoting others' goals, and maintaining social harmony" (Markus & Kitayama, 1991, p. 242). Empirical research has confirmed the conceptualization of independent and interdependent self-identities as two distinct constructs rather than as two ends of a single continuum (e.g., Singelis, 1994; Singelis, Triandis, Bhawuk, & Gelfand, 1995).

Drawing on the literatures just outlined, we reasoned that whether people are inclined to seek revenge against or to forgive a transgressor will be guided by their self-construal. On the one hand, the goal of revenge is to inflict pain and suffering on the offender as a means of elevating oneself to a superior position (Hampton, 1988; Nozick, 1981). Given that individuals with a strong independent self-construal derive positive self-esteem from achievement and from outperforming others, we reasoned that these individuals would be more motivated than those with a weak independent self-construal to seek revenge following a transgression so as to restore a positive self-view. On the other hand, the goal of forgiveness is to maintain or establish a positive relationship with the offender (e.g., McCullough, 2000). Given that individuals with a strong interdependent self-construal derive positive self-esteem from belonging and maintaining social harmony, we reasoned that these individuals would be more motivated than those with a weak interdependent self-construal to forgive the transgressor so as to restore a positive self-view.[1]

Recent theorizing on revenge and forgiveness supports our conceptualization of these reactions as means by which different people may attempt to repair self-esteem threat following the experience of injustice. For example, Vidmar (2001) proposed a six-stage model of the psychological dynamics of revenge in which he theorized that perceptions of injustice threaten or harm victims' self-regard and that revenge may serve to return self-regard to homeostasis. Kim and Smith (1993) also suggested that the enactment of revenge can restore victims' degraded feelings of self-worth (for similar ideas, see Aquino, Tripp, & Bies, 2006; Bies, Tripp, & Kramer, 1997; Bradfield & Aquino, 1999; Folger & Skarlicki, 1998).

Correspondingly, research on forgiveness has demonstrated a positive relation between forgiveness and positive self-regard. For example, Freedman and Enright (1996) found that women incest survivors felt better about themselves following an intervention promoting forgiveness toward their abuser, as compared to women who had not yet received the intervention. Similarly, Struthers, Dupuis, and Eaton (2005) found that a psychological intervention aimed at promoting forgiveness in the workplace enhanced the self-image of workers, as compared to a control intervention. Karremans, Van Lange, Ouwerkerk, and Kluwer (2003) found that forgiving, compared to not forgiving, was associated with greater levels of state self-esteem, life satisfaction, and positive affect when the victim's commitment to the offender was strong.

On the basis of the foregoing analysis, we conceptualize revenge and forgiveness as different means by which people may respond to injustice—depending on their self-construal—in an effort to restore self-esteem. Consequently, at the empirical level we should observe (a) a significant association between the strength of people's independent self-construal and revenge—the more they define the self as unique and different from others, the more motivated they should be to seek revenge—and (b) a significant association between the strength of people's interdependent self-construal and forgiveness—the more they define the self as interconnected with others, the more motivated they should be to forgive. These relations are depicted in Figure 2.1. As should be clear from the preceding summary, we conceptualize revenge and forgiveness as separate constructs (rather than as opposite ends of a single continuum) that are uniquely associated with the people's independent and interdependent selves, respectively. We have tested these ideas in a series of studies conducted in laboratory settings and in the workplace setting, with converging results. Next, we review the primary findings from some of the key studies.

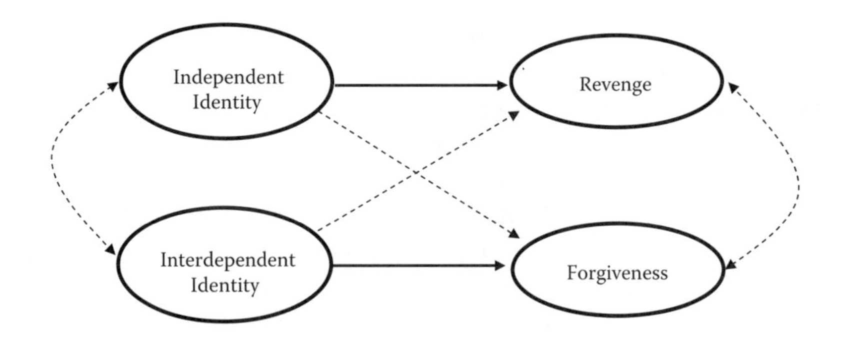

Figure 2.1 This model depicts the hypothesized relations between self-construal—independent and interdependent—and response—revenge versus forgiveness. The solid lines indicate significant relations; the dashed lines indicate nonsignificant relations.

Part II: Review of Our Research

Employee Study

We began with a critical incident study, in which working adults recounted an unfair experience and self-reported their reactions (see Bobocel & Zdaniuk, 2008; Zdaniuk & Bobocel, 2008). Participants were 101 employees (64 females and 37 males; M_{age} = 33.03, *SD* = 8.17) who were recruited at random from a database of University of Waterloo alumni to participate in a two-part Web survey of employee relations. At Time 1, we assessed the strength of employees' self-construals using the interdependent and independent self-construal scales developed and validated by Singelis and colleagues (Singelis et al., 1995).[2] Two example items from the interdependent self-construal scale are "I would sacrifice my self-interest for the benefit of the work group I am in" and "It is important for me to maintain harmony with co-workers" (1 = *strongly disagree*, 7 = *strongly agree*). Two example items from the vertical independent self-construal scale are "Winning is everything" and "When another person does better than I do, I get tense and aroused" (1 = *strongly disagree*, 7 = *strongly agree*).

Approximately 1 week later, participants were asked to describe a recent incident in which another person in their current organization treated them unfairly. Following their written account, participants completed a questionnaire containing measures to assess perceived unfairness of the event, as well as a number of potentially important contextual variables (discussion to follow). In addition, participants responded to multi-item scales taken from the literature to assess the extent to which they engaged in revenge and the extent to which they forgave the perpetrator. Forgiveness was assessed with McCullough and Hoyt's (2002) Benevolence Scale (α = .94, *M* = 4.16, *SD* = 1.54). Two items are "I forgave him/her for what he/she did" and "Despite what he/she did, I want us to 'bury the hatchet' and move forward with our relationship" (1 = *strongly disagree*, 7 = *strongly agree*). Revenge was assessed with McCullough et al.'s (1998) Transgression-Related Interpersonal Motivations (TRIM) Inventory (items were reworded to assess behavior; $\alpha = .73$, $M = 1.40$, $SD = 0.86$). Two items were "I got even with the offender" and "I did something to hurt the offender" (1 = *strongly disagree*, 7 = *strongly agree*).

Results As expected, overall participants perceived the workplace offense as relatively unfair ($M = 5.45$, $SD = 1.48$). Moreover, this perception was true regardless of the strength of their self-construals. That is, there was no significant correlation between perceived unfairness and the strength of employees' self-construals, either as independent ($r = -.08$) or interdependent ($r = -.19$, although in the latter case, there was an inverse trend). Finally, consistent with past research, the strength of participants' independent and interdependent self-views were not statistically intercorrelated ($r = .04$), and revenge and forgiveness were only weakly inversely related ($r = -.18$).

We conducted two simultaneous regression analyses to examine the hypothesized relations. In one, we regressed the two predictors (interdependent and independent self-construals) on revenge, and in another we regressed the same predictors on forgiveness. As expected, we found a significant positive, unique relation between independent self-construal and revenge ($\beta = .21$, $p < .05$), such that the more employees defined themselves as independent, the more they sought revenge. The relation between interdependent self-construal and revenge was nonsignificant ($\beta = -.01$). In contrast, in the analysis predicting forgiveness, we found a significant positive, unique relation between interdependent identity and forgiveness ($\beta = .24$, $p < .05$), such that the more employees defined themselves as interdependent, the greater was their forgiveness. The relation between independent self-construal and forgiveness was nonsignificant ($\beta = -.08$). These findings were entirely in line with our predictions.

As noted earlier, we had also measured a number of variables to rule out several possible third-variable explanations. Our goal was to demonstrate the incremental effect of employees' self-views after controlling for the possible effect of these alternative variables. On the basis of past research on predictors of revenge and forgiveness, we assessed (a) perceived severity of the offense, (b) time since the offense, (c) nature of the relationship with the transgressor, (d) whether the transgressor had attempted to redress the offense, (e) past unfair treatment by the transgressor, and (f) participant gender and age. Hierarchical regression analyses revealed that the hypothesized effects remained statistically significant after controlling for these variables, indicating that these other variables do not account for our findings.[3] Supplementary analyses also revealed that these variables do not moderate our central findings.

Examining Behavior

The advantage of the employee study is that participants were reporting on a transgression that had occurred in their own lives; therefore, the event was likely to be meaningful to them. The disadvantages, however, were that participants were recalling a past event and that they self-reported their reactions. Therefore, we conducted two complementary studies in the laboratory to directly examine people's responses to interactional unfairness.

Does Independence Predict Retaliation? In one study (see Zdaniuk & Bobocel, 2008), we examined the relation between the strength of one's independent identity and a behavioral index of retaliation. In the first phase, participants were treated unfairly, ostensibly by a fellow participant; in the second phase, participants had an opportunity to harm the transgressor by administering noise blasts, under the guise of an unrelated study on learning. In line with the employee study, we predicted that the stronger the participants' independent self-views, the more often they would harm the transgressor. We also included a control condition in which participants did not experience unfair treatment in the first phase, to rule out the possibility that independence predicts the delivery of noise blasts in this paradigm in the absence of the transgression.

Participants were introductory psychology students who participated for course credit. Early in the school term, they completed Singelis et al.'s (1995) self-construal scales, among a number of unrelated measures, as part of an online mass-testing questionnaire. About 3 weeks later, a random sample of these students was invited to the laboratory to participate in two ostensibly unrelated studies ($N = 81$, 49 females and 32 males; $M_{age} = 18.79$, $SD = 1.21$). Participants arrived in dyads and were led to believe that the purpose of the first study was to investigate the quality of group versus individual decision making. Participants then worked together on an involving decision-making task that required them to generate possible solutions to a workplace grievance.

After completing the task, participants were escorted to separate rooms and asked to evaluate each other's performance. In the injustice condition, the experimenter then ostensibly exchanged participants' evaluations. In reality, all participants received identical (fictitious) feedback, which was designed to be nonnormatively harsh and critical. In brief, participants read that their partner rated them as unlikeable and incompetent and as having put forth relatively less effort on the task. Drawing on past research on procedural and interactional justice (for reviews, see Bies, 2001; Miller, 2001), we expected that participants would perceive their evaluation to be both disrespectful and undeserved and therefore to be unfair. The results from an item included in what participants believed was a "poststudy" questionnaire indicated that overall participants did indeed perceive the evaluation as relatively unfair ($M = 5.62$, $SD = 1.23$; scale endpoints were 1 = *fair*, 7 = *unfair*). In the control condition, the experimenter did not exchange participants' evaluations.

After each participant ostensibly completed the first study, a different experimenter entered each participant's room and briefly introduced the second study.[4] Participants were told that the researchers were examining how aversive stimuli affect learning. Using a paradigm modeled on Milgram's obedience studies (e.g., Milgram, 1963; adapted from Bushman, 1995), the experimenters led the participants to believe that they had been randomly assigned to a "teacher" role and that other participants had been assigned to a "learner" role. In fact, all participants were assigned to the role of teacher, and the learner's responses were preprogrammed on a computer. Ostensibly, the learner would answer 10 verbal analogy questions, which were presented on a computer, and the teacher was to provide feedback by entering on the keyboard either *C* for correct responses or *I* for incorrect responses. On incorrect responses, the teacher could also deliver a burst of noise (which sounded like radio static). The experimenter then exited the participant's room, and the presumed learning trials began: The learner ostensibly responded incorrectly on 7 of 10 trials. The primary dependent variable was the number of times participants delivered a blast of noise in response to an incorrect trial.[5]

Results As before, preliminary analyses revealed that independence was not significantly correlated with perceived unfairness, indicating that participants in the injustice condition perceived the evaluation similarly regardless of their self-view. The primary regression analysis, however, revealed a significant interaction between independent self-construal and condition ($\beta = .44$, $p < .05$). As expected, independent self-construal predicted the frequency of noise blasts in the experimental condition, such that the stronger the participant's independent self-view, the more often they delivered a blast of noise to the transgressor ($\beta = .42$, $p < .01$). In contrast, there was no significant relation between independent self-construal and frequency of noise blasts in the control condition ($\beta = -.27$, $p > .10$).[6]

Finally, we examined whether *interdependent* self-construal had any relation to the frequency of noise blasts. Consistent with the results of the employee study, in which interdependence failed to predict self-reported revenge, interdependence was not significantly associated with the frequency of noise blasts in either condition (injustice condition, $\beta = .13$, $p > .10$; control condition, $\beta = .10$, $p > .10$). As in the employee study, the strength of participants' independent and interdependent self-construals were not significantly correlated ($r = -.03$).

Does Interdependence Predict Helping? Having conceptually replicated the relation between an independent identity and revenge that we observed in the employee study, we next focused on the relation between an interdependent identity and forgiveness (see Bobocel & Zdaniuk, 2008). Following past research that conceptualizes acts of good will toward the offender (especially when they also involve self-sacrifice) as a behavioral indicator of forgiveness (e.g., McCullough et al., 2000), we provided participants with an opportunity to help a transgressor and in so doing to forgo a reward.

As in the previous behavioral study, we first assessed students' self-construals in online mass testing using the measure of Singelis et al. (1995). About 3 weeks later, we invited a random sample of these students to the laboratory to participate in a study ostensibly examining cognitive processing strategies and decision-making ability (N = 73, 36 females and 37 males; M_{age} = 20.41, SD = 1.76). Each participant arrived at the session with another person whom they believed to be another student but who in fact was our confederate. It was explained that the two participants would first work on a task together, and then they would work individually. In between, they would complete an unrelated filler task purportedly to minimize possible "carry-over" effects. They were told that the filler task was an unrelated survey being conducted by another researcher in the department, for which they would receive $1 in exchange for participation. At this point, the confederate pulled out a few envelopes from her backpack in the presence of the participant and explained that she needed a few more surveys completed for a project in a leisure studies class; she asked the experimenter if it was possible for the participant to help her out. The experimenter simply responded that she would see what she could do, and she took a survey.

The participant and the confederate then worked together on a decision-making task. We developed a detailed script to ensure that the confederate responded in a standard fashion across all participants; moreover, she was blind to the hypotheses and to participants' self-views. In addition, the script ensured that the participant and confederate participated about equally, to bolster the perceived unfairness of the subsequent evaluation.

On completion of the decision-making task, the participant and confederate were taken to separate rooms and asked to evaluate each other's performance. As in the previous study, the experimenter then ostensibly exchanged the two evaluations. In reality, all participants received the overly harsh evaluation as before (presumably from the confederate).

As before, we expected participants to perceive the evaluation as unfair because it was disrespectful and in this case objectively undeserved.[7]

At this point, participants completed a short questionnaire, and then they were asked to complete the filler task. The experimenter also gave participants a second alternative, that is, if they preferred, they could complete the survey that the confederate asked about at the start of the session. The experimenter then left the participant alone with the two sealed envelopes (one was labeled "Psychology Department Survey" and contained a $1 coin). The primary dependent variable was whether participants volunteered to help the confederate by completing her survey and forgoing the $1 payment (0 = no, 1 = yes). (The confederate's survey required participants to respond to a series of questions about their preferred leisure activities; later examination of the surveys indicated that participants who chose to complete it did so authentically.)

Results Embedded in the questionnaire that participants completed after the group decision-making task were several items to assess perceived unfairness. Overall, participants perceived the evaluation to be relatively unfair ($M = 5.20$, $SD = 0.90$; scale endpoints were 1 = *fair*, 7 = *unfair*). Consistent with the earlier studies, perceived unfairness was not significantly associated with interdependence, indicating similar perceptions of the evaluation regardless of people's self-view. As expected, however, the primary regression analysis revealed a significant effect of interdependent self-construal on helping, such that the stronger the participant's interdependent identity, the more likely they were to help the offender ($\beta = .25$, $p < .05$).

Finally, we examined whether *independent* self-construal predicted helping (inversely). Consistent with the employee study, there was no significant effect ($\beta = -.15$, $p > .20$), although there was a slight trend. As before, the strength of participants' independent and interdependent self-construals were not significantly correlated ($r = .15$).

Manipulating Self-Identity

The preceding studies reveal a reliable predictive association between (a) the strength of people's independent self-construal and revenge or retaliation and (b) the strength of people's interdependent self-construal and forgiveness. The results were the same whether participants self-reported their behavior or we observed it in the laboratory. In these studies, we measured, rather than manipulated, self-construal. So we next conducted two parallel studies in which we experimentally raised the accessibility of either the independent self or the interdependent self in working memory and examined the effect on participants' reactions.

Priming the Independent Self In one study (see Zdaniuk & Bobocel, 2008), participants were 70 psychology students (38 females and 32 males; M_{age} = 19.36, *SD* = 1.69) who took part for credit in various psychology classes. The study was a two-group (independent prime versus control condition) between-subjects design. Participants were told that they would be completing a number of unrelated tasks being developed for subsequent use in various studies. Half of the participants—those in the independent prime condition—completed a scrambled-sentence task, which ostensibly assessed their language abilities. In reality, this task was intended to raise the accessibility of the independent self in working memory, without participant awareness (for validation, see Holmvall & Bobocel, 2008; for related manipulations, see Trafimow, Triandis, & Goto, 1991). Participants were asked to unscramble 15 five-word sentences, using four words in each. Ten sentences contained key words reflecting an independent self-construal (e.g., the sit sun like *I* → I like the sun; *mine* it the money is → the money is mine).

Following the priming task, or as their first task (in the control condition), participants were asked to think about a hypothetical situation depicting a violation of honor (Bies & Tripp, 2005) and to imagine how they would respond as the protagonist. Specifically, they read a short passage describing a situation in which a trusted coworker betrayed them by stealing their ideas on an important collaborative project. Participants then rated their motivation to seek revenge (using the measure of McCullough et al., 1998; 1 = *strongly disagree*, 7 = *strongly agree*).

Results As expected, ANOVA revealed a significant main effect of the prime, $F(1, 64) = 4.92$, $p < .05$, such that participants reported significantly greater revenge motivation in the independent prime condition ($M = 2.34$, $SD = 1.48$) as compared to the control condition ($M = 1.78$, $SD = 1.23$).

Priming the Interdependent Self In a parallel study (see Bobocel & Zdaniuk, 2008), we primed the interdependent self (versus a control condition) and assessed motivation to forgive. Participants were 60 psychology students (34 males and 26 females; M_{age} = 21.63, *SD* = 6.19) who took part for credit in a variety of courses. As previously, participants in the interdependent prime condition completed the 15-item scrambled sentence task; the 10 critical sentences, however, now contained key words relating to the interdependent self (e.g., the sit sun like *we* → *we* like the sun; *ours* it the money is → the money is *ours*; see Holmvall & Bobocel, 2008, for validation). After reading the same passage as in the preceding study, participants rated both (a) perceived unfairness (1 = *not at all unfair*, 7 = *very unfair*) and (b) motivation to forgive (using the McCullough & Hoyt, 2002, Benevolence Scale; 1 = *strongly disagree*, 7 = *strongly agree*).

Results We first examined whether the priming manipulation affected perceived unfairness. The two-group ANOVA was nonsignificant, $F(1, 64) < 1$. Thus, participants perceived the vignette to be equally unfair in the interdependent prime condition ($M = 6.35$, $SD = 0.85$) and the control condition ($M = 6.09$, $SD = 1.03$). As expected, however, we found a significant main effect of the priming manipulation on motivation to forgive, $F(1, 54) = 5.07$, $p < .05$. Participants reported significantly greater motivation to forgive the perpetrator in the interdependent prime condition ($M = 4.10$, $SD = 0.92$) as compared to the control condition ($M = 3.45$, $SD = 1.13$).

Summary So Far

Taken together, our data demonstrate that people's reactions toward transgressors are unquestionably guided by their self-view. In fact, we observed markedly different reactions as a function of self-view. The more people define themselves as independent and different from others, the greater their proclivity to seek revenge. In contrast, the more people define themselves as interdependent and connected to others, the greater their tendency to forgive.

Attempts to Repair a Transgression

Given that self-construals influence revenge and forgiveness, do they also influence how victims react to perpetrators' attempts to *repair* interpersonal transgressions? Recently, we have extended our research to investigate whether victims' self-construals moderate their reactions to the interpersonal apology. Although it is often said in both the research literature and the popular press that interpersonal apologies are an effective means by which an offender can foster forgiveness, the evidence is in fact mixed. Indeed, an apology can sometimes have a deleterious effect on victims' reactions (e.g., Conlon & Ross, 1997; Shaw, Wild, & Colquitt, 2003; Skarlicki, Folger, & Gee, 2004; for a review see Bobocel & Zdaniuk, 2005). Thus, research is needed to examine when, and for whom, interpersonal apologies will foster forgiveness following a transgression.

Given our earlier findings, it is conceivable that the victim's self-identity could play a role in determining the efficacy of interpersonal apologies. As noted earlier, those people with a strong interdependent self-identity gain positive self-regard from maintaining harmony within important relationships, and they are particularly attentive to others' needs. Accepting an apology is consistent with such communal orientations. In contrast, interpersonal apologies could be less effective for those with a stronger independent self-view. Those with a strong independent self-identity gain self-worth by asserting their autonomy and outperforming others when threatened. Accepting an apology is not consistent with these goals; in fact, accepting an apology could lead these individuals to feel submissive and possibly open to further exploitation. Thus, we reasoned that an interpersonal apology may have *greater* remedial value the stronger the victim's interdependent self-construal and *weaker* remedial value the stronger the victim's independent self-construal.[8]

We explored these ideas in two preliminary studies. In one study, we assessed students' interdependent and independent self-construals, using the same procedures as earlier. Three weeks later, a random subset of these students ($N = 60$, 32 females and 28 males; $M_{age} = 18.87$, $SD = 1.20$) was recruited through e-mail to participate in a study concerning interpersonal relations in the workplace. Participants read a story describing a derogation of status (Bies & Tripp, 2005) in which an employee was publicly ridiculed by a supervisor. Asked to imagine themselves as the employee, they then responded to a number of questions pertaining to the story. Of relevance here, they rated (a) perceived unfairness (1 = *not at all unfair*, 7 = *very unfair*) and (b) the extent to which an apology from the boss would remedy the situation (1 = *not at all*, 7 = *a great deal*).

Results Preliminary analyses revealed that overall participants perceived the event to be quite unfair ($M = 6.26$, $SD = 1.06$). As previously, independent and interdependent self-construals were not significantly interrelated ($r = -.03$). For the primary analysis, we simultaneously regressed interdependent and independent self-construals on participants' ratings of the remedial value of an apology. As expected, the results revealed a significant positive relation for interdependent self-construal ($\beta = .29$, $p < .05$) but a significant negative relation for independent self-construal ($\beta = -.28$, $p < .05$). Thus, the stronger the person's interdependent self-construal, the more value they saw in an apology, but the stronger the person's independent self-construal, the less value they saw in an apology.

In a second study using the critical incident method described earlier, students first completed the self-construal scales in mass testing. A week later, students from a random subset ($N = 100$, 66 females and 34 males) were asked to describe a recent incident in which another person treated them unfairly. After providing a written account of the event, they answered a series of questions. Among these, they first rated perceived unfairness of the event (1 = *not at all unfair*, 7 = *very unfair*), and they indicated whether they had received an apology from the perpetrator. Then they rated the extent to which an apology would (or did) make them forgive the offender (1 = *not at all*, 7 = *a great deal*). Finally, we asked participants to rate how accepting an apology would (or did) make them feel about themselves on two (7-point) bipolar scales with endpoints labeled *negative–positive* and *disrespected–respected*.

Results Overall participants perceived the event to be quite unfair ($M = 5.05$, $SD = 1.27$), and perceptions of unfairness were not related to the strength of participants' self-construals (interdependence $\beta = .02$, independence $\beta = .08$). Preliminary analyses also revealed that very few participants reported receiving an apology ($n = 16$), and self-construals did not predict responses to this question (interdependence $\beta = -.07$, independence $\beta = -.10$). Finally, unlike in our earlier studies, we found a weak positive relation between interdependent and independent self-construals ($r = .32$, $p < .01$). For the primary analyses, we conducted two simultaneous regression analyses—one on participants' ratings of the expected efficacy of an apology and another on participants' expected self-worth. Both analyses revealed a dissociative pattern as a function of self-construal, as predicted. The stronger the interdependent self-construal, the more an apology would foster forgiveness ($\beta = .20$, $p < .05$); in contrast, the stronger the independent self-construal, the *less* an apology would foster forgiveness ($\beta = -.22$, $p < .05$). Similarly, there was a marginally significant positive relation between interdependence and self-worth, such that the stronger the interdependent self-construal, the better participants expected to feel about themselves by accepting an apology ($\beta = .19$, $p = .09$). In contrast, the stronger the independent self-construal, the worse participants expected to feel ($\beta = -.25$, $p < .05$).

Part III: Summary and Implications

Our data suggest that people respond toward others who treat them unfairly in a manner that is consistent with how they construe the self. Those who define the self as interdependent and connected to others are motivated to forgive the perpetrator of an interpersonal transgression; in contrast, those who define the self as independent and different from others are motivated to seek revenge. Moreover, preliminary data suggest that people, as a function of their self-view, may react differently toward the perpetrator's attempt to redress the wrongdoing. In particular, an interpersonal apology may have greater remedial value the more that people define themselves as interdependent, but it may have the opposite effect the more that people define themselves as independent.

Theoretical Implications: The Role of Positive Self-Regard in Responding to Injustice

From a theoretical perspective, our findings support the idea that reactions to unfair treatment can be motivated by the victim's desire to restore positive self-regard following threat. Our studies shed light on this process by virtue of the moderating variables (interdependent and independent self-construal) that we examined. That is, we predicted that forgiveness would be more likely the stronger the victim's interdependent identity, precisely because such individuals derive positive self-esteem from maintaining social harmony. Similarly, we expected those with a stronger independent identity to be more likely to retaliate because they derive positive self-esteem from outperforming others (see Spencer, Zanna, & Fong, 2005, for a discussion of the role of "moderation-of-process" designs to test mediation).

Nevertheless, in two studies, we also examined the role of self-esteem maintenance in motivating participants' reactions more directly (i.e., using a "measurement-of-mediation" design; Spencer et al., 2005). For example, in the laboratory study in which we examined the association between independent self-identity and a behavioral index of retaliation, we measured participants' state self-esteem (using Heatherton & Polivy's [1991] measure of performance state self-esteem) at three points in the session: at the outset of the session to provide a baseline measure, immediately following the "decision-making study" in which participants in the injustice condition received the unfair evaluation (Phase A), and again on completion of the "aversive noise study" (Phase B). As expected, we found a significant interaction between condition and time on state self-esteem. In the injustice condition, participants' self-esteem declined following the transgression (compared to baseline) and then rose to baseline levels following the opportunity to retaliate. In contrast, in the control condition in which participants did not experience unfair treatment, self-esteem rose after the initial interaction between participants (Phase A) and remained at this same level after the aversive noise study (Phase B). Comparing the two conditions within each time, we found no significant difference in participants' self-esteem at baseline or after Phase B. The only significant difference occurred immediately following the experience of unfairness, such that self-esteem was significantly lower in the injustice condition than in the control condition. These data are consistent with our underlying assumption that the experience of unfair treatment can threaten people's feelings of positive self-regard.

In addition, to explore the idea that reactions to injustice may serve to restore self-worth, we examined—within the injustice condition—the correlation between the frequency with which participants delivered a noise blast to the perpetrator in Phase B and the degree to which their self-esteem recovered, as a function of their independent self-construal. There was a significant positive correlation among participants with a strong independent self-view ($r = .47$, $p < .01$) but not among participants with a weak independent self-view ($r = .17$, $p > .10$). Although these data are not definitive, they are consistent with the idea that retaliation among those with a strong independent self-construal was motivated by a need to preserve positive self-regard following threat.

In summary, our research program corroborates the ideas that the experience of injustice has important implications for people's identity and that researchers can elucidate important justice processes by considering differences in the source of people's feelings of positive self-regard.

Our results also have implications for theory and research on revenge and forgiveness more specifically. Although revenge and forgiveness constitute distinctly different reactions to perceived unfairness, our data are in line with the idea that they may be motivated by a common mechanism, namely, the restoration of positive self-regard. Thus, our findings are consistent with other research in social psychology that demonstrates a dual effect of self-enhancement motives. For example, research has demonstrated that self-threatened individuals can be more defensive and hostile and more likely to derogate or deceive others (e.g., Baumeister, Smart, & Boden, 1996; Fein & Spencer, 1997; Hodgins, Liebeskind, & Schwartz, 1996). Other research has demonstrated that self-threat can lead individuals to respond in more socially desirable ways, such as by conforming to others or doing favors for others (e.g., Leary, Twenge, & Quinlivan, 2006). Our findings suggest that depending on the source of people's self-esteem, their reaction toward a transgressor will differ dramatically.

Three caveats are worth noting. In the current research, we focused on reactions to violations of interpersonal codes of conduct (although employees in the field study did also describe procedural and distributive justice violations). Thus, it is not clear whether our findings would generalize to injustice that is more severe or far-reaching in scope, such as being the victim of discrimination or being unfairly laid off from one's job. Similarly, in all of our studies, the offender was someone with whom the participant shared a common fate (e.g., in the employee study, the offender was either a coworker or a supervisor; in the behavioral study, the offender was ostensibly a fellow student). It therefore is unclear whether the forgiving orientation of individuals with a strong interdependent self-identity that we observed in our research would generalize to perpetrators who are considered to be a part of an out-group (e.g., see Markus & Kitayama, 1991). Finally, in the present research, we focused on understanding how people react to the threat to self-worth that may arise from the experience of injustice. Injustice can, however, threaten several needs or motives. In particular, much research has confirmed Lerner's just-world theory (e.g., Lerner, 1974, 1977), which argues that injustice can threaten people's need to believe in a just world and that people are motivated to respond to injustice in ways that preserve or maintain their just-world beliefs (see Callan & Ellard and Hafer & Gosse, both in this volume). As noted by Hafer and Gosse, it remains a challenge for future research to determine what motive or motives people are trying to fulfill when they respond to injustice.

Self-Identity and Justice Processes

More broadly, our data contribute to a growing body of studies that examine the intersection between people's self-identity and justice processes (e.g., Brockner, Chen, Mannix, Leung, Skarlicki, 2000; De Cremer & Tyler, 2005; Johnson, Selenta, & Lord, 2006; Skitka, 2003; Tyler & Blader, 2003). For example, Skitka (2003) recently proposed an accessible identity model (AIM) of justice, in which she argued that justice reasoning should vary as a function of which aspect of people's identity (i.e., their material, social, or personal identity) is either chronically accessible or activated in a particular situation (also see Skitka et al., in this volume). Clayton and Opotow (2003) also considered how justice reasoning can differ depending on whether one conceives of oneself as an individual or as a representative of a broader group.

Empirical investigations have primarily focused on the possible moderating effects of self-identity. For example, Brockner and De Cremer and colleagues have demonstrated that the positive effect of procedural justice on attitudes and behaviors, demonstrated repeatedly in past research, is more pronounced for people with a strong, rather than weak, interdependent self-construal (Brockner, De Cremer, Van den Bos, & Chen, 2005; also see Brockner et al., 2000). Johnson et al. (2006) further argued that the positive effects of distributive, procedural, and interactional justice on specific employee reactions are differentially moderated by different levels of the self-concept. Consistent with their reasoning, they found an interaction between procedural justice and the collective self-concept on employee reactions, such that procedural justice had a stronger positive effect on reactions when people's collective identity was salient. Similarly, there was a joint effect of interactional justice and the relational self-concept, such that interactional justice had a stronger positive effect on reactions when people's relational identity was salient.

In prior research in our laboratory, Holmvall and Bobocel (2008) extended this line of inquiry by demonstrating that procedural fairness, in the context of receiving a negative outcome, has different meanings to people, depending on their self-construal. Consistent with past findings, we found that the stronger a person's interdependent self-view, the stronger the fair process effect, that is, the more *positively* they respond following fair (versus unfair) procedures. In this case, fair procedures are reassuring because they signal respect for the recipient, a pattern consistent with relational theories of justice (e.g., Tyler & Lind, 1992) and other social identity approaches (e.g., Wenzel, 2002). In contrast, and more novel, the stronger people's independent self-view, the more *negatively* they responded to fair (versus unfair) procedures. In this case, fair procedures are self-threatening because, drawing on attribution theory (e.g., Weiner, 1985), they imply that the negative outcome is deserved. Presumably, those with a strong independent self-view reacted defensively in an effort to maintain a positive view of the self. Holmvall and Bobocel also demonstrated that these effects of individual differences in the chronic accessibility of people's interdependent or independent selves can be mitigated by activating the alternate identity with a priming manipulation.

In summary, the current research adds to a growing body of work that suggests that investigators can shed considerable light on a number of justice processes by incorporating *differences* in how people define themselves. Whereas contemporary justice theory has emphasized relational and social identity concerns in justice processes (e.g., Lind & Tyler, 1988; Tyler & Lind, 1992; Wenzel, 2002), such concerns should be most relevant to people who have a strong interdependent self-view. Relational concerns should be of less relevance to those with a strong independent self-view. Thus, by examining justice processes through the lens of identity, investigators may learn more about the motives and processes underlying justice effects, as well as potentially reconcile previously inconsistent findings (see Holmvall & Bobocel, 2008).

Almost all of the empirical findings to date have focused on identity as a moderator of people's reactions to fair or unfair treatment. In line with Skitka's (2003) model, it would certainly be of interest to examine in future research whether identity influences how people form fairness judgments. Although in the current research we found no systematic relation between the strength of participants' self-construals (as either interdependent or independent) and their perceptions of unfairness, this may have been due to the fact that we asked participants to recall an unfair event in the field study or if we imposed unfair treatment on them in the laboratory studies. It remains to be examined therefore whether self-construal shapes people's perceptions of more ambiguous events.

Finally, it is also conceivable that the experience of fairness or unfairness could itself serve to raise the accessibility of victims' interdependent or independent selves, respectively. Consistent with this idea, in a recent laboratory study, Johnson and Lord (2007) found that unfair treatment activated the independent self, whereas fair treatment activated the interdependent self. In addition, the researchers found that self-identity mediated the effects of the manipulation of fairness on more distal reactions.

Finally, research in this vein should have implications for the growing literature on the cross-cultural study of justice (see Leung, 2005, for a review). As noted by Leung (2005), researchers do not yet have a coherent view of how culture shapes justice processes. Studying the effects of interdependent and independent self-construals *within* culture may help investigators understand differences *between* cultures that differ in their relative emphasis on collectivistic values (which promote an interdependent view of the self) and individualistic values (which promote an independent view of the self).

Conclusion

In this chapter, we explored how people respond to those who have treated them unfairly. Whereas undoubtedly there are numerous factors that contribute to victims' responses to transgressors, our research highlights the motivating role of the victim's self-identity and need to maintain positive self-regard. Taken together, the data suggest that how victims respond to those who have treated them unfairly is fundamentally shaped by how victims perceive themselves.

Notes

1. Singelis et al. (1995) distinguished between two facets of independent self-construal: *horizontal*, where the self is defined as autonomous and there is acceptance of equality between individuals, and *vertical*, where the self is defined as autonomous and there is acceptance of inequality between individuals. Vertical independence represents the blending of individualist values and achievement orientation with an emphasis on outperforming others (also see Triandis, 1996). The attainment of self-esteem from competing with and outperforming others is especially relevant for those with a strong versus weak *vertical* independent self-construal. Given this, we focus on the vertical dimension of independence in the present research. We do not make a distinction between horizontal and vertical *interdependence*, as this distinction has been found to be less important empirically (see Singelis et al., 1995).
2. We adapted the wording of the scale to fit the workplace context.
3. In the hierarchical analysis predicting revenge, there was a significant unique effect of perceived severity ($\beta = .28$, $p < .01$) in addition to independent self-construal. In the hierarchical analysis predicting forgiveness, there was a significant unique effect of perceived severity ($\beta = -.25$, $p < .05$) and of past unfair treatment by the transgressor ($\beta = -.21$, $p < .05$) in addition to interdependent self-construal.
4. Detailed instructions were presented to participants by audiotape and computer.
5. In each of the laboratory studies, we probed participants for suspicion about any aspect of the procedure using a funneled debriefing procedure (Bargh, Chen, & Burrows, 1996). The true purpose of the study and the deceptions involved were then explained fully using process debriefing (Ross, Lepper, & Hubbard, 1975).
6. In this study, we also examined whether our observed effect of independent self-construal remained significant after controlling for the effect of several individual difference variables that have been shown to predict revenge in past research. In mass testing, we included validated measures of moral identity,

self-esteem, agreeableness, consciousness, extraversion, and neuroticism. As expected, independent self-construal predicted behavior significantly regardless of whether these alternative variables were in the regression analysis.

7. We did not include a "no injustice" control condition in this study.
8. Note that in the employee study described earlier, we measured whether the offender had apologized or otherwise attempted to redress the offense. We found no significant relations between employees' self-construals and their responses to these questions; moreover, our primary results were not accounted for by differential behavior of the offender.

References

Adams, J. S. (1965). Inequity in social exchange. In L. Berkowitz (Ed.), *Advances in experimental social psychology* (Vol. 2, pp. 267–299). New York: Academic Press.

Aquino, K., Tripp, T. M., & Bies, R. J. (2001). How employees respond to personal offense: The effects of blame attribution, victim status, and offender status on revenge and reconciliation in the workplace. *Journal of Applied Psychology, 86*, 52–59.

Aquino, K., Tripp, T. M., & Bies, R. J. (2006). Getting even or moving on? Power, procedural justice, and types of offense as predictors of revenge, forgiveness, reconciliation, and avoidance in organizations. *Journal of Applied Psychology, 91*, 653–668.

Bargh, J. A., Chen, M., & Burrows, L. (1996). Automaticity of social behavior: Direct effects of trait construct and stereotype activation on action. *Journal of Personality and Social Psychology, 71*, 230–244.

Baumeister, R. F., Smart, L., & Boden, J. M. (1996). Relation of threatened egotism to violence and aggression: The dark side of high self-esteem. *Psychological Review, 103*, 5–33.

Bies, R. J. (2005). Are procedural justice and interactional justice conceptually distinct? In J. Greenberg & J. A. Colquitt (Eds.), *Handbook of organizational justice* (pp. 85–112). Mahwah, NJ: Lawrence Erlbaum.

Bies, R. J. (2001). Interactional (in)justice: The sacred and the profane. In J. Greenberg & R. Cropanzano (Eds.), Advances in organizational justice. CA: Stanford University Press.

Bies, R. J., & Moag, J. F. (1986). Interactional justice: Communication criteria of fairness. In R. J. Lewicki, B. H. Sheppard, & M. H. Bazerman (Eds.), *Research on negotiation in organizations* (Vol. 1, pp. 43–55). Greenwich, CT: JAI Press.

Bies, R. J., & Tripp, T. M. (2005). The study of revenge in the workplace: Conceptual, ideological, and empirical issues. In S. Fox & P. E. Spector (Eds.), *Counterproductive work behavior: Investigations of actors and targets* (pp. 65–81). Washington, DC: American Psychological Association.

Bies, R. J., Tripp, T. M., & Kramer, R. M. (1997). At the breaking point: Cognitive and social dynamics of revenge in organizations. In R. A. Giacalone & J. Greenberg (Eds.), *Antisocial behavior in organizations* (pp. 18–36). Thousand Oaks, CA: Sage.

Bobocel, D. R., & Zdaniuk, A. (2005). How can explanations be used to foster organizational justice? In J. Greenberg & J. A. Colquitt (Eds.), *Handbook of organizational justice* (pp. 469–498). Mahwah, NJ: Lawrence Erlbaum.

Bobocel, D. R., & Zdaniuk, A. (2008). *Forgiving injustice: The importance of perceiving ourselves as connected to others.* Manuscript submitted for publication.

Bradfield, M., & Aquino, K. (1999). The effects of blame attributions and offender likableness on forgiveness and revenge in the workplace. *Journal of Management, 25*, 607–631.

Brockner, J., Chen, Y., Mannix, E. A., Leung, K., & Skarlicki, D. P. (2000). Culture and procedural fairness: When the effects of what you do depend on how you do it. *Administrative Science Quarterly, 45*, 138–159.

Brockner, J., De Cremer, D., Van den Bos, K., & Chen, Y.-R. (2005). The influence of interdependent self-construal on procedural fairness judgments. *Organizational Behavior and Human Decision Processes*, *96*, 155–167.

Brockner, J., Heuer, L., Magner, N., Folger, R., Umphress, E., Van den Bos, K., Vermunt, R., Magner, M., & Siegel, P. (2003). High procedural fairness heightens the effect of outcome favorability on self-evaluations: An attributional analysis. *Organizational Behavior and Human Decision Processes, 91*, 51–68.

Brown, R. P. (2003). Measuring individual differences in the tendency to forgive: Construct validity and links with depression. *Personality and Social Psychology Bulletin, 29*, 759–771.

Bushman, B. J. (1995). Moderating role of trait aggressiveness in the effects of violent media on aggression. *Journal of Personality and Social Psychology, 69*, 950–960.

Clayton, S., & Opotow, S. (2003). Justice and identity: Changing perspectives on what is fair. *Personality and Social Psychology Review*, *4*, 298–310.

Conlon, D. E., & Ross, W. H. (1997). Appearances do account: The effects of outcomes and explanations on disputant fairness judgments and supervisory evaluations. *International Journal of Conflict Management*, *8*, 5–31.

De Cremer, D., Van Knippenberg, B., Van Knippenberg, D., Mullenders, D., & Stinglhamber, F. (2005). Rewarding leadership and fair procedures as determinants of self-esteem. *Journal of Applied Psychology*, *90*, 3–12.

De Cremer, D., & Tyler, T. (2005). Managing group behavior: The interplay between procedural justice, sense of self, and cooperation. In M. P. Zanna (Ed.). Advances in experimental social psychology (Vol. 37, pp. 151–218). London: Elsevier Academic Press.

Fein, S., & Spencer, S. J. (1997). Prejudice as self-image maintenance: Affirming the self through derogating others. *Journal of Personality and Social Psychology, 73*, 31–44.

Folger, R., & Skarlicki, D. P. (1998). A popcorn metaphor for employee aggression. In R. Griffin, A. O'Leary-Kelly, & J. Collins (Eds.), *Dysfunctional behavior in organizations: Violent and deviant behavior* (pp. 43–81). Stamford, CT: JAI Press.

Freedman, S. R., & Enright, R. D. (1996). Forgiveness as an intervention goal with incest survivors. *Journal of Consulting and Clinical Psychology*, *64*, 983–992.

Hampton, J. (1988). Forgiveness, resentment and hatred. In J. G. Murphy & J. Hampton (Eds.), *Forgiveness and mercy* (pp. 35–87). New York: Cambridge University Press.

Heatherton, T. F., & Polivy, J. (1991). Development and validation of a scale for measuring state self-esteem. *Journal of Personality and Social Psychology*, *60*, 895–910.

Hodgins, H. S., Liebeskind, E., & Schwartz, W. (1996). Getting out of hot water: Facework in social predicaments. *Journal of Personality and Social Psychology*, *71*, 300–314.

Holmvall, C. M., & Bobocel, D. R. (2008). What fair procedures say about me: Self-construals and reactions to procedural justice. *Organizational Behavior and Human Decision Processes*, *105*, 147–168.

Johnson, R. E., & Lord, R. G. (2007). *The implicit effects of (un)fairness on self-concept: Unconscious shifts in identity levels.* Paper presented at the annual meeting of the Society for Industrial and Organizational Psychology, New York.

Johnson, R. E., Selenta, C., & Lord, R. G. (2006). When organizational justice and the self-concept meet: Consequences for the organization and its members. *Organizational Behavior and Human Decision Processes*, *99*, 175–201.

Karremans, J. C., & Van Lange, P. A. M. (2004). Back to caring after being hurt: The role of forgiveness. *European Journal of Social Psychology*, *34*, 207–227.

Karremans, J. C., Van Lange, P. A. M., & Holland, R. W. (2005). Forgiveness and its associations with prosocial thinking, feeling, and doing beyond the relationship with the offender. *Personality and Social Psychology Bulletin*, *31*, 1315–1326.

Karremans, J. C., Van Lange, P. A. M., Ouwerkerk, J. W., & Kluwer, E. S. (2003). When forgiving enhances psychological well-being: The role of interpersonal commitment. *Journal of Personality and Social Psychology*, *84*, 1011–1026.

Kim, S. H., & Smith, R. H. (1993). Revenge and conflict escalation. *Negotiation Journal*, *9*, 37–43.

Koper, G., Knippenberg, D. V., Bouhuijs, F., Vermunt, R., & Wilke, H. (1993). Procedural fairness and self-esteem. *European Journal of Social Psychology*, *23*, 313–325.

Krause, N., & Ellison, C. G. (2003). Forgiveness by God, forgiveness by others, and psychological well-being in late life. *Journal for the Scientific Study of Religion*, *42*, 77–93.

Leary, M. R., Twenge, J. M., & Quinlivan, E. (2006). Interpersonal rejection as a determinant of anger and aggression. *Personality and Social Psychology Review*, *10*, 111–132.

Lerner, M. J. (1974). The justice motive: "Equity" and "parity" among children. *Journal of Personality and Social Psychology*, *29*, 539–550.

Lerner, M. J. (1977). The justice motive: Some hypotheses as to its origins and forms. *Journal of Personality*, *45*, 1–52.

Leung, K. (2005). How generalizable are justice effects across cultures? In J. Greenberg & J. A. Colquitt (Eds.), *Handbook of organizational justice* (pp. 555–586). Mahwah, NJ: Lawrence Erlbaum.

Lind, E. A. (2001). Fairness heuristic theory: Justice judgments as pivotal cognitions in organizational relations. In J. Greenberg & R. Cropanzano (Eds.), *Advances in organizational justice* (pp. 56–88). Lexington, MA: New Lexington.

Lind, E. A., & Tyler, T. R. (1988). *The social psychology of procedural justice*. New York: Plenum.

Markus, H. R., & Kitayama, S. (1991). Culture and the self: Implications for cognition, emotion, and motivation. *Psychological Review, 98*, 224–253.

McCullough, M. E. (2000). Forgiveness as human strength: Theory, measurement, and links to well-being. *Journal of Social and Clinical Psychology, 19*, 43–55.

McCullough, M. E. (2001). Forgiving. In C. R. Snyder (Ed.), *Coping with stress: Effective people and processes* (pp. 93–113). New York: Oxford University Press.

McCullough, M. E., & Hoyt, W. T. (2002). Transgression-related motivational dispositions: Personality substrates of forgiveness and their links to the Big Five. *Personality and Social Psychology Bulletin, 28*, 1556–1573.

McCullough, M. E., Pargament, K. I., & Thorensen, C. E. (Eds.). (2000). *Forgiveness: Theory, research, and practice.* New York: Guilford.

McCullough, M. E., Rachal, K., Sandage, S. J., Worthington, E. L., Jr., Brown, S. W., & Hight, T. L. (1998). Interpersonal forgiving in close relationships: II. Theoretical elaboration and measurement. *Journal of Personality and Social Psychology, 75*, 1586–1603.

Milgram, S. (1963). Behavioral study of obedience. *Journal of Abnormal and Social Psychology, 67*, 371–378.

Miller, D. (2001). Disrespect and the experience of injustice. *Annual Review of Psychology, 52*, 527–553.

Nozick, R. (1981). *Philosophical explanations.* Cambridge, MA: Harvard University Press.

Rose, A. J., & Asher, S. R. (1999). Children's goals and strategies in response to conflicts within a friendship. *Developmental Psychology, 35*, 69–79.

Ross, L., Lepper, M. R., & Hubbard, M. (1975). Perseverance in self-perception and social perception: Biased attributional processes in the debriefing paradigm. *Journal of Personality and Social Psychology, 32*, 880–892.

Schafer, R. B. (1988). Equity/inequity, and self-esteem: A reassessment. *Psychological Reports, 63*, 637–638.

Shaw, J. C., Wild, E., & Colquitt, J. A. (2003). To justify or excuse? A meta-analytic review of the effects of explanations. *Journal of Applied Psychology, 88*, 444–458.

Singelis, T. M. (1994). The measurement of independent and interdependent self-construals. *Personality and Social Psychology Bulletin, 20*, 580–591.

Singelis, T. M., Triandis, H. C., Bhawuk, D. P. S., & Gelfand, M. J. (1995). Horizontal and vertical dimensions of individualism and collectivism: A theoretical and measurement refinement. *Cross-Cultural Research, 29*, 240–275.

Skarlicki, D. P., Folger, R., & Gee, J. (2004). When social accounts backfire: The exacerbating effects of a polite message or an apology on reactions to an unfair outcome. *Journal of Applied Social Psychology, 4*, 322–341.

Skitka, L. J. (2003). Of different minds: An accessible identity model of justice reasoning. *Personality and Social Psychology Review, 7*, 286–297.

Smith, H. J., & Tyler, T. R. (1997). Choosing the right pond: How group membership shapes self-esteem and group-oriented behavior. *Journal of Experimental Social Psychology, 33*, 146–170.

Smith, H. J., Tyler, T. R., Huo, Y. J., Ortiz, D. J., & Lind, E. A. (1998). The self-relevant implications of the group-value model: Group membership, self-worth, and treatment quality. *Journal of Experimental Social Psychology, 34*, 470–493.

Spencer, S. J., Zanna, M. P., & Fong, G. T. (2005). Establishing a causal chain: Why experiments are often more effective than meditational analyses in examining psychological process. *Journal of Personality and Social Psychology, 89*, 845–851.

Struthers, C. W., Dupuis, R., & Eaton, J. (2005). Promoting forgiveness among co-workers following a workplace transgression: The effects of social motivation training [*Special issue*]. *Canadian Journal of Behavioral Sciences, 37*, 299–308.

Tesser, A. (2000). On the confluence of self-esteem maintenance mechanisms. *Personality and Social Psychology Review, 4*, 290–299.

Toussaint, L., & Webb, J. R. (2005). Theoretical and empirical connections between forgiveness, mental health and well-being. In E. L. Worthington, Jr. (Ed.), *Handbook of forgiveness* (pp. 349–362). New York: Routledge.

Trafimow, D., Triandis, H. C., & Goto, S. G. (1991). Some tests of the distinction between the private self and the collective self. *Journal of Personality and Social Psychology, 60*, 649–655.

Triandis, H. C. (1996). The psychological measurement of cultural syndromes. *American Psychologist, 51*, 407–415.

Triandis, H. C. (2001). Individualism–collectivism and personality. *Journal of Personality, 69*, 907–924.

Tyler, T. R., & Blader, S. L. (2003). The group engagement model: Procedural justice, social identity, and cooperative behavior. *Personality and Social Psychology Review, 7*, 349–361.

Tyler, T. R., & De Cremer, D. (2005). Process-based leadership: Fair procedures and reactions to organizational change. *Leadership Quarterly, 16*, 529–545.

Tyler, T. R., Degoey, P., & Smith, H. (1996). Understanding why the justice of group procedures matters: A test of the psychological dynamics of the group-value model. *Journal of Personality and Social Psychology, 70*, 913–930.

Tyler, T. R., & Lind, E. A. (1992). A relational model of authority in groups. In M. Zanna (Ed.), *Advances in experimental social psychology* (Vol. 25, pp. 115–191). San Diego, CA: Academic Press.

Van den Bos, K., Lind, E. A., & Wilke, H. A. M. (2001). The psychology of procedural and distributive justice viewed from the perspective of fairness heuristic theory. In R. Cropanzano (Ed.), *Justice in the workplace: From theory to practice* (Vol. 2, pp. 49–66). Mahwah, NJ: Lawrence Erlbaum.

Vidmar, N. (2001). Retribution and revenge. In J. Sanders and V. L. Hamilton (Eds.), *Handbook of justice research in the law* (pp. 31–63). New York: Kluwer Academic.

Walster, E., Walster, G., & Berscheid, E. (1978). *Equity: Theory and research*. Boston: Allyn & Bacon.

Weiner, B. (1985). An attributional theory of achievement motivation and emotion. *Psychological Review, 92*(4), 548–573.

Wenzel, M. (2002). What is social about justice? Inclusive identity and group values as the basis of the justice motive. *Journal of Experimental Social Psychology, 38*, 205–218.

Witvliet, C. V. O., Ludwig, T. E., & Van der Laan, K. L. (2001). Granting forgiveness or harboring grudges: Implications for emotion, physiology, and health. *Psychological Science, 121*, 117–123.
Zdaniuk, A., & Bobocel, D. R. (2008). *Who is most likely to avenge an injustice, and why? The role of the independent self.* Manuscript submitted for publication.

CHAPTER 3

Beyond Blame and Derogation of Victims

Just-World Dynamics in Everyday Life

MITCHELL J. CALLAN
University of Western Ontario

JOHN H. ELLARD
University of Calgary

Abstract: Melvin Lerner's (1980) just-world theory proposes that people need to believe that the world is a just place where people get what they deserve. Over forty years of just-world research has compellingly demonstrated that people reinterpret their experiences to sustain a commitment to justice, with victim derogation and blame being the most researched examples of this process. Lerner and his colleagues' original conceptualization of the justice motive as an organizing principle in people's lives suggests the potential for manifestations of the concern for justice beyond victim derogation and blame. To that end, in this chapter the authors discuss a series of studies where they examined how the desire for justice—both for self and others—influences the ways people respond to and reconstruct more everyday occurrences of injustice. Specifically, in a first series of experiments (Studies 1 to 4), the authors explored people's reactions to the fates of others in terms of their causal explanations for events and what they remembered about the past. In Studies 5 to 7, the authors demonstrated that people's personal deservingness concerns can influence their desire to gamble and wanting of consumer goods.

In our everyday lives, we are often confronted with, and threatened by, the pain, suffering, and misfortunes of others. Indeed, many children live in poverty, people lose their loved ones to disease and disaster, and many otherwise innocent victims—including ourselves—are humiliated, bullied, discriminated against, and laid off. Melvin Lerner's (1977, 1980; Lerner, Miller, & Holmes, 1976) just-world theory posits that such episodes of undeserved suffering and misfortune are threatening because they violate the sense that the world is basically a just place where people generally get what they deserve. According to Lerner (1977), the "belief in a just world" (BJW) serves an adaptive function in that it enables people to commit to long-term goal pursuits with confidence that their investments for the future will result in the outcomes they deserve (see also Hafer, 2000b; Hafer, Begue, Choma, & Dempsey, 2005). Evidence that one's world is not a just place constitutes a threat to the viability of the BJW and motivates cognitive and behavioral reactions aimed at restoring a sense of justice. Indeed, from both experimental and individual difference approaches, a long tradition of just-world research has compellingly demonstrated that people often reinterpret their experiences of people and events to sustain a commitment to justice and deserving (Dalbert, 2001; Furnham, 2003; Hafer & Begue, 2005; Lerner, 1980).

Victim derogation and blame have been, and continue to be, the most frequently investigated strategies people employ to restore their faith in justice. In an early investigation, Lerner and Simmons (1966) found that observers of an ostensible shock victim derogated her character when they were unable to intercede on her behalf and when they believed her suffering would persist in a second experimental phase. Derogating the victim presumably enabled Lerner and Simmons's participants to maintain the sense that the victim *deserved* her fate precisely because of the sort of person she was construed to be: Bad things happen to bad people.

Dozens of studies since Lerner and Simmons's (1966) classic work have documented how people's concerns with justice can lead to the rejection of innocent victims. Indeed, a majority of the post-1980 just-world studies reviewed by Hafer and Begue (2005) involved assessments of reactions to victims. This continuing research focus of victim derogation and blame is, of course, completely warranted and understandable, as the human capacity for rejecting innocent victims is a troubling yet important social issue worthy of investigation. Indeed, derogating innocent victims and blaming them for their plight might, for example, serve as justifications for limiting social support for the poor (Cozzarelli, Wilkinson, & Tagler, 2001; Napier, Mandisodza, Andersen, & Jost, 2006) and reducing empathy for victims of illness (Gruman & Sloan, 1983).

At the same time, however, research that is heavily focused on one type of reaction to injustice can potentially limit our understandings of the various ways the theme of justice appears in people's lives. Lerner's (1977, 1980) original conceptualization of the justice motive as an organizing principle in people's lives suggests the potential for manifestations of just-world beliefs beyond victim derogation and blame. As Lerner and Whitehead (1980) noted,

> The concern with justice seems to provide a central and guiding theme in our lives, as well as in our societies. People set their goals and evaluate their own fates, seem capable of sacrificing almost all other resources and values in our society, all in the service of justice. (p. 228)

If the BJW represents a core set of functional assumptions that are essential for people's goal-seeking activities, then one might expect to find reactions to injustices occurring to self and others that are not limited to the rejection of victims. The research we report in this chapter is intended to add to the growing interest among just-world researchers in manifestations of the commitment to a just world other than victim derogation and blame. Important work on justice concerns and victimization of course continues (see Hafer & Begue, 2005), including investigations of reactions to perpetrators (Ellard, Miller, Baumle, & Olson, 2002). At the same time, other scholars have explored the implications of just-world beliefs for other aspects of daily life (see Ross & Miller, 2002). For instance, Dalbert's (2001) research has demonstrated that the BJW functions as a personal resource in fostering subjective well-being. Hafer (2000b) and Callan, Shead, and Olson (2009) explored the important links between the need to believe in a just world, long-term goal commitments, and the processes of delaying gratification. Callan, Powell, and Ellard (2007) found that the perceived tragedy of a victim's fate varied as a function of his or her physical attractiveness. These and other studies have extended the reach of just-world research by expanding the domain of inquiry to one's *own* experiences (see also Wood, Heimpel, Manwell, & Whittington, in press) and to phenomena other than derogation and blame. The studies reported in this chapter adopted this approach of seeking to understand how the need to believe in a just world influences previously unexamined aspects of daily experience.

Overview of Research

The research we describe attempts to better understand how the desire for justice, both for self and others, influences the ways people respond to, construe, and reconstruct events unfolding around them. Across seven studies, we examined different potential strategies and mechanisms for sustaining a BJW. Although these mechanisms are quite different from one another (e.g., gambling versus memory reconstruction), we argue that they share the common purpose of serving the need to believe in a just world. In the first four studies, we examined people's reactions to the fates of others in terms of (a) their causal explanations for events and (b) what they remembered about the past. The first two studies investigate the just-world underpinnings of immanent justice reasoning: causally linking someone's fortuitous bad experiences to prior immoral behavior. The next two studies focus on memory and, specifically, whether reconstructing the past can serve achieving justice in the present. This research similarly exploits fortuitous outcomes by examining whether people misremember the value of a lottery prize as a function of the moral worth of the recipient. In all of these studies, the case for just-world processes relies on examining whether the hypothesized effect in causal reasoning and recall is most apparent following a just-world threat manipulation in an unrelated context.

The last three studies shift focus to the role of justice concerns in reactions to one's own fate, in this case feeling relatively deprived. We wondered whether the experience of feeling unfairly deprived would give rise to impulsive and potentially costly behavior. Accordingly, we manipulated our research participants' justice concerns by leading them to believe they had less discretionary income than similar others and then assessed their desire to gamble or their desire for consumer goods.

Studies 1 and 2: The BJW and Immanent Justice Reasoning

Immanent justice reasoning, first identified in children by Piaget (1932/1965), occurs when our *causal* understanding of bad experiences reflects the dictates of justice. Thus, negative experiences are the result of prior bad behavior or immoral character, even when plausible causal linkages are missing. Although immanent justice reasoning was once considered a form of moral reasoning limited only to younger children, Kushner (1981) detailed in his best-selling book *When Bad Things Happen to Good People* a number of examples of *adults* engaging in immanent justice reasoning, such as a fellow who came to understand his friend's recent paralysis after being shot during a robbery gone bad as the result of his pride, arrogance, and insensitivity toward others.

Raman and Winer (2002, 2004) offered a number of explanations for why adults might engage in immanent justice reasoning, including the notion that adults have the capacity for multifocused thinking that enables them to entertain a number of different causal explanations for events (e.g., rational, scientific accounts), including immanent justice ones. In collaboration with Jennifer Nicol (Callan, Ellard, & Nicol, 2006), we examined a justice motive account of immanent justice reasoning in adults.

We reasoned that adults' immanent justice accounts of events might be a just-world-based desire for "deservingness" such that bad or good outcomes would more or less automatically be assumed to reflect prior bad or good deeds, respectively. In the first study, we examined whether people might be more inclined to draw a causal connection between a fortuitous negative outcome and the value of a person's prior behavior when the person's prior behavior was perceived as bad versus good. The second study manipulated just-world threat to see if participants' tendencies to reason in immanent justice terms were more apparent when their BJW was temporarily challenged.

Study 1: Immanent Justice Reasoning

Participants in this study were led to believe they were contributing to research concerned with the effects of different media portrayals of news events (i.e., print versus Web portrayals) on people's reactions to the news. Participants were given two articles (one target and one filler) constructed to look like they were printed from an online magazine. They then completed a series of questionnaires that included a number of filler items (e.g., "How informative was this article?") and our primary dependent measures. The target article reported the story of David, a pedestrian struck by a car. Participants received one of two versions of "David's Story," each designed to create a good or bad impression of David. In one version, participants learned that David was having an extramarital affair with a female travel agent and was planning a trip to Mexico with her. In the other condition, participants learned that David's prior dealings with the travel agent involved purchasing a surprise vacation to Mexico for his family. In each condition, participants learned that David was badly injured when he was struck by a car as he was walking across the street.

Our primary interest in Study 1 was to see whether participants would *causally* relate David's accident to his prior behavior. Manipulation check items confirmed that participants perceived David as more of a bad person and more deserving of the accident when he cheated with the agent ($p < .001$) than when he did not cheat ($p < .001$). Most important, to assess immanent justice reasoning, we asked the participants to rate the extent to which they believed David's accident was the result of his dealings with the travel agent on a scale ranging from 1 (*not at all*) to 7 (*a great deal*). The results showed that participants were more willing on average to endorse an immanent justice account of David's accident when he cheated with the agent (M = 3.05, SD = 1.64) than when he did not cheat (M = 1.61, SD = 0.70), $p = .001$, $d = 1.14$. These results supported our just-world theory account of immanent justice reasoning, as participants were more likely to engage in immanent justice reasoning when the value of a person's fortuitous outcome, in this case a serious accident, conformed to participants' deservingness expectations: Bad people deserve bad outcomes.

Study 2: Just-World Threat and Immanent Justice Reasoning

In Study 2, we examined whether justice motivation aroused in one context would influence immanent justice reasoning in an unrelated context. Research has demonstrated that justice concerns produced in one context can influence justice-restoring reactions in unrelated contexts (e.g., Correia & Vala, 2003; Kay, Gaucher, Napier, Callan, & Laurin, 2008; Lerner, Goldberg, & Tetlock, 1998). Our just-world threat manipulation was modeled after Lerner and Simmons's (1966) manipulation involving the continued versus ended suffering of an innocent victim, and it is consistent with research demonstrating the influence of perceived suffering on observers' reactions to injustice (e.g., Callan et al., 2007; Hafer, 2000b; Starzyk & Ross, 2008). We also sought to extend understanding of immanent justice reasoning by seeing whether it would be apparent for both bad and good outcomes. Would, for instance, participants be more inclined to assume that prior virtuous behavior caused a good outcome after their just-world beliefs had been threatened?

Method. Participants were informed that the study was about how perceptions of a person's emotional cues are affected by the medium of communication (video versus Web pages). They then viewed a series of video clips from *People Like Us* (Fisher & Fisher, 1992) depicting a young woman, Kerry, discussing her emotionally trying experiences living with HIV. In the video presentation, Kerry discusses, among other things, how being HIV positive has negatively affected her life. Prior to watching the video, participants learned that the young woman contracted HIV when a condom broke during intercourse with a person she knew. Following the video presentation, the experimenter gave the participants a brief summary of an ostensible second video of the young woman that was designed to either affirm or threaten participants' need to believe in a just world. In the just-world threat condition, participants learned that Kerry's antiviral medications were ineffective at suppressing the virus and that she was suffering a great deal (and would continue to suffer). In the low threat condition, participants learned that her medications were effective and that her suffering had ended. The results of two separate validation studies confirm that this manipulation produces heightened concern with justice. First, we found that knowledge of the woman's prolonged (versus ended) suffering produced attentional biases toward justice words during a modified Stroop task (see Callan et al., 2006; cf. Hafer, 2000b). Second, Kay et al. (2008) recently found that participants perceived the progression of the young woman's suffering as more unfair when her medications were ineffective versus when they were effective in eliminating her suffering. Following the video presentation, participants were asked to read two articles titled "Charitable Gold" and "Accidental Dream" that were ostensibly printed from the Web. In "Charitable Gold," participants read about a virtuous and charitable older couple who won a $315 million lottery prize. In "Accidental Dream," participants read about an aspiring model named Sarah who was involved in an automobile accident that left her facially disfigured. Participants learned that Sarah often berated her fellow classmates for flaws in their physical appearance. After reading the articles, participants were asked to complete a series of questionnaires that included our primary dependent measures of immanent justice reasoning. For "Charitable Gold," participants were asked to rate the extent to which they believed the couple's lottery winning was the result of their past generosity. For "Accidental Dream," participants were asked to rate the extent to which Sarah's disfiguring accident was the result of her past superficiality and poor treatment of her classmates. Participants were also asked to provide a rationale for each of the immanent justice ratings.

Results and Discussion. For the "Accidental Dream" article, immanent justice reasoning was significantly higher in the high ($M = 3.43$) than in the low ($M = 2.63$) just-world threat conditions ($p < .05$, $d = 0.45$). Furthermore, participants' coded rationales for their immanent justice responses varied systematically as a function of the prior just-world threat. As shown in Figure 3.1, the percentage of rationales with a theme of "justice" (e.g., "She got what she deserved") versus a theme of "chance" or randomness (e.g., "I think it was a random event") were significantly greater in the high (i.e., prolonged suffering) than in the low (i.e., ended suffering) just-world threat conditions ($p < .05$). For the "Charitable Gold" article, the just-world threat

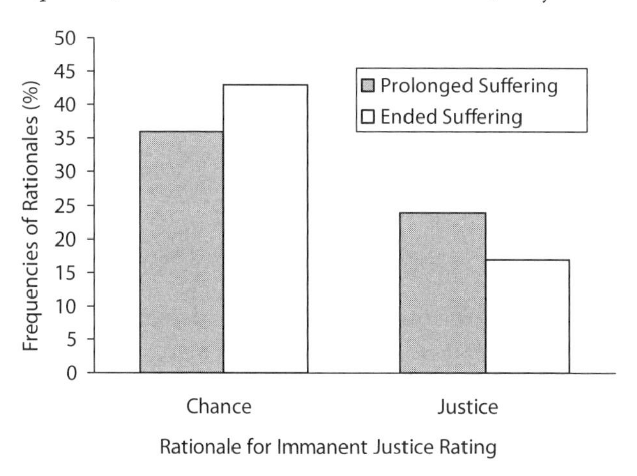

Figure 3.1 Percentage of rationales for immanent justice ratings with themes of chance or justice by suffering status conditions (Study 2).

manipulation did not influence participants' immanent justice ratings or rationales. Although there are likely a number of reasons why an effect of just-world threat on immanent justice reasoning was not observed for the positive outcome scenario (see Callan et al., 2006), one possible reason is that within the "Charitable Gold" scenario, participants learned that the winning couple gave their money to charity, which may have remedied their concerns with the young woman's suffering status and the need to further engage in immanent justice reasoning.

At least in the realm of bad experiences, our undergraduate research participants, schooled in the workings of the natural world enough to have recognized a fluke event for what it was, nonetheless displayed a willingness to understand its occurrence in just-world terms. The apparent ease with which a majority of participants in the critical conditions were able to endorse an immanent justice account points to arguably one of the most straightforward and literal mechanisms for maintaining one's BJW: Assume that (at least) bad things happen for a reason, and the reason is to be found in the prior misdeeds of the unfortunate victim. That people, without any apparent sense of irony or inconsistency, are able to endorse nonrational, superstitious causal accounts that so transparently violate everyday naturalistic understanding of events is intriguing but not entirely surprising given what we have learned about moral intuition (Haidt, 2001). Thus, everyday encounters with the misfortune of others appear to be readily integrated into a worldview that is not only morally coherent in terms of matters such as blame, derogation, and responsibility but also *causally* coherent. Our immanent justice findings also highlight the extent to which the BJW functions as an expectancy, linking past injustice with ongoing daily experience. When we notice evidence of unresolved injustice that is remembered, the just-world expectation encourages attention to outcomes that address the injustice. Thus, daily events are assimilated into a continuing just-world narrative. In the following studies, the linking of past with present in the service of just-world needs was examined directly through two studies that focused on just-world biases in recall.

Studies 3 and 4: Justice Motivation and Memory Reconstruction

A long tradition of just-world research has demonstrated that people often alter their *perceptions* of others and events to maintain their BJW (see Hafer & Begue, 2005; Lerner & Miller, 1978). Thus, we know that victim derogation and blame are mechanisms for sustaining one's BJW. Along with Aaron Kay and Nicolas Davidenko (Callan, Kay, Davidenko, & Ellard, in press), we examined whether people might similarly *misremember* specific details of the past to maintain a sense of justice in the present. Drawing on research demonstrating that motivational concerns (e.g., self-esteem; Sanitioso, Kunda, & Fong, 1990) can influence memory reconstruction (see Hirt, Lynn, Payne, Krackow, & McCrea, 1999; Kunda, 1999), we investigated whether people might misremember the value of a lottery prize depending on (a) whether the winner was a good or bad person and (b) a subsequent experience of just-world threat. We proposed that the need to believe in a just world operates as a frame of reference for inferring the past, such that memorial distortions serve to maintain an ongoing sense of the world as a just and orderly place (cf. McDonald & Hirt, 1997). In the current context, knowledge that a bad person won a lottery is inconsistent with the deservingness expectation that good things happen to good people and thus might influence recall of a lottery prize in a way that renders its value more consistent with what a bad person deserves. Moreover, to the extent that memory reconstruction of events serves a compensatory strategy of maintaining a BJW, a threatened sense of justice might further influence memory reconstruction of a lottery prize given to an undeserving person.

Study 3: Lottery Recipient's Moral Worth and Lottery Prize Recall

Participants read an article ostensibly printed from an online magazine about a man named Roger who won an $18.42 million lottery prize. Participants read one of two versions of the article, each designed to create a different impression of the sort of person that Roger was (see Callan et al., 2006). In one version, Roger was characterized as a good person, who, according to the waitress at a local diner, always left big tips, smiled at everyone, and never complained about the food or service. In a second, bad person portrayal, participants learned that Roger never tipped, never smiled, and always complained about the food and service. After reading a distracter article about the health benefits of coffee consumption and completing a number of filler questions (e.g., "How informative was this article?"), participants were asked to recall the value of the lottery prize by providing any value between $17.49 and $20.49 million.

Consistent with our hypothesis, the results showed that participants in the "bad winner" condition recalled a significantly smaller lottery prize on average (M = $18.18 million) than did participants in the "good winner" condition (M = $18.46 million), $p < .01$, $d = 1.02$ (M difference = $280,000). This finding is compatible with the idea that participants in the "bad winner" condition reconstructed their memory of the lottery prize in a way that rendered its value as more consistent with what a bad person deserves.

Study 4: Just-World Threat and Lottery Prize Recall

In this study, we employed the lottery scenario with Roger winning $18.42 million, but before asking participants to remember the value of the lottery prize, we introduced the same just-world threat scenario of the HIV victim, Kerry, employed in Study 2. Participants reviewed the lottery article and learned that Roger was either a good person or a bad person. They then watched the video presentation of Kerry discussing her experiences living with HIV. As in Study 2, the experimenter informed participants that Kerry's medications were either effective or ineffective in reducing her suffering. Participants were then asked to recall the value of the lottery prize within the given values of \$15.09 and \$21.75 million. Applying the same logic about the just-world-enhancing effects of the HIV victim manipulation, we expected knowledge of the young woman's suffering to influence participants' recall of the lottery prize. We hypothesized that just-world threat would enhance the earlier observed just-world-biasing recall of a bad person's lottery winnings.

Figure 3.2 portrays the expected interaction between the moral worth of the lottery recipient and the suffering status of the young woman on participants' recall of the lottery prize ($p = .03$). Within the "bad winner" condition, participants who learned that the young woman continued to suffer recalled a smaller lottery prize than did participants who learned her suffering ended ($p < .01$, $d = 1.09$; M difference = \$831,200). The suffering status manipulation did not significantly influence recall within the "good winner" condition ($p = .83$; M difference = \$59,900). This finding indicates that when an appropriate relation already exists between the value of person and the value of his or her outcome (good outcome, good

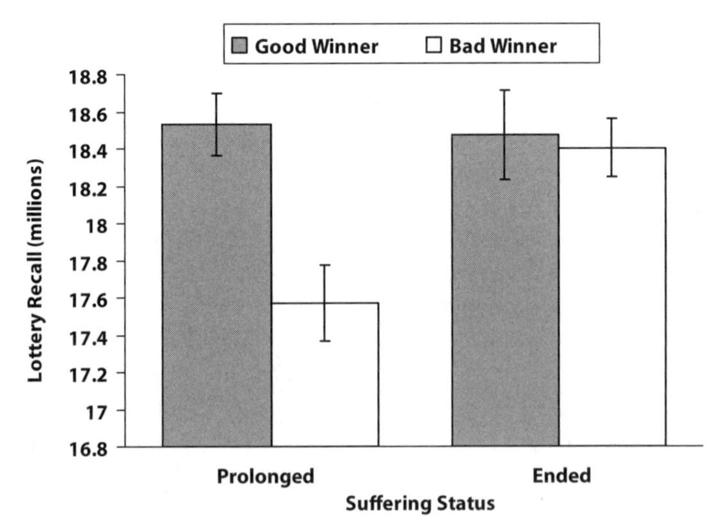

Figure 3.2 Mean lottery recall as a function of the young woman's suffering status and the lottery winner's moral worth (Study 4). The correct lottery value was \$18.42 million. Error bars show standard errors.

person), an interceding just-world threat will not influence memory reconstruction. This finding is consistent with research suggesting that accurate recall generally occurs when an outcome matches one's expectancy for the outcome (see Hirt et al., 1999). Instead, justice-motivated recall occurred only when the lottery prize was undeserved. Nevertheless, significant differences in recall of the lottery were not found between the "good winner" and "bad winner" conditions when participants' just-world beliefs were affirmed (i.e., when her medications were effective), which, at first blush, seems inconsistent with the Study 3 findings. Although the conditions under which recall was made between Studies 3 and 4 were quite different (indeed, see Callan et al., in press, for a replication of Study 3), this finding suggests the possibility that participants might have felt sanguine about the bad person winning the lottery if their just-world beliefs were subsequently reaffirmed by the knowledge that an innocent victim was emancipated from her suffering.

The results of these studies point to the importance of conducting investigations into how misremembering the past supports perceived justice in the present. To that end, and with our colleagues, we have also demonstrated that people will selectively remember their own prior good and bad deeds as a function of recently experienced good and bad outcomes (Callan et al., in press) and distort their memories of a victim's physical attractiveness as a function of his or her suffering status (Callan et al., 2007).

Studies 5 and 6: Gambling as a Search for Justice

With Will Shead and David Hodgins (Callan, Ellard, Shead, & Hodgins, 2008), we explored the idea that gambling urges and gambling behavior might be motivated, in part, by people's justice concerns. We reasoned that gambling might serve a justice-seeking function for some people, as gambling presumably offers an easy means of pursuing "deserved" outcomes (e.g., money, status) that are, for some people, not attainable through more conventional means or self-improvement efforts (Olson, Roese, Meen, & Robertson, 1995). Drawing on relative deprivation theory and research (Crosby, 1976, 1982), we believe gambling offers the prospect of an immediate strategy to reduce perceived unfair deprivation, perhaps because the prospect of winning money obviates the larger problem of whether partaking in behaviors to improve one's lot, such as improving one's occupational qualifications, will be successful in the shorter term.

Interestingly, correlational and anecdotal evidence indicates that gambling behavior may be influenced by personal relative deprivation. Research has demonstrated, for instance, that poverty rates correlate positively with lottery ticket sales (Blalock, Just, & Simon, 2007), and gambling activities increased during the Great Depression (Cross, 2000). Gambling opportunities are also more readily available in more economically disadvantaged geographical regions (Wheeler, Rigby, & Huriwai, 2006). In Studies 5 and 6, we sought evidence for the idea that feeling relatively deprived (either self-reported or experimentally induced) increases urges to gamble and gambling behavior.

Study 5: Self-Reports of Personal Relative Deprivation and Urges to Gamble

We first sought evidence for a link between gambling and personal relative deprivation by conducting two online surveys at the University of Calgary (Sample A; $n = 130$) and the University of Western Ontario (Sample B; $n = 170$). Undergraduate students who reported gambling in some form at least twice in the previous year were recruited to complete the Problem Gambling Severity Index (PGSI; Ferris & Wynne, 2001), a measure of gambling urges (Raylu & Oei, 2004), and a 4-item Personal Relative Deprivation Scale (PRDS) constructed by the authors. The PRDS assesses both the *belief* that one is worse off than others and the associated *feelings* (e.g., "When I think about what I have compared to others, I feel deprived"; "I feel resentful when I see how prosperous other people seem to be"; see Smith & Ortiz, 2001). Given that both self-esteem and personality dimensions have been linked to problem gambling and personal relative deprivation in previous research, in Sample B we also included Rosenberg's (1965) Self-Esteem Scale and the 10-item Personality Inventory (Gosling, Rentfrow, & Swann, 2003) as control measures.

Consistent with our hypothesis, the PRDS significantly correlated with each of the gambling measures (see Table 3.1). Higher personal relative deprivation was related to stronger urges to gamble and more severe gambling problems. Moreover, in Sample B, multiple regression analyses revealed that the PRDS predicted gambling urges and problem gambling over and above self-esteem and Big Five personality dimensions ($p < .05$).

The results of this study provided important evidence in support of the hypothesis that the desire to gamble is linked to personal relative deprivation concerns. Indeed, current concerns associated with personal deprivation predicted "in the moment" desires to gamble. Moreover, these data offer evidence of *individual*-level concerns associated with relative deprivation that had not previously existed in the literature.

Study 6: Personal Relative Deprivation and Gambling Behavior

Given the correlational nature of our first gambling study, we followed up with a study that manipulated personal relative deprivation. The manipulation involved having participants directly compare their discretionary income to the discretionary income of "similar others." The discretionary income of similar others varied between experimental conditions,

Table 3.1 Intercorrelations Among Measures Employed in Study 5

Measure	1	2	3
Sample A (n = 130)			
Problem Gambling Severity Index	(.84)		
Gambling Urges Scale	.51*	(.89)	
Personal Relative Deprivation Scale	.28*	.25*	(.74)
Sample B (n = 170)			
Problem Gambling Severity Index	(.81)		
Gambling Urge and Craving Items	.52*	(.88)	
Personal Relative Deprivation Scale	.24*	.22*	(.64)

Note: Higher values indicate more of each construct (e.g., stronger urges to gamble). Alpha reliabilities are presented along the diagonal.
*$p < .01$.

such that some participants learned that it was relatively high ($759), whereas other participants learned that it was relatively low ($244). This comparison information appeared in the context of a "Demographics Questionnaire," in which participants first reported their income, spending, and discretionary income. They then completed a bogus "Normative Discretionary Income Index" (NDI-Index) where they subtracted their average monthly discretionary income from the average monthly discretionary income of similar others, in this case, "students taking psychology courses." To facilitate the perceived sophistication of the measure, the NDI-Index was used in further calculations to arrive at an adjusted NDI-Index score using point values representing $100 of discretionary income in their calculations. For example, a participant with $366 of discretionary income was asked to use a point value of 4 ($366 rounded up to $400 for a point value of 4) and subtract the obtained point value from 8 or 2 (representing the "relatively deprived" or "not deprived" conditions, respectively). Thus, on average, participants in the "relatively deprived" condition learned that they had less than the similar others, whereas participants in the "not deprived" condition learned that they had a similar level of discretionary income.

Manipulation Validation Studies In two validation studies, we assessed the extent to which the relative deprivation manipulation produces the intended concern with justice and resentment with one's relative standing on discretionary income. In a first study, we found that participants who learned that they had less discretionary income than other psychology students demonstrated selective attention toward justice words versus neutral words in a modified Stroop task (cf. Hafer, 2000a). Specifically, reaction times for justice words and neutral words were significantly different in the "relatively deprived" condition (M = 580 ms and M = 563 ms, respectively, p = .01) but not in the "not deprived" condition (M = 550 ms and M = 552 ms, respectively, p = .45). In a second validation study, we found that participants in the "relatively deprived" condition reported feeling more resentful ($p < .05$, d = 0.69) and thought that their current-level discretionary income was less than they deserved ($p < .01$, d = 0.91) than participants in the "not deprived" condition. Furthermore, consistent with the notion that relative deprivation stems from a violated sense of justice (Crosby, 1976; Olson, 1986), perceived deservingness significantly mediated the effect of the deprivation manipulation on resentment (Sobel's Z = 2.36, $p < .05$). The results of these two validation studies demonstrate that our manipulation of personal relative deprivation produces concern with justice in terms of both the activation of concepts relevant to the goal of maintaining a BJW and the related concern with deservingness.

Personal Deprivation and Gambling Behavior Confident that our manipulation would induce the intended feelings of deprivation and injustice, we proceeded with a study described to participants as a study of "gambling beliefs and attitudes and decision making during a gambling game." The participants first completed measures of problem gambling, including the PGSI. Within the same questionnaire package, participants completed the "Demographics Questionnaire" that included our manipulation of personal relative deprivation. Once participants completed the questionnaires, we gave them a $20 bill and the opportunity to play a card-cutting gambling game (see Breen & Zuckerman, 1999). The experimenter informed the participants that the game was real; thus, they could win more than the $20 or lose the money. The game was played on a computer, and participants could bet up to $10 at a time on whether cards drawn from a standard deck of cards would be low (2 to 7) or high (9 to ace). If participants opted to *not* play the game, they kept the $20 and were thanked for their time. If they opted to play, they "bought into" the game by giving the experimenter back the $20. Participants who played the game were given whatever money they won or $5 if they lost it all during the game. Our primary dependent measure was whether participants played the game. The gambling task, however, enabled us to also examine the number of bets players made during the game.

We hypothesized that a greater percentage of participants in the relatively deprived condition would play the game than participants in the "not deprived" condition. We analyzed participants' decisions to play using a logistic regression analysis where we regressed decisions to play (1 = player, 0 = nonplayer) onto our manipulation variable (1 = deprived, –1 = not deprived), scores on the PGSI, and their interaction term. Results revealed that participants who scored higher on the PGSI were marginally more likely to gamble ($p = .08$), which is consistent with the notion that problem gamblers are more likely to gamble. Consistent with our main hypothesis, a significantly greater percentage of participants in the "relatively deprived" condition (88%) opted to play the gambling game than participants in the "not deprived" condition (60%; $p = .04$). The interaction term did not achieve statistical significance ($p = .35$). Furthermore, our manipulation of relative deprivation did not significantly affect within-game gambling behavior among participants who chose to play the game, with participants in the "relatively deprived" condition betting as many times as participants in the "not deprived" condition (M bets = 39 and 36, respectively).

The results of these studies revealed that feelings of being unfairly deprived predicted participants' gambling urges and influenced their decisions to play a real gambling game. Our analysis of the psychology of these situations is that gambling offers the prospect of people getting what salient social comparison information tells them they deserve. The gambling findings highlight as well a particular behavioral option for managing the justness of one's own world. We suspect that the sort of transitory reminder that similar others may be doing materially better is not uncommon in a materialistically oriented society such as ours. Maintaining the sense that one's own world is just is thus an ongoing life project with such experiences stimulating compensatory responses that can include an impulse to gamble one's way to deserved prosperity. How else might people respond to threats to their personal deserving and BJW? Our search for an answer led us to consider and investigate another behavioral possibility, again in the realm of economic behavior: seeking justice through buying.

Study 7: Justice and Wanting Material Goods

People acquire material goods for all manner of reasons. In collaboration with Leigh Henderson (Ellard, Henderson, & Callan, 2007), we endeavored to see if those reasons include the impulse to buy to assuage feeling relatively deprived. We reasoned that people may be moved to acquire possessions in an attempt to demonstrate to themselves, if not others, that they deserve the same outcomes as similar others. Whatever else our possessions do for us, they can serve as manifest evidence of our material relative standing. Thus, acquisitiveness in this analysis is less about status display (Veblen, 1902) than about minimizing feeling personally deprived and the associated implications for the justness of one's world. The great irony of course is that excessive buying can make one's relative standing *worse* in the long run, thus we might expect justice-motivated buying to be less characteristic of people with a strong commitment to delay of gratification.

The research strategy employed to investigate this idea parallels the approach taken in our investigation of gambling. After gathering relevant background information, we followed the same personal relative deprivation manipulation in this study, with a measure of participants' self-reported wanting of a consumer item they had identified.

Method Participants were told the study was concerned with students' thoughts and behaviors regarding their own and others' spending. In an initial session, participants completed a number of premeasures, including Ray and Najman's (1986) Deferment of Gratification Scale (DGS; e.g., "Are you good at planning things way in advance?"; "Do you fairly often find that it is worthwhile to wait and think things over before deciding"; "Can you tolerate being kept waiting for things fairly easily most of the time?") and the ability subscale of Gibbons and Buunk's (1999) Iowa-Netherlands Comparison Orientation Measure (INCOM; e.g., "I am not the type of person who often compares my outcomes with others"). The DGS measures individual differences in people's abilities to delay gratification and was included as a potential moderator of the effect of relative deprivation on wanting material goods, with the expectation that the effect of deprivation would be most apparent for participants relatively low in delay of gratification. The INCOM assesses individual differences in propensity to socially compare, another variable we thought might serve as a moderator, in this case on the assumption that those most moved to respond to social-comparison-based relative deprivation will be those most given to social comparison.

In a second session, participants completed a "Demographics Questionnaire" that included the personal relative deprivation manipulation, which as before had participants learning that their discretionary income was either less than or about the same as that of similar others. Participants then completed a "Purchases Intentions Inventory" form. In the form, participants were asked to report major purchases they made that cost over $100 in the past year that they deemed as "wants" and "needs." They were then asked to think of an item over $100 they were considering purchasing that was clearly a "want." To determine our primary dependent measure, we then asked participants to rate how badly they wanted the item (1 = *a little bit*, 5 = *it's a "must have"*).

Results and Discussion The wanting item was regressed onto relative deprivation (effect coded), DGS scores (mean centered), INCOM scores (mean centered), and their cross-product interaction terms (including the three-way interaction). Taken together, the predictor variables accounted for 24% of the variance in wanting the consumer good. Shown in Table 3.2, the results of the analysis, contrary to our expectation, revealed no main effect of the personal relative deprivation manipulation ($p = .75$). Relative deprivation, however, did separately interact with delay of gratification and social comparison proclivity (the three-way interaction was not

Table 3.2 Unstandardized Regression Coefficients From Multiple Regression Analyses Predicting Wanting of a Consumer Good, Study 7

	Regression Coefficient	
Predictor Variable ($R^2 = .24$)	***B***	***SE***
PRD Manipulation (PRD)	.03	0.09
Delay of Gratification (DG)	–.03	0.02
Comparison Orientation (CO)	–.24	0.18
PRD × DG	–.04*	0.02
PRD × CO	.35*	0.16
DG × CO	.06	0.04
PRD × DG × CO	.06	0.04

Note: PRD = personal relative deprivation. Dependent variable is ratings of wanting the identified consumer good.
*$p < .05$.

significant). Delay of gratification significantly predicted wanting in the "relatively deprived" condition (see Figure 3.3; $B = -.08$, $SE = 0.03$, $p = .01$) but not in the "not deprived" condition ($B = .02$, $SE = 0.02$, $p = .34$). This pattern suggests that the ability to delay gratification might buffer the effects of personal relative deprivation on wanting consumer goods, as the "relatively deprived" participants with low DGS scores were the most given to wanting the item they identified (cf. Dalbert, 2001).

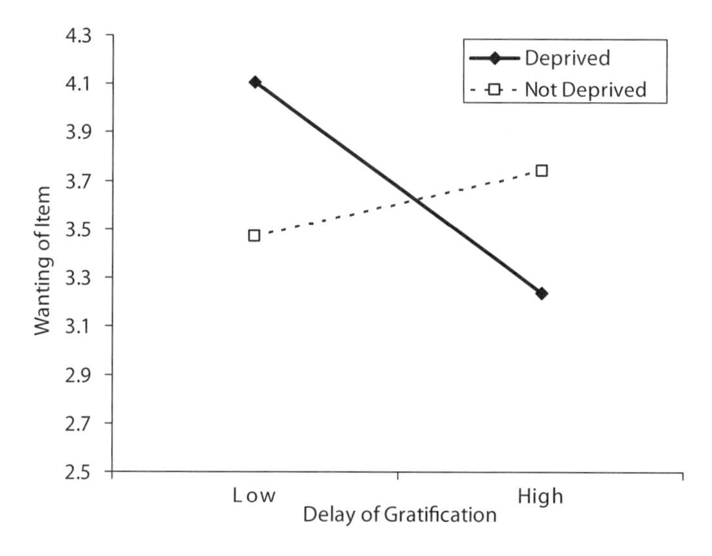

Figure 3.3 Wanting of a consumer good as a function of delay of gratification and personal relative deprivation (Study 7).

Follow-up analysis of the social comparison orientation and relative deprivation interaction revealed unexpectedly that INCOM scores significantly predicted wanting the item in the "not deprived" condition ($B = -.68$, $SE = 0.25$, $p = .01$) but not in the "relatively deprived" condition ($B = .11$, $SE = 0.26$, $p = .68$). Looking at the interaction pattern shown in Figure 3.4 a different way, however, indicates that inducing feelings of deprivation resulted in participants higher in social comparison orientation (1 *SD* above the mean) wanting the item more compared to when they were not deprived (*M* difference = 0.53, $p = .08$), with the difference being marginally significant. For participants lower in social comparison orientation (1 *SD* below the mean), the relative deprivation manipulation did not significantly affect wanting (*M* difference = −0.40, $p = .18$). This pattern of results is consistent with the idea that those most given to social comparisons might be more responsive to social comparison information relevant to feeling relatively deprived—in this case, enhanced wanting a desired material good. Given the unexpected relation between social comparison orientation and wanting the consumer good in the "not deprived" condition, however, the findings for social comparison orientation should be viewed with caution. Indeed, although social comparison orientation is conceptually relevant to our manipulation of personal relative deprivation, it is not clear at this point how social comparison orientation relates to wanting consumer goods following an experience of personal relative deprivation.

Taken together, the results of our investigation of wanting consumer goods as justice-seeking behavior were consistent with our expectations.

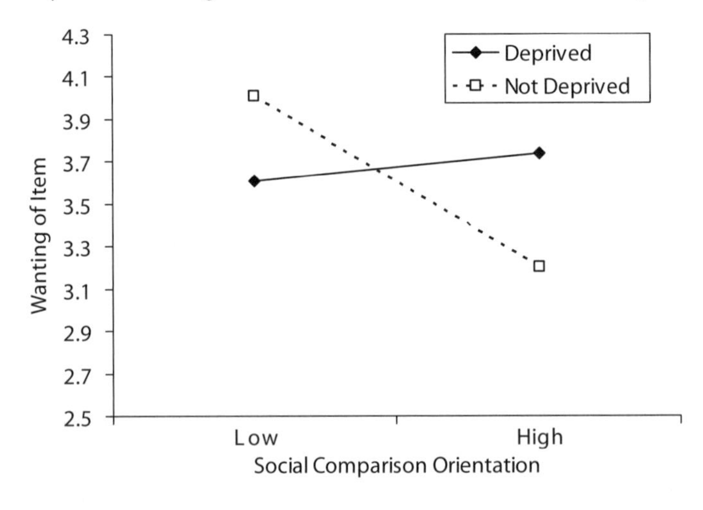

Figure 3.4 Wanting of a consumer good as a function of social comparison orientation and personal relative deprivation (Study 7).

When participants were presented with salient information of being personally deprived relative to similar others, the wanting of a desired material good increased for those individuals most sensitive to social comparison information and for those less skilled at delay of gratification. The delay of gratification findings are particularly intriguing from a just-world perspective given the important role delay of gratification is assumed to play in the development and maintenance of just-world beliefs (Callan et al., 2009; Hafer, 2000b; Lerner, 1977). It appears that people with a strong commitment to delay of gratification may be more sensitive to the potential long-term threat that impulsive acquisitions pose to relative deprivation and personal deserving. Although the specific processes remain to be studied, it may be that commitment to delayed gratification results in a more rational, less impulsive orientation to just-world-threatening information, and as such delay of gratification buffers transitory reminders of one's relative standing.

General Discussion

Lerner (1980) conceptualized the BJW as a central organizing theme in people's lives, apparent in the "more or less articulated assumptions that underlie the way people orient themselves to their environment" (p. 9). As such, the BJW provides the enabling psychological context for people to pursue their life goals with a measure of confidence. Given the broad functional significance of the belief, much of our research, including the studies reported here, has been guided by the premise that the processes involved in sustaining the BJW include more than victim derogation and blame. Although victim derogation and blame provide powerful and compelling evidence for the importance of a just world to people, the ongoing task of sustaining the BJW in response to less dramatic everyday reminders of how imperfectly just the world is undoubtedly relies on a variety of psychological processes and behaviors. In much the same way that researchers have come to understand how sustaining a favorable view of the self is achieved through a diversity of processes (e.g., biased self-justification, memory distortion, attributional bias, unrealistic optimism, self-serving information processing; cf. Tavris & Aronson, 2007; Taylor, 1989), we expect the variety of strategies involved in maintaining one's BJW are also more varied than we currently know.

People find justice in daily life through their construal of events and through their actions. The studies we report cover both domains. Our investigations of immanent justice reasoning focused on construal of causal forces. Although we have no doubt that our participants had as firm a grasp on conventional naturalistic understandings of cause and effect as anyone else, we also found that they were willing to construe cause and effect in nonnaturalistic but justice-sustaining terms. Particularly following a salient reminder of unjust suffering, causal assessments appeared to reflect the view that there are causal forces at work governed by the first principle of a just world: People get what they deserve.

Evidence from our studies on memory bias similarly suggests that our ongoing need to believe in a just world benefits from biased recollections of the past. Thus, justice can be added to the list of present concerns that inform and bias our memories. More important, the immanent justice reasoning and memory studies highlight the reality-distorting nature of construal in the service of a specific motivationally anchored belief. As such, we now know more about the psychological-sustaining mechanisms for the BJW as a *functional illusion*. The conceptual parallels between positive illusions for the self and the just-world illusion suggest interesting directions for research. For instance, the positive illusion work has identified unrealistic optimism as a mechanism that allows people to retain a positive perspective of the self (Taylor, 1989). Unrealistic optimism is arguably as important for being able to believe the world is just insofar as optimism is about the probability of good and bad things happening in the future. To the extent that most people retain a positive view of the self, to have an unrealistic expectation of relatively more good things and fewer bad things happening to the self compared to others is at the same time useful for believing the world is just (Ellard, 1982).

Our work on gambling and wanting consumer goods as behavioral manifestations of just-world motivation resonates with the early Lerner and Simmons (1966) finding that people are indeed moved to *act* to redress injustice and indulge in psychological adaptations such as victim derogation when they are unable to do so. Notwithstanding that important finding, relatively little attention has been devoted to what people actually do in their daily lives in the service of their need to believe in a just world. Gambling and wanting were examined here as responses people can have to threats to *personal* deserving, but more needs to be learned about how people behaviorally manage such threats. For instance, some behaviors may anticipate threat by proactively avoiding social relationships and contexts that invite the sort of social comparisons our participants were required to make with peers. Gambling and wanting consumer goods, then, are likely only two of many justice-seeking responses people engage in. For instance, given the apparent role of the BJW in the "what is beautiful is good" stereotype (Callan et al., 2007; Dion & Dion, 1987), are people spurred to spend on beauty-enhancing products and designer clothing in an attempt to demonstrate they are as beautiful as they deserve to be? If people experience physical attractiveness as an outcome, then it ought to play a role in people's sense of personal deservingness just like any other outcome. This notion might suggest that part of the motivation people have for buying beauty products is to have an outcome (beauty) that will place them socially in a way that fits what they think they deserve. Relatedly, to the extent that social status is experienced by people as an outcome (Ellard & Bates, 1990), threats to personal deservingness might also be expected to increase status-striving behavior. This behavior could take form in a variety of ways, including seeking social roles with higher status, engaging in dominance behaviors or competitive pursuits, or making conspicuous consumption acquisitions. In an individualistic, status-sensitive society, there are no shortages of behavioral options for addressing unfair deprivation with status moves.

Finally, our gambling and wanting consumer goods findings provide an interesting perspective on the dynamics of legitimizing social inequality (see Jost, Banaji, & Nosek, 2004; Kay et al., 2007). Paradoxically, to the extent that gambling and spending represent short-term responses to unfair deprivation, they have the potential to be system-sustaining behaviors because both can serve to undermine the longer term prospects of those already disadvantaged (e.g., by chasing gambling losses). In this way, justice-seeking behavior can contribute to injustice if people are overspending or gambling to achieve the outcomes they feel they deserve.

References

Blalock, G., Just, D. R., & Simon, D. H. (2007). Hitting the jackpot or hitting the skids: Entertainment, poverty, and the demand for state lotteries. *American Journal of Economics and Sociology, 66,* 545–570.

Breen, R. B., & Zuckerman, M. (1999). "Chasing" in gambling behavior: Personality and cognitive determinants. *Personality and Individual Differences, 27,* 1097–1111.

Callan, M. J., Ellard, J. H., & Nicol, J. E. (2006). The belief in a just world and immanent justice reasoning in adults. *Personality and Social Psychology Bulletin, 32,* 1646–1658.

Callan, M. J., Ellard, J. H., Shead, N. W., & Hodgins, D. (2008). Gambling as a search for justice: Examining the role of personal relative deprivation in gambling urges and gambling behavior. *Personality and Social Psychology Bulletin, 34,* 1514–1529.

Callan, M. J., Kay, A. C., Davidenko, N., & Ellard, J. H. (in press). *The effects of justice motivation on memory for self- and other-relevant events.* Journal of Experimental Social Psychology.

Callan, M. J., Powell, N. G., & Ellard, J. H. (2007). The consequences of victim's physical attractiveness on reactions to injustice: The role of observers' belief in a just world. *Social Justice Research, 4,* 433–456.

Callan, M. J., Shead, N. W., & Olson, J. M. (2009). Foregoing the labor for the fruits: The effect of just world threat on the desire for immediate monetary rewards. *Journal of Experimental Social Psychology, 45,* 246–249.

Correia, I., & Vala, J. (2003). When will a victim be secondarily victimized? The effect of observer's belief in a just world, victim's innocence, and persistence of suffering. *Social Justice Research, 16*, 379–400.

Cozzarelli, C., Wilkinson, A. V., & Tagler, M. J. (2001). Attitudes toward the poor and attributions for poverty. *Journal of Social Issues, 57*, 207–227.

Crosby, F. (1976). A model of egoistical relative deprivation. *Psychological Review, 83*, 85–113.

Crosby, F. (1982). *Relative deprivation and working women.* New York: Oxford University Press.

Cross, G. (2000). *An all-consuming century: Why commercialism won in modern America.* New York: Columbia University Press.

Dalbert, C. (2001). *The justice motive as a personal resource.* New York: Plenum.

Dion, K. L., & Dion, K. K. (1987). Belief in a just world and physical attractiveness stereotyping. *Journal of Personality and Social Psychology, 52*, 775–780.

Ellard, J. H. (1982). *The belief in a just world: Dimension or style.* Unpublished master's thesis, University of Waterloo, Ontario, Canada.

Ellard, J. H., & Bates, D. D. (1990). Evidence for the role of the justice motive in status generalization process. *Social Justice Research, 4*, 115–134.

Ellard, J. H., Henderson, L. C., & Callan, M. J. (2007). *The effect of personal deprivation on spending desires and intentions: A hypothesized mechanism of overspending.* Unpublished manuscript.

Ellard, J. H., Miller, C. D., Baumle, T., & Olson, J. M. (2002). Just world processes in demonizing. In M. Ross & D. T. Miller (Eds.), *The justice motive in everyday life* (pp. 350–362). Cambridge, UK: Cambridge University Press.

Ferris, J., & Wynne, H. J. (2001). *The Canadian Problem Gambling Index final report.* Ottawa, Canada: Canadian Centre on Substance Abuse.

Fisher, J. D., & Fisher, W. A. (1992). *People like us* [Videotape]. Storrs: Department of Psychology, University of Connecticut.

Furnham, A. (2003). Belief in a just world: Research progress over the past decade. *Personality and Individual Differences, 34*, 795–817.

Gibbons, F. X., & Buunk, B. P. (1999). Individual differences in social comparison: Development of a scale of social comparison orientation. *Journal of Personality and Social Psychology, 76*(1), 129–142.

Gosling, S. D., Rentfrow, P. J., & Swann, W. B., Jr. (2003). A very brief measure of the Big Five personality domains. *Journal of Research in Personality, 37*, 504–528.

Gruman, J. C., & Sloan, R. P. (1983). Disease as justice: Perceptions of the victims of physical illness. *Basic and Applied Social Psychology, 4*, 39–46.

Hafer, C. L. (2000a). Do innocent victims threaten the belief in a just world? Evidence from a modified Stroop task. *Journal of Personality and Social Psychology, 79*, 165–173.

Hafer, C. L. (2000b). Investment in long term goals and commitment to just means drive the need to believe in a just world. *Personality and Social Psychology Bulletin, 26*, 1059–1073.

Hafer, C. L., & Begue, L. (2005). Experimental research on just world theory: Problems, developments, and future challenges. *Psychological Bulletin, 131*, 128–166.

Hafer, C. L., Begue, L., Choma, B. L., & Dempsey, J. L. (2005). Belief in a just world and commitment to long-term deserved outcomes. *Social Justice Research, 18*, 429–444.

Haidt, J. (2001). The emotional dog and its rational tail: A social intuitionist approach to moral judgment. *Psychological Review, 108*, 814–834.

Hirt, E. R., Lynn, S. J., Payne, D. G., Krackow, E., & McCrea, S. M. (1999). Expectancies and memory: Inferring the past from what must have been. In I. Kirsch (Ed.), *How expectancies shape experience* (pp. 93–124). Washington, DC: American Psychological Association.

Jost, J. T., Banaji, M. R., & Nosek, B. A. (2004). A decade of system justification theory: Accumulated evidence of conscious and unconscious bolstering of the status quo. *Political Psychology, 25*(6), 881–920.

Kay, A. C., Gaucher, D., Napier, J., Callan, M. J., & Laurin, L. (2008). God and the government: Testing a compensatory control mechanism for the support of external systems. *Journal of Personality and Social Psychology, 95*, 18–35.

Kay, A. C., Jost, J. T., Mandisodza, A. N., Sherman, S. J., Petrocelli, J. V., & Johnson, A. L. (2007). Panglossian ideology in the service of system justification: How complementary stereotypes help us to rationalize inequality. In M. P. Zanna (Ed.), *Advances in experimental social psychology* (Vol. 38, pp. 305–358). San Diego, CA: Academic Press.

Kunda, Z. (1999). *Social cognition: Making sense of people*. Cambridge, MA: MIT Press.

Kushner, H. S. (1981). *When bad things happen to good people*. New York: Avon Books.

Lerner, J. S., Goldberg, J. H., & Tetlock, P. E. (1998). Sober second thought: The effects of accountability, anger, and authoritarianism on attributions of responsibility. *Personality and Social Psychology Bulletin, 24*, 563–574.

Lerner, M. J. (1977). The justice motive: Some hypotheses as to its origins and forms. *Journal of Personality, 45*, 1–32.

Lerner, M. J. (1980). *The belief in a just world: A fundamental delusion*. New York: Plenum.

Lerner, M. J., & Miller, D. T. (1978). Just world research and the attribution process: Looking back and ahead. *Psychological Bulletin, 85*, 1030–1051.

Lerner, M. J., Miller, D. T., & Holmes, J. G. (1976). Deserving and the emergence of forms of justice. In L. Berkowitz & E. Walster (Eds.), *Advances in experimental social psychology* (Vol. 9, pp. 133–162). New York: Academic Press.

Lerner, M. J., & Simmons, C. H. (1966). Observer's reaction to the "innocent victim": Compassion or rejection? *Journal of Personality and Social Psychology, 4*, 203–210.

Lerner, M. J., & Whitehead, L. A. (1980). Procedural justice viewed in the context of justice motive theory. In G. Mikula (Ed.), *Justice and social interaction: Experimental and theoretical contributions from psychological research* (pp. 219–256). Bern, Austria: Huber.

McDonald, H. E., & Hirt, E. R. (1997). When expectancy meets desire: Motivational effects in reconstructive memory. *Journal of Personality and Social Psychology, 72*, 5–23.

Napier, J. L., Mandisodza, A. N., Andersen, S. M., & Jost, J. T. (2006). System justification in responding to the poor and displaced in the aftermath of Hurricane Katrina. *Analyses of Social Issues and Public Policy, 6*, 57–73.

Olson, J. M. (1986). Resentment about deprivation: Entitlement and hopefulness as mediators of the effects of qualifications. In J. M. Olson, C. P. Herman, & M. P. Zanna (Eds.), *Relative deprivation and social comparison: The Ontario Symposium* (Vol. 4, pp. 57–77). Hillsdale, NJ: Lawrence Erlbaum.

Olson, J. M., Roese, N. J., Meen, J., & Robertson, D. J. (1995). The preconditions and consequences of relative deprivation: Two field studies. *Journal of Applied Social Psychology, 25*, 944–964.

Piaget, J. (1965). *The moral judgment of the child.* London: Kegan, Paul, Trench, Trubner. (Original work published 1932)

Raman, L., & Winer, G. A. (2002). Children's and adults' understanding of illness: Evidence in support of a coexistence model. *Genetic, Social, and General Psychology Monographs, 128*, 325–355.

Raman, L., & Winer, G. A. (2004). Evidence of more immanent justice responding in adults than children: A challenge to traditional developmental theories. *British Journal of Developmental Psychology, 22*, 255–274.

Ray, J. J., & Najman, J. M. (1986). The generalizability of deferment of gratification. *Journal of Social Psychology, 126*(1), 117–119.

Raylu, N., & Oei, T. P. (2004). The Gambling Urge Scale: Development, confirmatory factory analysis, and psychometric properties. *Psychology of Addictive Behaviors, 18*, 100–105.

Rosenberg, M. (1965). *Society and the adolescent self-image.* Princeton, NJ: Princeton University Press.

Ross, M., & Miller, D. T. (Eds.). (2002). *The justice motive in everyday life.* New York: Cambridge University Press.

Sanitioso, R., Kunda, Z., & Fong, G. T. (1990). Motivated recruitment of autobiographical memory. *Journal of Personality and Social Psychology, 59*, 229–241.

Smith, H. J., & Ortiz, D. J. (2001). Is it just me? The different consequences of personal and group relative deprivation. In I. Walker & H. J. Smith (Eds.), *Relative deprivation: Specification, development, and integration* (pp. 91–118). New York: Cambridge University Press.

Starzyk, K. B., & Ross, M. (2008). A tarnished silver lining: Victim suffering and support for reparations. *Personality and Social Psychology Bulletin, 34*, 366–380.

Tavris, C., & Aronson, E. (2007). *Mistakes were made (but not by me).* New York: Harcourt.

Taylor, S. E. (1989). *Positive illusions: Creative self-deception and the healthy mind.* New York: Basic Books.

Veblen, T. (1902). *The theory of the leisure class: An economic study of institutions.* New York: Macmillan.

Wheeler, B. W., Rigby, J. E., & Huriwai, T. (2006). Pokies and poverty: Problem gambling risk factor geography in New Zealand. *Health and Place, 12*, 86–96.

Wood, J. V., Heimpel, S. A., Manwell, L. A., & Whittington, E. J. (2009). This mood is familiar and I don't deserve to feel better anyway: Mechanisms underlying self-esteem differences in motivation to repair sad moods. *Journal of Personality and Social Psychology, 96*, 363–380.

CHAPTER 4

Preserving the Belief in a Just World

When and for Whom Are Different Strategies Preferred?

CAROLYN L. HAFER and LEANNE GOSSE

Brock University

Abstract: According to just-world theory, people have a need to believe in a just world where individuals get what they deserve. Instances of injustice, such as innocent suffering, will threaten this need and, in turn, will motivate people to try to restore their belief in a just world. Although researchers have suggested different strategies people might use to restore their belief in a just world in the face of threat, little research has addressed the question, "When and for whom are different just-world preservation strategies preferred?" In the present chapter the authors review research that, at least indirectly, suggests answers to this question. They begin with a description of different just-world preservation strategies, then discuss several potential situational predictors of strategy preference, including variables that likely influence the effortfulness, availability, and effectiveness of just-world preservation strategies, as well as the usefulness of various strategies in addressing other needs. Potential individual difference predictors are then reviewed, including ideological beliefs, justice beliefs, demographic variables, and coping style. Finally, the authors discuss several implications as well as considerations for future research that arise from their review.

On May 12, 2008, a massive earthquake hit Sichuan province in southwestern China. The quake killed, injured, or displaced tens of thousands of individuals (Anna, 2008), individuals who might easily be regarded as innocent victims. How do people respond to such apparently undeserved outcomes, whether it is the massive suffering caused by the Sichuan earthquake or the smaller scale misfortune experienced by a single individual? According to just-world theory (Lerner, 1980; Lerner, Miller, & Holmes, 1976), instances of undeserved suffering can threaten a fundamental human need to believe that the world is a just place in which people get what they deserve. This threat can motivate people to attempt to preserve their belief in a just world through any of a number of strategies, from helping victims to restore justice to reinterpreting victims' characters as undesirable and, therefore, their fate as deserved. How actors choose to maintain their belief in a just world can have very different, often profound, consequences for the victims, the actors, and perhaps society as a whole. Thus, it is important to understand the predictors of how people cope with threats to their need to believe in a just world. Unfortunately, although this issue was raised in the early literature (e.g., Lerner & Miller, 1978), it has not been at the forefront of research on just-world theory.

The aim of this chapter is to prompt future research on predictors of just-world preservation strategies by reviewing existing relevant literature, including some recent work from our own laboratory, and discussing possible issues and hypotheses for further research. We first review the various ways in which people might attempt to restore their belief in a just world in the face of threat. Then we divide our discussion of potential predictors of these strategies into two sections: situational predictors and individual difference predictors. Finally, we examine some further considerations and implications arising from our discussion.

Strategies for Maintaining a Belief in a Just World

A belief in a just world performs important functions in people's lives, such as fostering trust that one will be treated fairly and encouraging investment in long-term goals (see Dalbert, 2001; Hafer, 2000b; Hafer, Bègue, Choma, & Dempsey, 2005). Thus, when the belief is threatened by contrary evidence, people are unlikely to give it up but instead engage in strategies for preserving the belief.

Lerner discussed in his 1980 book several strategies people might use to deal with threats to the need to believe in a just world (see also Hafer & Bègue, 2005). Lerner first proposed a set of "rational" strategies, so named because they involve a conscious recognition of injustice. These strategies include attempts to prevent injustice before it occurs. For example, a variety of programs, such as Head Start in the United States (Zigler & Muenchow,

1992), are aimed at ensuring children from low-income families receive the benefits from education that they "deserve" (e.g., Virginia Head Start Association, 2007) before they enter the more formal school system.

Compensating victims of injustice or otherwise helping victims also fits within the rational strategies. Several studies on belief in a just world have investigated how people might be motivated to help victims of injustice by donating to their cause (e.g., Miller, 1977), supporting financial compensation through formal legal procedures (e.g., Haynes & Olson, 2006), placing them in more positive circumstances (e.g., Lerner & Simmons, 1966), volunteering time (e.g., DePalma, Madey, Tillman, & Wheeler, 1999; Miller, 1977), and so on. Finally, Lerner also proposed that one can deal rationally with threats to the need to believe in a just world by accepting that one cannot fix all the injustice one is exposed to in the world; therefore, one sets priorities and rules for whom and when to help.

Lerner (1980) next proposed several "nonrational" or defensive strategies, which can be characterized by an aversion to consciously recognizing or exposing oneself to injustice. These defensive tactics include, first, avoiding or withdrawing from situations involving injustice (e.g., Novak & Lerner, 1968; Pancer, 1988). For example, one might change the television channel to withdraw from ads featuring disturbing images of starving children, or one might avoid visiting an acquaintance who has contracted a terminal illness through little apparent fault of his or her own. In addition to investigating physical avoidance or withdrawal, some researchers have investigated one's tendency to respond to just-world threats by psychologically distancing oneself from victims of injustice. For example, an observer of injustice might reason that the victim is dissimilar to him or her (e.g., Drout & Gaertner, 1994; Hafer, 2000b, Studies 1 and 2), which implies that at least the observer's own immediate world is just, if not the world to which the victim belongs.

Cognitive distortions or reinterpretations that make the situation appear less unfair also fall under the nonrational strategies. One can reinterpret the cause of the injustice, the character of the victim, or the nature of the outcome, all in the service of maintaining a belief in a just world. Many authors, for example, have shown evidence that blaming certain victims of sexual assault (e.g., "She was raped because she was wearing provocative clothes") is a reinterpretation of the *cause* of suffering that is motivated by a need to restore a belief in a just world (e.g., Jones & Aronson, 1973; Karuza & Carey, 1984). Lerner and Simmons's (1966) classic experiment showed that people will sometimes denigrate the *character* of a victim of unjust suffering to restore a sense of justice. Character denigration can take two forms: (a) blaming aspects of a victim's character for his or her fate (e.g., Hafer et al., 2005; Karuza & Carey, 1984) or (b) "general character derogation," that is, generally perceiving the victim as an undesirable person who is, therefore, deserving of negative outcomes (e.g., Hafer, 2000a; Haynes & Olson, 2006; Lerner & Simmons, 1966). To reinterpret the *nature* of the outcome, one might, for example, defensively refer to the benefits of innocent suffering (e.g., "It will make him a stronger person").

Aside from rational and nonrational strategies, Lerner (1980) described two "protective strategies." These strategies embody a way of thinking about the world that, especially if they coexist, prevents a belief in a just world from ever being challenged. First, one can view justice in an ultimate sense. That is, whatever injustices are occurring at present, one can believe that people will get their just deserts in the long run. Ultimate justice, often meted out in an afterlife, is a ubiquitous part of religious ideology (see Wilson, 2003), but we suspect that many nonreligious individuals also believe in a form of long-term justice (see Maes, 1998). The second protective strategy can be described as a multiple-world view (see Hafer & Bègue, 2005): Instances of another's unjust suffering are relegated to a different world that is seen as operating on different principles than the just world within which one lives. Psychological distancing, referred to earlier in this section, can be seen as a mild form of this protective strategy.

An additional strategy for reducing threats to the need to believe in a just world, referred to by Lerner (1980) as the "penultimate strategy," involves false cynicism. People pretend to themselves and to others that they do not believe that the world is, overall, a just place. Consequently, they hide their attempts to maintain a sense of justice behind various charades. For example, people might make charitable donations under a veneer of self-interest, such as receiving a consumer product in return for their donation (Holmes, Miller, & Lerner, 2002).

Another category of strategies for dealing with threats to one's need to believe in a just world is responses to the perpetrator of injustice. Though this category was not highlighted by Lerner (1980), several researchers have investigated tactics in this category, including punishment through the courts (e.g., Kleinke & Meyer, 1990; Wyer, Bodenhausen, & Gorman, 1985) or military action in the case of international threats (Kaiser, Vick, & Major, 2004). In addition, Ellard, Miller, Baumle, and Olson (2002) found some evidence for a perpetrator-oriented strategy that does not involve punishment. These authors argued that perpetrators of grave injustices are sometimes "demonized," that is, they are seen as simply "evil," with no further explanation of their actions deemed necessary. Demonizing might help preserve an overall belief in a just world by allowing one to perceive the injustice as a highly unusual occurrence perpetrated by a highly unusual individual, which does not bear on any comprehensible laws that describe the world or human nature. The event, therefore, need not be dwelled on further.

In summary, researchers have discussed many different strategies for dealing with threats to the need to believe in a just world. Although some have alluded to conditions under which these strategies are more or less likely to be implemented, few researchers have directly tested the predictors of different strategies. In the following two sections, we draw on past research and our own ideas to propose examples of situational and individual difference determinants of how people cope with just-world threat.

Situational Predictors of Just-World Preservation Strategies

Theorists in related areas of social psychology (e.g., Adams, 1965; Festinger, 1957; Walster, Berscheid, & Walster, 1973) have suggested a number of principles that help determine how people will choose to cope with threats to certain needs (see also Kruglanski, 1996). For example, people will gravitate toward coping mechanisms that require less effort, that are available, that are effective, and that help fulfill other coexisting needs. The extent to which various strategies conform to these criteria will depend, in part, on the situation. Thus, we divided our discussion of situational predictors of mechanisms for preserving a belief in a just world into those that likely influence each of the choice criteria just mentioned.

Effortfulness and Availability

People might often prefer the just-world preservation strategy that requires the least effort, whether that effort involves psychological resources (e.g., cognition) or nonpsychological resources (e.g., time, physical energy, or material assets). Several situational variables could affect the degree of effort needed to engage in particular strategies for maintaining a belief in a just world in the face of threat.

With respect to psychological resources, reality constraints might render certain strategies less tenable (Lerner, 1980). In a classic study by Jones and Aronson (1973), for example, participants were exposed to a female victim of sexual assault who was either married, a virgin, or divorced. Participants exposed to one of the first two victims tended to blame her more for her fate compared to participants exposed to the divorcée. According to Jones and Aronson, participants found it difficult to derogate the character of the married woman or virgin, who were seen as more respectable than was the divorcée, and thus they resorted to blaming the victim's behavior to protect a belief in a just world (see also Smith, Keating, Hester, & Mitchell, 1976). Presumably, by reasoning that the respectable victim was responsible for her negative fate in this one instance, participants were able to see her suffering as somewhat deserved without having to reinterpret her overall character as less respectable than it seemed. Jones and Aronson (1973) did not assess alternative methods for resolving the threat the victim posed to participants' need to believe in a just world. Their reasoning, however, implies that participants who did not blame the victim's behavior might have blamed or generally derogated the victim's character to maintain their belief in a just world.

Another set of studies involves manipulations of cues for one or more just-world preservation strategies (see Hafer & Bègue, 2005, Table 2), based on the rationale that the presence of cues for a particular reaction makes that strategy less psychologically effortful. One of the few studies on belief in a just world to directly address people's choice of preservation strategy, conducted by Haynes and Olson (2006), falls within this category. These researchers had participants respond to a supposedly real news story about a car accident victim. Results supported the authors' argument that people will tend to choose the just-world preservation strategy that is easiest, given existing cues in the situation (see also, for example, Karuza & Carey, 1984; Kay, Jost, & Young, 2005). Specifically, participants appeared to defensively blame the victim's behavior or generally derogate the victim's character when cues to these strategies existed. Yet, when both the victim's character and behavior were beyond reproach, participants dealt with the threat that the victim posed to their need to believe in a just world by supporting compensation for the victim. Although many of the studies that manipulate cues to blame or derogation suffer from conceptual difficulties, and results are inconsistent (see Hafer & Bègue, 2005), the notion that people may gravitate toward just-world preservation strategies for which there are already cues in the situation is worthy of further investigation.

We know of no studies that have directly investigated situational predictors of the nonpsychological effortfulness of various strategies for maintaining a belief in a just world. We suggest, however, one idea for further research. The medium through which people are exposed to innocent suffering might affect the material effortfulness of, and therefore the preference for, particular just-world preservation strategies. For example, if an observer is exposed directly to innocent suffering (e.g., a victim of injustice approaches an individual in person), more physical effort will be required to avoid the victim than if observers are exposed to a similar victim through the media, such as on television (where one can easily change the channel or leave the room) or in the newspaper (where one can easily flip the page). The more direct a person's exposure to a just-world threat, therefore, the less likely he or she might be to withdraw from the victim and the more likely he or she might be to help or cognitively distort the situation. Withdrawing from or avoiding victims of injustice in the first place becomes more difficult the more media coverage a particular example of unjust suffering receives. Interestingly, our reasoning in this paragraph suggests that the cognitive distortions and helping behavior uncovered in laboratory experiments, where participants are exposed to portrayals of innocent suffering that they *cannot* avoid or from which they *cannot* withdraw, may overestimate the presence of these strategies. In real-life situations, avoidance or withdrawal might frequently be preferred tactics, as they are often possible and require very little effort.

Some situational variables affecting effortfulness, if taken to the extreme, essentially influence whether specific strategies are available in the first place rather than the ease with which they can be employed. For example, reality constraints on a particular strategy may be so strong that it is rendered impossible: Observers of innocent victims who died might believe it is impossible to engage in compensation and thus would necessarily resort to other means of reducing the threat to their need to believe in a just world, such as withdrawing, reinterpreting the cause of the misfortune, punishing a perpetrator, and so on.

Effectiveness

Not only might people prefer to maintain their belief in a just world with strategies that are less effortful, they might also prefer strategies that are greater in effectiveness. Although we know of no studies directly testing situational predictors of the effectiveness of and, therefore, preference for various just-world preservation strategies, we will review a few investigations that are indirectly relevant to this issue.

Miller (1977) argued that helping innocent victims will be a less attractive option when help does not effectively eliminate the threat that the situation poses to people's need to believe in a just world. Miller further reasoned that helping will be seen as less effective in situations where the victim's need appears persistent (versus temporary) or when the victim is perceived as only one of many (versus an isolated incident). In support of his arguments, Miller (1977) found that people with a strong belief in a just world volunteered less time and money in response to an appeal to assist a needy family when the appeal made the family appear to be one of many such families rather than a single case (Study 1). Strong believers in a just world also donated less money to needy families when the appeal implied chronic rather than temporary need (Study 2).

Some just-world research implies that the cost to the observer influences a strategy's effectiveness at alleviating the threat of innocent suffering and, thus, the likelihood that observers will employ the strategy. Lerner et al. (1976) theorized that people will help innocent victims to the extent that they do not perceive their own deserved resources to be in jeopardy. If the situation requires observers to give too much to alleviate another's unjust suffering, then observers become victims of injustice themselves, and the threat to their need to believe in a just world persists. Supporting this reasoning, Holmes et al. (2002) found that participants donated more to a worthy cause when donations could be framed within an economic transaction in which money was exchanged for a consumer product that was clearly not of intrinsic value to participants. One explanation for this finding is that participants were more comfortable giving to innocent victims if the donation could be masked by a veneer of self-interest, which allowed participants to at least pretend that their behavior had no bearing on their own just deserts (for alternative explanations of these findings, see Holmes et al., 2002; Simpson, Irwin, & Lawrence, 2006).

We do not know if the participants in the Miller (1977) or Holmes et al. (2002) studies resorted to other ways of maintaining their belief in a just world when helping was presumably seen as ineffective, because alternative strategies were not measured. Their research, however, does imply that the choice of coping mechanism in the face of threat to one's need to believe in a just world could be affected by variables that influence the effectiveness of those mechanisms.

Competing Motives

The need to maintain one's belief in a just world will not always be the sole motive aroused by situations involving undeserved fates. Often, social situations raise multiple concerns (e.g., Batson, Klein, Highberger, & Shaw, 1995; Insko, Smith, Alicke, Wade, & Taylor, 1985; Jost, Burgess, & Mosso, 2001). People's choice of just-world preservation strategy might be influenced by these other concerns in that people might prefer, when confronted with innocent suffering, a response that satisfies both the need to maintain a belief in a just world and one or more additional motives (see Kruglanski, 1996). Several situational variables could influence the extent to which other needs are also salient in cases of undeserved outcomes.

The extent to which a person identifies with an innocent victim's situation might arouse ego-defensive concerns that help determine the choice of just-world preservation strategy. If people can easily see themselves in the victim's situation, they might be less inclined to engage in strategies that downgrade the victim (e.g., blame and derogation), as these strategies might imply that they are similarly unworthy or irresponsible (e.g., Sorrentino & Hardy, 1974). Foley and Pigott (2000), for example, reasoned that women would identify more with a female sexual assault victim (whose attacker had already been convicted) than would male participants and, thus, would choose to compensate the victim to maintain their belief in a just world rather than blame her. Men, in contrast, were hypothesized to be more likely to respond to the victim's suffering by devaluing her rather than by suggesting compensation. Consistent with the researchers' reasoning, women with a strong belief in a just world thought the victim should receive more compensation than did women with a weak belief in a just world, but women with a strong belief versus a weak belief in a just world did not differ in the extent to which they held the victim responsible for her fate. The hypothesis for men was not supported. Perhaps men engaged in a strategy that was not assessed by Foley and Pigott, such as seeing benefits in the victim's suffering. Lerner (1980) summarized research suggesting that people sometimes avoid or withdraw from, rather than devalue, innocent victims with whom they identify as a way of maintaining their belief in a just world. Some participants in Foley and Pigott's (2000) experiment might have responded in this manner.

Aside from raising ego-defensive concerns, identification with the victim of injustice might motivate observers to altruistically increase the other's welfare (see Batson et al., 1995) as well as to restore their belief in a just world. Helping the victim could fulfill both these needs and, therefore, might be preferred over cognitive distortions. Perhaps, for example, women with a strong belief in a just world in the Foley and Pigott (2000) study were motivated by altruism, rather than ego-defensive concerns, as well as their need to believe in a just world.

Whether one is the observer or victim of unjust suffering might influence one's choice of just-world preservation strategy for a similar reason as the degree to which a person identifies with a victim. If one is the victim, responses that devalue the self might be less likely because they work against one's ego-defensive motives, though the literature is not without evidence that people sometimes unrealistically blame themselves for their own suffering (e.g., Bulman & Wortman, 1977; Davis, Lehman, Silver, Wortman, & Ellard, 1996; Downey, Silver, & Wortman, 1990). As an alternative to devaluation, metaphorical withdrawal might be possible through substance abuse or other methods of escaping from one's suffering. As far as we know, physically maladaptive responses to just-world threats have not been investigated. More adaptive alternatives to self-devaluation might include seeing benefits in suffering (see Park, Cohen, & Murch, 1996), seeking compensation, and so on. Note that whereas victims of injustice might gravitate away from devaluation, observers of injustice might prefer such a response, especially blaming the victim's behavior (see Karuza & Carey, 1984), because it not only allows them to restore their belief in a just world but also fulfills their need to avoid harm to themselves by suggesting that they are in control of their fate (see Shaw & McMartin, 1977).

Whether an individual is the perpetrator rather than merely the observer of injustice might also raise concerns that affect how the individual chooses to resolve a threat to his or her need to believe in a just world. To reduce guilt and avoid social censure, for example, a perpetrator might gravitate toward coping mechanisms that downgrade the victim. Chaikin and Darley (1973) found evidence of such a process. They manipulated whether people expected to be in a similar role as the perpetrator or the victim of an undeserved outcome, as well as whether the outcome had mild or extremely negative consequences for the victim. Participants who expected to be in the perpetrator's role reported disliking the victim more than participants in any other condition did. Downgrading the victim may have been a preferred method for restoring a belief in a just world by these participants because the tactic simultaneously deflected potentially negative evaluations away from the self (see Lerner & Miller, 1978, for a discussion of the interplay between the need to believe in a just world and the desire to reduce guilt). Though not assessed by Chaikin and Darley (1973), perhaps participants who expected to be in the victim's role took an alternative route to satisfying their need to believe in a just world.

A challenge to future research will be to show whether a given response to innocent suffering is employed because it fulfills a need to believe in a just world and a coexisting need or because it fulfills only one of these motives. We return to this issue later in our chapter.

Individual Difference Predictors of Just-World Preservation Strategies

Not only are situational variables likely to predict how people choose to preserve their belief in a just world in the face of threat, but individual differences might also be important predictors. The work on individual differences within the just-world literature has focused almost exclusively on variations in the strength of people's belief in a just world (for reviews, see Furnham, 2003; Furnham & Procter, 1989). These variations are assessed with explicit self-report measures that ask people to rate the extent to which they agree or disagree with a number of statements (e.g., "I am confident that justice always prevails over injustice," from Dalbert, 2001). There are various possible interpretations of just-world scales (see Hafer & Bègue, 2005). At least in experimental studies, however, just-world scale scores are most often used as an indirect indicator of people's motivation to restore a belief in a just world in the face of threat (as in several studies already noted in this chapter), rather than an indicator of how people choose to respond to just-world threats. Presumably, the stronger one's belief in a just world is, the greater the discrepancy is between an unjust event and one's belief, thus, the more threat experienced and the greater motivation to engage in some strategy for reducing the discrepancy in a way that preserves a belief in a just world, such as denigrating the character of an innocent victim.

Some just-world researchers have examined other individual differences in the context of predicting reactions to victims. These authors have usually presumed that the individual difference variable, similar to the interpretation of just-world scales in the previous paragraph, influences the degree of threat that the actor feels in the first place (e.g., Sorrentino, Hancock, & Fung, 1979). Alternative explanations, however, could involve individual differences in the type of just-world preservation strategy employed rather than the extent of just-world threat experienced. Our discussion of individual difference predictors of just-world preservation strategies is divided into four types of variables: ideological beliefs, justice beliefs, demographics, and coping style.

Ideological Beliefs

Sorrentino (Sorrentino et al., 1979; Sorrentino & Hardy, 1974) has investigated individual differences in religiosity and right-wing authoritarianism—two related ideological variables (e.g., Altemeyer, 1996)—within experiments designed to expose some participants to just-world threat. In conditions where a belief in a just world would presumably be most threatened, participants low in religiosity or authoritarianism evaluated an innocent victim more negatively than did those high in these characteristics. Perhaps individuals high in these ideological variables used different methods of resolving the threat to their need to believe in a just world that were not assessed in the studies. Lea and Hunsberger (1990), for example, suggested that general character derogation may be a less viable option for people high in Christian orthodoxy, or for whom religion is temporarily salient, because religious norms generally involve benevolence. Religious people (or people for whom religion is made salient) might prefer to deal with threats to their need to believe in a just world by reasoning that innocent victims will reap their just deserts in the long-run. Individuals high in right-wing authoritarianism, given their tendency to think in terms of clearly delineated ingroups and outgroups (Altemeyer, 1996; Duckitt, 2005), might prefer to deal with just-world threats by placing innocent victims in a separate world than their own. Alternatively, authoritarians' punitiveness (see Altemeyer, 1996) might lead them to focus on punishment of offenders rather than on strategies that are more victim-focussed.

As discussed in an earlier section, a belief that justice will occur in the long run or that innocent victims belong to a different world can exist as chronic worldviews that ward off just-world threat in the first place, rather than as more reactive strategies that are conjured up only in defensive response to particular threats. Perhaps, then, as suggested by Sorrentino (e.g., Sorrentino, 1981), religious individuals and right-wing authoritarians do not feel threatened by events that might be seen as unjust. Rather, their ideological beliefs protect their belief in a just world by guiding their interpretation of situations such that the situations are not threatening to the need to believe in a just world in the first place. Indeed, one function of ideological belief systems may be to protect one from various psychological threats (see Jost & Hunyady, 2002). In future, researchers could investigate the extent to which individual differences in ideological beliefs protect one from experiencing just-world threat in the first place or provide a readily available defense mechanism that can be employed against specific instances of just-world threat (see Hafer & Bègue, 2005). In any case, researchers should take into account individual difference measures of the strength of participants' belief in a just world in future research on ideological beliefs and just-world preservation strategies, given that these ideological variables tend to be related to scores on just-world scales (Furnham, 2003; Furnham & Procter, 1989).

Justice Beliefs

A few researchers have discussed individual differences in justice beliefs that might influence one's preference for just-world preservation strategies. We already mentioned that individual differences in the strength of one's overall belief in a just world are not interpreted as determinants of strategy preference. Individual differences in particular *forms* of belief in a just world, however, might be more relevant in this regard.

Maes (e.g., 1998; Maes & Schmitt, 1999) delineated two kinds of belief in a just world. A belief in "immanent justice" (see Piaget, 1932/1968) is an assumption that current outcomes are deserved on the basis of past positive or negative behavior or character; thus, current outcomes are seen as necessarily fair. A belief in "ultimate justice," as noted earlier, is an assumption that people will get what they deserve in the long run. Maes (1998) measured illness-specific versions of immanent justice (e.g., "Hardly anyone becomes seriously ill without having deserved it") and ultimate justice (e.g., "In the long run, the injustice imposed by illnesses receive appropriate reparation"). He found that the strength of these beliefs was related to how people responded to cancer patients, at least some of whom might be seen as innocent victims. The higher respondents scored on immanent justice beliefs (controlling for ultimate justice), the more they generally derogated the character of patients and blamed them for their illness. In contrast, the higher respondents scored on ultimate justice beliefs (controlling for immanent justice), the more they endorsed helping responses such as charitable work and support of expanded medical services. Furthermore, a belief in ultimate justice in the realm of illness was related to a belief that a cure for cancer could be found, which implies that unjust suffering of cancer patients will eventually be eradicated. We do not know whether these differential reactions, as noted for religiosity and right-wing authoritarianism, reflect defensive responses to victims who threatened the participants' need to believe in a just world or reflect the fact that a belief in immanent or ultimate justice are worldviews that protected participants from experiencing threats to their need to believe in a just world in the first place.

Future research on the predictors of just-world preservation strategies, as well as immanent versus ultimate justice, could investigate other ways of believing in a just world, for example, a belief that the world ought to be just and that one must and can make it so (see Mohiyeddini & Montada, 1998) versus a belief in a just world that is governed primarily by forces external to the self, such as God. The former might predict helping responses or punishment of the perpetrator in reaction to just-world threat, whereas the latter might predict less action-oriented coping mechanisms, such as conjuring up thoughts of ultimate justice or rationalizing a victim's plight to make it seem more deserved.

Demographics

In one of the few studies to directly investigate individual difference predictors of reactions to just-world threat, Drout and Gaertner (1994) argued for sex differences in how observers would respond to certain victims. They exposed participants to a female victim of sexual assault or a nonvictim and also measured the strength of participants' belief in a just world. As predicted, men with a strong belief in a just world were more likely to help the victim (on a task unrelated to her victimization) compared to the nonvictim, presumably because men are socialized to help female victims and thus preferred to restore their belief in a just world in this manner. Female participants showed no difference in helping but, in contrast, psychologically distanced themselves from the victim more than from the nonvictim (although this effect was not moderated by the strength of their belief in a just world). Perhaps women identified more with the victim, and therefore, both ego-defensive and just-world concerns were aroused (see Foley & Pigott, 2000). Distancing themselves from the victim allowed them to address both concerns. Drout and Gaertner's findings are open to many interpretations; their reasoning, however, raises the possibility of sex differences in how people choose to respond to threats to their need to believe in a just world.

Aside from sex, we know of no other demographic variables that have been examined as predictors of different just-world preservation strategies. Some ideas for future research, however, are culture, age, and socioeconomic status (SES). Lerner (1980, 2003) proposed that as children become adults, they will largely replace blame and general character derogation with the protective and penultimate strategies we discussed earlier. The former more "primitive" reactions, however, will still occur when people are emotionally aroused by an instance of injustice or they are forced by the situation to respond automatically. With respect to culture, aside from the demographic variables of nationality or ethnicity, researchers could investigate more psychological cultural variables, such as Hofstede's (1984) dimensions of culture (e.g., masculinity) or interdependent versus independent self-construals (see Markus & Kitayama, 1991). An individual's SES might also predict strategy choice. Perhaps, for example, individuals with lower SES are less likely to engage in strategies that tax their already constrained resources compared to individuals with higher SES (e.g., seeking compensation through the legal system). As noted for ideological beliefs, some demographic variables are related to the strength of people's self-reported belief in a just world (for reviews, see Furnham, 2003; Furnham & Procter, 1989); thus, researchers should take this relation into account in their studies.

Coping Style

A natural place to look for individual difference predictors of responses to just-world threat is the coping literature. Several modes of coping have been suggested that presumably have relevance for many kinds of threat, including threat to one's need to believe in a just world. We have recently investigated repressive coping style and its relation to people's attempts to preserve their belief in a just world. This research will be described in detail next.

According to Weinberger, Schwartz, and Davidson (1979), people with a repressive coping style or "repressors" score high on measures of defensiveness (e.g., the Social Desirability Scale; Crowne & Marlowe, 1960) and low on measures of trait anxiety (e.g., the Manifest Anxiety Scale; Taylor, 1953). In general, these individuals are motivated to avoid negative affect (for a review, see Myers et al., 2008). Specifically relevant to our investigation, Boden and Baumeister (1997) found evidence that repressors sometimes try to accomplish this goal by thinking positive thoughts when facing negative emotional stimuli. Furthermore, repressors often respond in an optimistic fashion to potentially threatening negative events (e.g., developing cancer; Myers & Reynolds, 2000). These findings led us to predict that repressors would be more likely than nonrepressors to react to a victim who threatens their need to believe in a just world by thinking of the victim's suffering in a positive light, for example, by thinking about benefits to suffering or about reaping deserved outcomes in the long run. We tested this hypothesis in two studies (Hafer & Gosse, 2008).

In Study 1, participants were classified as repressors or nonrepressors with scales used by Weinberger et al. (1979). In a later session, they watched a videotape of a young woman who had contracted a sexually transmitted disease after a condom had broken and who suffered from depression since hearing of her diagnosis (see Hafer, 2000b, Study 2). This victim can be seen as innocent in that she behaved responsibly (i.e., engaged in safer sexual practices) yet still contracted a disease. Information that followed the video manipulated how threatening the innocent victim was to participants' need to believe in a just world. In the high-threat condition, participants were told that the victim's suffering had ended, whereas in the low-threat condition, participants were told her suffering continued. Participants' tendencies to see the victim's suffering in a positive light were assessed by averaging their ratings for three questionnaire items, for example, how likely they thought the victim's situation would make her a better person.

Consistent with our hypothesis, when the victim's suffering had ended (i.e., when there was presumably little threat to people's need to believe in a just world), there was no significant difference between the repressors and nonrepressors in their tendency to view her suffering in a positive light. When the victim's suffering continued (i.e., which presumably posed a strong threat to people's need to believe in a just world), however, the repressors were significantly more likely to view her suffering in a positive light than were nonrepressors.

In Study 2, we used similar video stimuli but a different manipulation of threat to the need to believe in a just world. In the high-threat condition, the victim was innocent (as in Study 1) in that she contracted the disease after a condom had broken, whereas in the low-threat condition, she was more responsible for her condition in that she contracted the disease after not using a condom. In addition to assessing the tendency to see the victim's suffering in a positive light, we also assessed participants' tendencies to respond negatively toward the victim, for example, by interpreting her character as undesirable.

Results were again consistent with our reasoning. When the victim was responsible (low-threat condition), there was no significant difference between repressors and nonrepressors in their tendency to see the victim's suffering in a positive light; moreover, the responsible victim was seen as equally desirable by both repressors and nonrepressors. When the victim was innocent (high-threat condition), however, repressors had a greater tendency to see the victim's suffering in a positive light, and nonrepressors, as compared to repressors, saw the innocent victim as having a significantly less desirable character (see Thornton, 1992).

We also conducted a pilot test for Study 2 in which we found that both repressors and nonrepressors showed a greater attentional bias toward justice-related words when the victim was innocent than when she was responsible (see Hafer, 2000a). These results suggest that repressors in Study 2 did not merely respond more positively and less negatively to the innocent victim (compared to nonrepressors) because justice concerns were less salient to them. Instead, the differences between repressors and nonrepressors to the presumably high-threat victim were more likely the result of individual differences in strategies for dealing with just-world threat. Repressors interpreted the high-threat victim's suffering in a positive light, whereas nonrepressors downgraded her character. Both strategies presumably helped participants to maintain a belief in a just world in the face of continued undeserved suffering.

Further Considerations and Implications

Although we suggested a number of situational and individual difference variables that might influence one's choice of just-world preservation strategy, our ideas are only preliminary. Very little research *directly* addresses when and for whom various strategies will be preferred. Interestingly, researchers have raised a similar point with regard to the cognitive dissonance literature (e.g., Hardyck & Kardush, 1968). As these theorists' critiques lead to further developments in the area (e.g., for a recent review of cognitive dissonance research, see Olson & Stone, 2005), we hope that this chapter will also lead to an expansion of research on just-world theory. In these final sections, we review further considerations for research on the predictors of strategy choice, including additional research questions as well as methodological points. We also discuss some implications of the ideas raised in this chapter.

Considerations for Future Research

The situational and individual difference variables we mentioned in the previous two sections are not meant to be exhaustive of potential predictors. One additional situational predictor that might be examined in future research is whether the target of injustice is a victim or beneficiary of the injustice. Although we focused exclusively in this chapter on people's responses to unjust suffering—that is, situations in which individuals get less than they deserve—cases in which people obtain more than they deserve can also, theoretically, threaten the need to believe in a just world (e.g., Ellard & Bates, 1990). In addition to other situational variables, there are likely other choice criteria that are influenced by aspects of the situation that we did not discuss earlier, such as the extent to which a strategy provides immediate versus delayed fulfillment of the need to believe in a just world or the extent to which the strategy restores one's belief in a just world only in the short term versus indefinitely (see Kruglanski, 1996). With respect to individual differences not mentioned earlier, basic personality traits could be examined in future studies. Although researchers have occasionally investigated the relation between personality and people's scores on just-world scales (e.g., Wolfradt & Dalbert, 2003), they have not examined personality traits as predictors of how people might prefer to restore their belief in a just world when confronted with contradictory evidence.

Not only are strategies for protecting a belief in a just world in the face of threat likely influenced by situational and individual difference variables, but some tactics will often be preferred *across* situations and persons, perhaps because they tend to be low in effort, relatively available, effective, and so on in most circumstances and for most people. For example, we already suggested that avoidance and withdrawal might be relatively popular in daily life because these tactics usually require little effort. One might also predict that protective and penultimate strategies, if in the form of a general worldview, will usually be preferred over more reactive strategies, given that they can effectively stop most potential just-world threats from occurring at all.

Researchers could also examine the predictors of whether people will use multiple strategies versus a single strategy to maintain their belief in a just world (see Hafer & Bègue, 2005). Studies that have examined the use of multiple strategies have sometimes found evidence for the simultaneous use of many ways of coping with a threat to the need to believe in a just world. Others, including our own research on repression described in this chapter, have found a hydraulic effect, in which the use of one tactic seems to reduce the likelihood of another also being employed (cf. Bordieri, Sotolongo, & Wilson, 1983; Correia, Vala, & Aguiar, 2001; Haynes & Olson, 2006; Jones & Aronson, 1973; Kenrick, Reich, & Cialdini, 1976; Lerner & Simmons, 1966). Perhaps variables that influence the effectiveness of particular just-world preservation strategies are important in this regard: When a strategy is very effective, it might be less likely that others will also be used.

Future research on the predictors of responses to just-world threat should include measures or manipulations of theoretical mediators to help uncover underlying mechanisms. We discussed several mediators in previous sections of this chapter. For example, Drout and Gaertner (1994) theorized that their sex effects might be mediated by social norms (for men) or identification with the victim (for women). Furthermore, we suggested in our discussion of situational predictors that the effect of certain situational variables might work through their influence on specific choice criteria. Individual difference effects might also at times be mediated by these criteria. For example, certain stable tendencies might make some responses to just-world threat easier or perhaps more effective at maintaining a belief in a just world.

Moderator variables, as well as mediators, could be investigated. For instance, situational and individual difference factors might interact to influence people's preferences for how they maintain a belief in a just world (for an analogous example from the cognitive dissonance literature, see Zanna & Aziza, 1976). The Drout and Gaertner (1994) study discussed earlier raises this possibility. When an innocent victim is female, and people have the opportunity to help the victim on an unrelated task, men might prefer to help (at least certain kinds of help), whereas women might prefer to engage in psychological distancing (cf. Foley & Pigott, 2000); when an innocent victim is male, another pattern might emerge.

Future research directly addressing predictors of just-world preservation strategies will need to consider several methodological and conceptual issues. First, researchers should not only manipulate or measure the predictor variable of interest but also assess at least two potential coping tactics. Assessing more than one tactic will help test whether participants in different situations or possessing different characteristics are all attempting to cope with a threat to their need to believe in a just world, merely through different means. Our second study of repressive coping style described earlier employed such a design. Specifically, we assessed not only the predictor of interest—repressive coping style—but also two types of just-world preservation strategy: seeing injustice in a positive light and negatively evaluating the victim of injustice (e.g., generally derogating her character). With this design we hoped to show that repressors and nonrepressors have different ways of responding to threats to their need to believe in a just world.

Measures or manipulations indicative of variations in just-world threat will also help assess whether different responses are methods of achieving the identical goal of maintaining one's belief in a just world. For example, the situational or individual difference predictors of interest should be crossed with a manipulation of just-world threat (such as those used in our studies on repressive coping style), with the expectation that evidence of the differential coping mechanisms would not occur in the low-threat condition. Alternatively, or in addition, researchers could include a dependent variable indicative of justice-related threat. Work by Hafer (2000a) and others (Callan, Ellard, & Nicol, 2006, Study 3; Correia, Vala, & Aguiar, 2007) has shown that a modified Stroop task can be used as a subtle indicator that justice concerns have been raised in a particular context (see also Kay & Jost, 2003, Studies 3 and 4). The procedures noted in the past two paragraphs will also help address two issues we mentioned previously: whether those who react to an innocent victim in a particular way are guided by their need to believe in a just world and alternative motives or are acting to fulfill only one of these motives and whether different reactions to innocent victims occur because some individuals (or those in particular situations) experienced little threat to begin with.

Implications of Situational and Individual Difference Predictors of Just-World Preservation Strategies

The idea that people respond differently to just-world threat depending on the situation and their personal characteristics can help explain some past failures to support hypotheses based on just-world theory (Hafer & Bègue, 2005). Perhaps the situation was such that no one strategy was dominant; thus, individuals gravitated toward different coping mechanisms, depending on which matched their personal characteristics. Or perhaps the strategy assessed by the researchers was not the response of choice, given the situation. With a good a priori understanding of when and for whom different strategies will be preferred, researchers can constrain the experimental situation so that one particular coping mechanism is dominant, or they can assess multiple strategies and measure or manipulate those variables known to affect one's preference. One should keep in mind, however, that constraining the experimental situation too much might force participants to employ a strategy they would not pursue outside of the laboratory. Future research that assesses multiple responses under more naturalistic conditions would be informative.

Another methodological implication is that researchers assessing multiple responses to just-world threat should pay attention to the ordering of their measures. If strategies for maintaining a belief in a just world can work in a hydraulic fashion, then participants might merely employ the tactic that is offered to them first. A few just-world researchers have explored such order effects, with conflicting results (cf. Haynes & Olson, 2006; Kenrick et al., 1976; Lincoln & Levinger, 1972); thus, further work on this issue is warranted.

Our discussion in this chapter also has practical value. We noted in our introductory paragraph that the predictors of how people choose to maintain their belief in a just world in the face of threat are important to investigate because that choice can have significant implications for those involved. We now return to the event with which we opened this chapter—the 2008 earthquake in Sichuan, China. Media coverage of this event exhibited a vast array of responses. Much was written, for example, about the great outpouring of help extended to the victims, by both the Chinese and the international community (e.g., Bennett & MacArtney, 2008; Yardley & Barboza, 2008). In contrast, explicit and implicit messages that the Chinese deserved their fate—for example, because of their poor human rights record—were readily available (e.g., Jane, 2008; Jolly, 2008). Others described positive consequences of the incident, including the political expediency of the event as China tried to improve its image leading up to the 2008 Olympics, or the invaluable experience gained in civic behavior (Ni, 2008; Yardley & Barboza, 2008; York, 2008). Aside from responses found in the media, some individuals likely changed the television channel or turned the page of the newspaper when confronted with images of the suffering victims. The consequences of these different reactions are wide-ranging—from emotional and psychological consequences to observers, to groups' decisions to redistribute millions of dollars in resources, to the life and death of victims.

Of course, there are many motivations behind the varying opinions and behaviors with respect to the Sichuan earthquake. If even some of these responses are partly motivated by a need to believe in a just world, however, then it becomes of great practical importance to understand when and for whom different strategies for preserving the belief are preferred. Given the practical and research implications of the questions raised in this chapter, we believe the predictors of just-world preservation strategies deserve greater research efforts in the future.

References

Adams, S. J. (1965). Inequity in social exchange. In L. Berkowitz (Ed.), *Advances in experimental social psychology* (Vol. 2, pp. 267–299). New York: Academic Press.

Altemeyer, B. (1996). *The authoritarian specter.* Cambridge, MA: Harvard University Press.

Anna, C. (2008, May 27). Families hit by disaster to get exemption from one-child policy. *Globe and Mail*, p. A14.

Batson, C. D., Klein, T. R., Highberger, L., & Shaw, L. L. (1995). Immorality from empathy-induced altruism: When compassion and justice conflict. *Journal of Personality and Social Psychology, 68*, 1042–1054.

Bennett, R., & MacArtney, J. (2008, May 22). Burma victims lose aid battle as cash flows into China on a flood of sympathy: Disaster response. *The Times*, p. 2.

Boden, J. M., & Baumeister, R. F. (1997). Repressive coping: Distraction using pleasant thoughts and memories. *Journal of Personality and Social Psychology, 73*, 45–62.

Bordieri, J. E., Sotolongo, M., & Wilson, M. (1983). Physical attractiveness and attributions for disability. *Rehabilitation Psychology, 28*, 207–215.

Bulman, R. J., & Wortman, C. B. (1977). Attributions of blame and coping in the "real world": Severe accident victims react to their lot. *Journal of Personality and Social Psychology, 35*, 351–363.

Callan, M. J., Ellard, J. H., & Nicol, J. E. (2006). The belief in a just world and immanent justice reasoning in adults. *Personality and Social Psychology Bulletin, 32*, 1646–1658.

Chaikin, A. L., & Darley, J. M. (1973). Victim or perpetrator? Defensive attribution of responsibility and the need for order and justice. *Journal of Personality and Social Psychology, 25*, 268–275.

Correia, I., Vala, J., & Aguiar, P. (2001). The effects of belief in a just world and victim's innocence on secondary victimization, judgements of justice and deservingness. *Social Justice Research, 14*, 327–342.

Correia, I., Vala, J., & Aguiar, P. (2007). Victim's innocence, social categorization, and the threat to the belief in a just world. *Journal of Experimental Social Psychology, 43*, 31–38.

Crowne, D. P., & Marlowe, D. (1960). A new scale of social desirability independent of psychopathology. *Journal of Consulting Psychology, 24*, 349–354.

Dalbert, C. (2001). *The justice motive as a personal resource: Dealing with challenges and critical life events.* New York: Kluwer Academic/Plenum.

Davis, C. G., Lehman, D. R., Silver, R. C., Wortman, C. B., & Ellard, J. H. (1996). Self-blame following a traumatic event: The role of perceived avoidability. *Personality and Social Psychology Bulletin, 22*, 557–567.

DePalma, M. T., Madey, S. F., Tillman, T. C., & Wheeler, J. (1999). Perceived patient responsibility and belief in a just world affect helping. *Basic and Applied Social Psychology, 21*, 131–137.

Downey, G., Silver, R. C., & Wortman, C. B. (1990). Reconsidering the attribution-adjustment relation following a major negative event: Coping with the loss of a child. *Journal of Personality and Social Psychology, 59*, 925–940.

Drout, C. E., & Gaertner, S. L. (1994). Gender differences in reactions to female victims. *Social Behavior and Personality, 22*, 267–278.

Duckitt, J. (2005). Personality and prejudice. In J. F. Dovidio, P. Glick, & L. A. Rudman (Eds.), *On the nature of prejudice: Fifty years after Allport* (pp. 395–412). Malden, MA: Blackwell.
Ellard, J. H., & Bates, D. D. (1990). Evidence for the role of the justice motive in status generalization processes. *Social Justice Research, 4*, 115–134.
Ellard, J. H., Miller, C. D., Baumle, T., & Olson, J. M. (2002). Just world processes in demonizing. In M. Ross & D. T. Miller (Eds.), *The justice motive in everyday life* (pp. 350–362). Cambridge, UK: Cambridge University Press.
Festinger, L. (1957). *A theory of cognitive dissonance.* Stanford, CA: Stanford University Press.
Foley, L. A., & Pigott, M. A. (2000). Belief in a just world and jury decisions in a civil rape trial. *Journal of Applied Social Psychology, 30*, 935–951.
Furnham, A. (2003). Belief in a just world: Research progress over the past decade. *Personality and Individual Differences, 34*, 795–817.
Furnham, A., & Procter, E. (1989). Belief in a just world: Review and critique of the individual difference literature. *British Journal of Social Psychology, 28*, 365–384.
Hafer, C. L. (2000a). Do innocent victims threaten the belief in a just world? Evidence from a modified Stroop task. *Journal of Personality and Social Psychology, 79*, 165–173.
Hafer, C. L. (2000b). Investment in long-term goals and commitment to just means drive the need to believe in a just world. *Personality and Social Psychology Bulletin, 26*, 1059–1073.
Hafer, C. L., & Bègue, L. (2005). Experimental research on just-world theory: Problems, developments, and future challenges. *Psychological Bulletin, 131*, 128–167.
Hafer, C. L., Bègue, L., Choma, B. L., & Dempsey, J. L. (2005). Belief in a just world and commitment to long-term deserved outcomes. *Social Justice Research, 18*, 429–444.
Hafer, C. L., & Gosse, L. (2009). *Predicting alternative strategies for preserving a belief in a just world: The case of repressive coping style.* Manuscript submitted for publication.
Hardyck, J. A., & Kardush, M. (1968). A modest modish model for dissonance reduction. In R. P. Abelson, E. Aronson, W. J. McGuire, T. M. Newcomb, M. J. Rosenberg, & P. H. Tannenbaum (Eds.), *Theories of cognitive consistency: A sourcebook* (pp. 684–692). Chicago: Rand McNally.
Haynes, G. A., & Olson, J. M. (2006). Coping with threats to just-world beliefs: Derogate, blame, or help? *Journal of Applied Social Psychology, 36*, 664–682.
Hofstede, G. (1984). *Culture's consequences.* London: Sage.
Holmes, J. G., Miller, D. T., & Lerner, M. J. (2002). Committing altruism under the cloak of self-interest: The exchange fiction. *Journal of Experimental Social Psychology, 38*, 144–151.
Insko, C. A., Smith, R. H., Alicke, M. D., Wade, J., & Taylor, S. (1985). Conformity and group size: The concern with being right and the concern with being liked. *Personality and Social Psychology Bulletin, 11*, 41–50.
Jane. (2008, May 18). *Los Angeles Times.* Retrieved from http://www.latimes.com/news/nationworld/world/la-fg-china-earthquake-gb,1,6020570.graffitiboard?slice=15&limit=10

Jolly, D. (2008, May 30). Dior drops actress from ads after China remarks. *New York Times*, p. C8(L).

Jones, C., & Aronson, E. (1973). Attribution of fault to a rape victim as a function of respectability of the victim. *Journal of Personality and Social Psychology, 26*, 415–419.

Jost, J. T., Burgess, D., & Mosso, C. O. (2001). Conflicts of legitimation among self, group, and system: The integrative potential of system justification theory. In J. T. Jost & B. Major (Eds.), *The psychology of legitimacy: Emerging perspectives on ideology, justice, and intergroup relations* (pp. 363–388). New York: Cambridge University Press.

Jost, J. T., & Hunyady, O. (2002). The psychology of system justification and the palliative function of ideology. *European Review of Social Psychology, 13*, 111–153.

Kaiser, C. R., Vick, S. B., & Major, B. (2004). A prospective investigation of the relationship between just-world beliefs and the desire for revenge after September 11, 2001. *Psychological Science, 15*, 503–506.

Karuza, J., Jr., & Carey, T. O. (1984). Relative preference and adaptiveness of behavioral blame for observers of rape victims. *Journal of Personality, 52*, 249–260.

Kay, A. C., & Jost, J. T. (2003). Complementary justice: Effects of "poor but happy" and "poor but honest" stereotype exemplars on system justification and implicit activation of the justice motive. *Journal of Personality and Social Psychology, 85*, 823–837.

Kay, A. C., Jost, J. T., & Young, S. (2005). Victim-derogation and victim-enhancement as alternate routes to system justification. *Psychological Science, 16*, 240–246.

Kenrick, D. T., Reich, J. W., & Cialdini, R. B. (1976). Justification and compensation: Rosier skies for the devalued victim. *Journal of Personality and Social Psychology, 34*, 654–657.

Kleinke, C. L., & Meyer, C. (1990). Evaluation of rape victim by men and women with high and low belief in a just world. *Psychology of Women Quarterly, 14*, 343–353.

Kruglanski, A. W. (1996). Motivated social cognition: Principles of the interface. In E. T. Higgins & A. W. Kruglanski (Eds.), *Social psychology: Handbook of basic principles* (pp. 493–520). New York: Guilford.

Lea, J. A., & Hunsberger, B. E. (1990). Christian orthodoxy and victim derogation: The impact of the salience of religion. *Journal for the Scientific Study of Religion, 29*, 512–518.

Lerner, M. J. (1980). *The belief in a just world: A fundamental delusion*. New York: Plenum.

Lerner, M. J. (2003). The justice motive: Where psychologists found it, how they lost it, and why they may not find it again. *Personality and Social Psychology Review, 7*, 388–399.

Lerner, M. J., & Miller, D. T. (1978). Just world research and the attribution process: Looking back and ahead. *Psychological Bulletin, 85*, 1030–1051.

Lerner, M. J., Miller, D. T., & Holmes, J. G. (1976). Deserving and the emergence of forms of justice. In L. Berkowitz & E. Walster (Eds.), *Advances in experimental social psychology* (Vol. 9, pp. 133–162). New York: Academic Press.

Lerner, M. J., & Simmons, C. H. (1966). Observer's reaction to the "innocent victim": Compassion or rejection? *Journal of Personality and Social Psychology, 4*, 203–210.

Lincoln, A., & Levinger, G. (1972). Observers' evaluations of the victim and the attacker in an aggressive incident. *Journal of Personality and Social Psychology, 22*, 202–210.

Maes, J. (1998). Immanent justice and ultimate justice: Two ways of believing in justice. In L. Montada & M. J. Lerner (Eds.), *Responses to victimizations and belief in a just world* (pp. 9–40). New York: Plenum.

Maes, J., & Schmitt, M. (1999). More on ultimate and immanent justice: Results from the research project "Justice as a problem within reunified Germany." *Social Justice Research, 12*, 65–78.

Markus, H., & Kitayama, S. (1991). Culture and the self: Implications for cognition, emotion, and motivation. *Psychological Review, 98*, 224–253.

Miller, D. T. (1977). Altruism and threat to a belief in a just world. *Journal of Experimental Social Psychology, 13*, 113–124.

Mohiyeddini, C., & Montada, L. (1998). Belief in a just world and self-efficacy in coping with observed victimization. In L. Montada & M. J. Lerner (Eds.), *Responses to victimizations and belief in a just world* (pp. 41–54). New York: Plenum.

Myers, L. B., Burns, J. W., Derakshan, N., Elfant, E., Eysenck, M. W., & Phipps, S. (2008). Current issues in repressive coping and health. In A. Vingerhoets, I. Nyklícek, & J. Denollet (Eds.), *Emotion regulation: Conceptual and clinical issues* (pp. 69–86). New York: Springer Science + Business Media.

Myers, L. B., & Reynolds, D. (2000). How optimistic are repressors? The relationship between repressive coping, controllability, self-esteem and comparative optimism for health-related events. *Psychology and Health, 15*, 677–687.

Ni, C. (2008, May 30). China eager to show it can handle Olympics and crisis: Officials say the quake has offered a rare opportunity for the Beijing Games host to practice its crisis management skills. *Los Angeles Times*, p. 1.

Novak, D. W., & Lerner, M. J. (1968). Rejection as a consequence of perceived similarity. *Journal of Personality and Social Psychology, 9*, 147–152.

Olson, J. M., & Stone, J. (2005). The influence of behavior on attitudes. In D. Albarracín, B. T. Johnson, & M. P. Zanna (Eds.), *The handbook of attitudes* (pp. 226–249). Mahwah, NJ: Lawrence Erlbaum.

Pancer, S. M. (1988). Salience of appeal and avoidance of helping situations. *Canadian Journal of Behavioural Science, 20*, 133–139.

Park, C. L., Cohen, L. H., & Murch, R. L., (1996). Assessment and prediction of stress-related growth. *Journal of Personality, 64*, 71–105.

Shaw, J. I., & McMartin, J. A. (1977). Personal and situational determinants of attribution of responsibility for an accident. *Human Relations, 30*, 95–107.

Simpson, B., Irwin, K., & Lawrence, P. (2006). Does a "norm of self-interest" discourage prosocial behavior? Rationality and quid pro quo in charitable giving. *Social Psychology Quarterly, 69*, 296–306.

Smith, R. E., Keating, J. P., Hester, R. K., & Mitchell, H. E. (1976). Role and justice considerations in the attribution of responsibility to a rape victim. *Journal of Research in Personality, 10*, 346–357.

Sorrentino, R. M. (1981). Derogation of an innocently suffering victim: So who's the "good guy"? In J. P. Rushton & R. M. Sorrentino (Eds.), *Altruism and helping behavior* (pp. 267–283). Hillsdale, NJ: Lawrence Erlbaum.

Sorrentino, R. M., Hancock, R. D., & Fung, K. (1979). Derogation of an innocent victim as a function of authoritarianism and involvement. *Journal of Research in Personality, 13*, 39–48.

Sorrentino, R. M., & Hardy, J. E. (1974). Religiousness and derogation of an innocent victim. *Journal of Personality, 42*, 372–382.

Taylor, J. A. (1953). A personality scale of manifest anxiety. *Journal of Abnormal and Social Psychology, 48*, 285–290.

Thornton, B. (1992). Repression and its mediating influence on the defensive attribution of responsibility. *Journal of Research in Personality, 26*, 44–57.

Virginia Head Start Association. (2007). *About Virginia Head Start.* Retrieved June 19, 2008, from http://www.headstartva.org/about/index.htm

Walster, E., Berscheid, E., & Walster, G. W. (1973). New directions in equity research. *Journal of Personality and Social Psychology, 25*, 151–176.

Weinberger, D. A., Schwartz, G. E., & Davidson, R. J. (1979). Low-anxious, high-anxious, and repressive coping styles: Psychometric patterns and behavioral and physiological responses to stress. *Journal of Abnormal Psychology, 88*, 369–380.

Wilson, A. (Ed.). (2003). *World scripture: A comparative anthology of sacred texts.* St. Paul, MN: Paragon House.

Wolfradt, U., & Dalbert, C. (2003). Personality, values and belief in a just world. *Personality and Individual Differences, 35*, 1911–1918.

Wyer, R. S., Jr., Bodenhausen, G. V., & Gorman, T. F. (1985). Cognitive mediators of reactions to rape. *Journal of Personality and Social Psychology, 48*, 324–338.

Yardley, J., & Barboza, D. (2008, May 20). Many hands, not held by China, aid in quake. *New York Times*, p. A1.

York, G. (2008, May 17). "Shock of consciousness" sweeps autocratic China in wake of temblor. *Globe and Mail*, p. A1.

Zanna, M. P., & Aziza, C. (1976). On the interaction of repression-sensitization and attention in resolving cognitive dissonance. *Journal of Personality, 44*, 577–593.

Zigler, E., & Muenchow, S. (1992). *Head Start: The inside story of America's most successful educational experiment.* New York: Basic Books.

CHAPTER 5

From Moral Outrage to Social Protest

The Role of Psychological Standing

DALE T. MILLER, DANIEL A. EFFRON, and SONYA V. ZAK

Stanford University

Abstract: The thesis of this chapter is that the decision to protest requires not only that people experience outrage but that they also feel entitled to act on their outrage. The authors call this feeling of entitlement *psychological standing*. One important determinant of a person's standing to protest an injustice is the extent to which he or she is materially affected by it. The more one is materially affected by the source of outrage, the more standing one has to protest it. When people lack a material stake in an issue, they can nonetheless feel that they have the standing to protest if they observe other nonvested individuals protesting or if they perceive themselves as having a moral stake in the issue. Having a personal characteristic or history that justifies to others why one feels such outrage can also provide one with standing. Not just any connection to an issue will suffice, however. Having committed a particular transgression in the past or simply being a member of a group that has committed (or continues to commit) that transgression deprives one of the standing to protest that particular transgression. Finally, having a material stake in an issue's outcome is not always sufficient to license protest. Victims lack the standing to retaliate against a transgressor when others who have been more victimized by the transgression choose to turn the other cheek. The authors conclude the chapter by showing that the concept of standing, in addition to permitting unique predictions, offers an alternative frame for viewing previous findings.

Keywords: social protest, entitlement, legitimacy, moral mandate, psychological standing

When people are confronted with circumstances they judge to be unfair, they commonly experience some form of anger or moral outrage (Mikula, 1993; Miller, 2001; Solomon, 1990). How they respond to their outrage varies: Sometimes it leads them to protest the perceived injustice, and sometimes not (Jackman, 1994; Olson & Hafer, 2001). Clarifying when and for whom moral outrage provokes social protest is the goal of this chapter. In particular, we will show that the decision to protest requires not only that people experience outrage but that they also feel entitled to act on their outrage. We call this feeling of entitlement *psychological standing*, and in this chapter we illustrate how its absence can constrain—and, conversely, how its presence can liberate—social protest.

Our analysis begins with a consideration of an empirical anomaly: The personal cost inflicted by a perceived injustice influences less how one *feels about it* than how one *acts toward it* (Green & Cowden, 1992). More specifically, whereas those affected and unaffected by a controversial policy or action are often equally outraged by it, those directly affected by it are much more likely to protest it. As an example, consider American Whites' attitudes toward school busing as a means of achieving racial integration during the 1960s and 1970s. This policy was unpopular among Whites in the targeted areas when it was implemented, and this was as true for those Whites who had no school-age children, and hence were not personally inconvenienced by it, as for those who did (Sears, Hensler, & Speer, 1979). On the other hand, it was Whites with school-age children who dominated the membership of antibusing organizations (Green & Cowden, 1992).

The finding that personal relevance does not dispose people to feel more negatively toward a controversial social policy but renders them more likely to act on their outrage is well established. Regan and Fazio (1977), for example, found that Cornell University undergraduates were equally outraged by a campus housing shortage whether or not they were personally inconvenienced by it, but those directly affected were much more likely to participate in social action to alleviate it. Similarly, Sivacek and Crano (1982) found that Michigan college students affected by a proposal to raise the state drinking age were no more opposed to it than were those old enough that their ability to drink would not be hindered by the new drinking age; those younger than the drinking age, however, were more likely than those older than it to get involved with a local group protesting the change.

The clear conclusion from this research is that being outraged by an action or policy in which one does not have a stake is insufficient to prompt one to take social action. Green and Cowden (1992) offered an account for why outrage alone does not prompt social action. They proposed that the opportunity to protest an injustice (in contrast to merely feeling moral outrage about it) leads the sympathetic actor to ask "Is it worth it?" They argued that the answer to this question (especially if money, time, or effort is required) is more likely to be affirmative when the actor has a stake in the issue. Stated more generally, one may not need to be directly affected by a policy or action to be outraged by it, but one does require a level of motivation that only having a stake in the issue can provide to convert a sense of outrage into an act of protest.

Miller and Ratner (Miller, 1999; Miller & Ratner, 1996, 1998; Ratner & Miller, 2001) offered an alternative account for why a sense of outrage is insufficient to produce social protest. They emphasized not the diminished motivation of would-be actors who lack a material stake but their diminished feelings of entitlement to act on their outrage. They contended that not everyone outraged by an act or policy will feel equally comfortable acting on their outrage. The actor faced with behavioral involvement in a social cause must ask not only "Is it worth it?" but also "Is it appropriate?" Concluding that it is not their "place" to protest will inhibit people from acting, irrespective of how unjust they find the state of affairs. In other words, what often prevents people from protesting is not a lack of motivation to protest but rather their feeling that they lack the legitimacy to do so.

The difference between Miller and Ratner's and Green and Cowden's accounts is captured by the distinction between avoidance and approach motivation. As Lewin (1951) noted, people's reluctance to undertake a particular action can be due either to weak motivation to do so (the absence of approach motivation) or to strong inhibitions against doing so (the presence of avoidance motivation). In explaining the greater tendency of vested actors than nonvested actors to protest, Green and Cowden pointed to the greater approach motivation afforded by being vested, and Miller and Ratner pointed to the greater avoidance motivation afforded by not being vested.

The idea that not everyone who believes something is unjust is equally entitled to protest is captured in the legal system by the concept of legal standing. According to the law, one cannot bring a suit for judicial review merely because one feels an injustice has occurred; one must additionally be able to show that one has been materially affected by it. Even if one is morally outraged by the policies or actions of another, without demonstrable personal injury, one does not have sufficient grounds for bringing legal action. So, for example, Whites who could not show that they were materially affected by the school busing policy or college students who could not show that they would be materially affected by the change in the drinking age could not gain access to the courts—for they, unlike their counterparts with a vested interest, would lack standing.

Psychological Standing

In this section we argue that the lack of psychological standing, defined as the subjective feeling of entitlement or legitimacy to perform a particular action, inhibits people from expressing their outrage through social protest. At least four factors determine the extent to which people feel entitled or disentitled to protest. These factors are the extent to which one has either a material stake or a moral stake in an issue, the extent to which one can link an issue in which one does not have a stake to another in which one does, and the extent to which one has a personal characteristic or history that explains one's outrage. We consider each factor in turn.

Psychological Standing From Material Stakes

One important determinant of a person's psychological standing to protest an injustice is the extent to which he or she is materially affected by it. Initial empirical support for the claim that a lack of standing inhibits protest among those not materially affected by the injustice came from a study by Ratner and Miller (2001). These researchers presented Princeton undergraduates with a circumstance designed to offend their sense of justice—the proposed decision to shift government funds from a worthy cause to an unworthy one. After reading of the injustice ("Proposition 174"), participants were given the opportunity to indicate their opposition to it and also their willingness to protest it by assisting an organization called Princeton Opponents of Proposition 174. The researchers manipulated the participants' material stake in the cause by varying whether the threatened "worthy" cause would exclusively benefit their sex or the opposite sex.

Ratner and Miller correctly anticipated that although participants would be equally opposed to the proposed action irrespective of whether it exclusively affected their sex or the opposite sex, these comparably strong feelings of opposition would more likely translate into behavioral opposition among those with a material stake (i.e., vested participants) than among those without a stake (i.e., nonvested participants). Specifically, whereas 94% of the vested students signed a petition and 50% agreed to write a statement, only 78% of the nonvested students signed the petition, and only 22% agreed to write a statement.

Ratner and Miller (2001) further reasoned that if the higher rate of protest among vested participants than nonvested participants in this study was due to the nonvested participants' lack of psychological standing rather than their lack of incentive, then granting standing to the nonvested participants should equalize the rates of protest between both groups. They tested this hypothesis by including another set of conditions that employed a different name for the group that participants had the opportunity to help. The rationale for this manipulation was that the name Princeton Opponents of Proposition 174 did not explicitly exclude the unvested gender, but it also did not extend to students the standing they needed to feel comfortable acting. In this other set of conditions, therefore, Ratner and Miller called the group Princeton Men and Women Opposed to Proposition 174, reasoning that the inclusiveness of this name would provide even the nonvested gender with the standing needed to feel comfortable participating. If this organization welcomed both male and female advocates, then both males and females must be entitled to advocate, even if one gender lacked a self-interested stake in the proposition. Stated differently, the inclusive name was expected to license members of the nonvested gender to act on their feelings. It communicated that a self-interested stake was not necessary for them to have psychological standing.

The results supported these predictions: When the group bore the inclusive name, nonvested participants were just as likely as vested participants to protest by signing the petition and writing the statement. Ratner and Miller argued that the fact that the group label explicitly acknowledged the appropriateness of nonvested actors' joining the protest group disinhibited them from joining. Without a sufficient stake, even passionate actors may feel inhibited from acting unless their inhibitions are reduced by a framing that grants them standing to protest in some other way.

Although Ratner and Miller contended that the inclusive group label—Princeton Men and Women Opposed to Proposition 174—elicited more protest behavior from nonvested actors because it legitimated their participation, there is another possibility. Namely, the inclusive label may have generated greater rates of protest behavior among the nonvested actors not because it liberated them to protest but because it increased their felt obligation to protest. The fact, however, that there was a stronger (positive) correlation between attitudinal opposition and commitment to social protest among the nonvested participants in the inclusive condition than in the exclusive condition favors the disinhibition hypothesis over the obligation hypothesis. That is, the correlational pattern suggests that nonvested actors felt freer to act on their attitudes; where the disinhibition account suggests they should feel most free, but the obligation account suggests they should feel least free.

It is not necessary that the cost of the injustice be highly significant for concerns about standing to influence protest behavior. Consider the case of an illegitimate intrusion into an existing queue (line). When someone enters a queue at some point other than the end, the "first come first served" justice principle is violated, and protest commonly ensues. To examine how standing affects the inclination of queue members to speak up in response to an unjust line intrusion, Zak and Miller (2008) conducted a field study at a large national retailer location. The procedure they used to create a violation of queue norms was as follows: One confederate (the buffer) took a place in line; when he reached the third position from the front and at least the third position from the back, another confederate (the intruder) asked to cut in line. The buffer always allowed the intruder to cut in line, though the researchers manipulated whether he allowed the intruder ahead of or behind his own position.

Following the staging of the intrusion, the pattern of protest behavior among queue members was noted. There was a strong tendency for people ahead of and behind the intrusion point to speak up more when the buffer let the intruder in the line behind him as opposed to ahead of him. Most relevant to the present analysis, however, there was a strong tendency for those behind the intruder to speak up more than those ahead of the intruder. That people in front of the intrusion point spoke up less frequently than those behind may simply indicate that they were less outraged by it. After all, they were not going to be delayed by the intrusion. But it is also possible that their reluctance to speak up was due, at least partly, to their lacking the standing to do so. Support for the latter claim comes from the finding that those not materially affected by the intrusion tended to speak up only once one or more of the materially affected victims spoke up. The idea that people in front of the intrusion cared less about this violation than those behind the intrusion cannot by itself easily account for this finding. Their lack of standing, on the other hand, can explain both why they were less inclined to speak up and why when they did speak up, it was only after those bearing a material cost did.

Psychological Standing From Moral Stakes

Ratner and Miller's (2001) research suggests that having a self-interested stake grants people the psychological standing to protest. How else might people obtain psychological standing? Once again, the legal system provides a concept to which there may be a psychological analogue. In international law, as in domestic law, not all parties are entitled to bring legal action against perpetrators. The United States, for example, does not have the jurisdiction to prosecute Canadian citizens who commit crimes in Canada against other Canadians, in part because the United States has no material stake in preventing such crimes. The concept of *universal jurisdiction*, however, provides an exception. Under universal jurisdiction, all countries are permitted to apprehend and prosecute the so-called *hostis humani generis* (i.e., "the enemies of all people") because all countries are thought to have a stake in preventing such crimes. Because all countries have a material stake in keeping the seas free from piracy, for example, the United States would be allowed to apprehend and prosecute pirates who attack a British ship outside of American waters. More important, international law extends universal jurisdiction to matters in which countries have only a *symbolic* or *moral stake*. Such is the case with offenses dubbed "crimes against humanity." For example, although genocide committed in a foreign country by foreign nationals against their own people does not affect the material interests of the United States, the egregious nature of such a crime offends the international community's basic moral values and thus vests the United States and other nations with the entitlement to prosecute those responsible (Randall, 1988).

Individuals may similarly feel entitled to protest against policies they perceive as morally reprehensible, even if these individuals lack a self-interested stake in these policies' outcomes. In other words, perceiving oneself as having a moral stake in an issue may grant one the psychological standing to protest, even if one lacks a material stake.

Historical and contemporary acts of protest lend anecdotal support to this idea. Consider the Freedom Riders of the early 1960s who risked (and in many cases experienced) incarceration and bodily harm to protest racial discrimination in the Deep South. Although the many White members of the Freedom Riders may have lacked a material stake in the civil rights movement, they no doubt perceived civil rights as a moral issue. Consider also the observations that many heterosexuals vocally protest efforts to legalize or prohibit same-sex marriage, or that many men vocally protest abortion or efforts to restrict it, despite these groups' apparent lack of a self-interested stake in these issues. Notably, however, these are issues that the public discourse (on both sides of each debate) "moralizes" (i.e., characterizes as relevant to moral values).

Although it is possible that moralization of these issues motivated these nonvested groups to protest by increasing their feelings of obligation, moralization may have also licensed members of these groups to act on their preexisting outrage. This licensing explanation is consistent with the United Nations' application of universal jurisdiction to acts of piracy: "Parties have the right, but not the obligation, to assume jurisdiction over piratical acts with which they have no connection" (Randall, 1988, p. 792). In other words, a perceived moral stake may give them the entitlement to protest rather than obligate them to do something they would rather avoid.

If moral stakes grant people psychological standing, then nonvested actors who feel comfortable protesting an issue should moralize the issue to a greater extent than vested actors, whose self-interested stake in the issue already grants them standing. Effron and Miller (2008) tested this prediction in a sample of undergraduates who all supported legalized abortion. The researchers measured participants' attitudes about abortion and how comfortable they would feel engaging in different behaviors in protest of antiabortion legislation (e.g., by attending a demonstration, signing a petition, attending a meeting of people who shared their attitude about abortion). They also measured participants' "moral mandates" (Skitka, Bauman & Sargis, 2005), that is, how much participants felt that their attitudes about abortion reflect their "core moral values and convictions." Effron and Miller reasoned that because of women's greater self-interested stake in the abortion issue, women would feel more comfortable protesting than men—unless men expressed a relatively strong moral mandate, in which case men and women would feel equally comfortable. The results supported this reasoning: Among male and female participants who shared the same strong attitudes about abortion and felt equally comfortable protesting antiabortion laws, men had a significantly stronger moral mandate than did women. Controlling for moral mandates revealed that if men and women had both moralized the issue to the same extent, then men would have felt less comfortable protesting than women. One interpretation of these correlational results is that for people who lack a material stake to feel just as comfortable speaking up as those with a material stake, they must have a moral stake in the issue.

In a second study, Effron and Miller (2008) manipulated whether participants had a moral stake in an issue and again measured how comfortable they would feel publicly expressing their privately held attitude. In a paradigm closely paralleling the previously described study by Ratner and Miller (2001), they told participants about Proposition 174, which would shift funding from a worthy cause to an unworthy one. To manipulate a self-interested stake, they randomly assigned participants to learn that the worthy cause benefited either only their own gender (vested condition) or only the other gender (nonvested condition). Orthogonal to this manipulation, they manipulated whether participants read a moralizing passage. Participants randomly assigned to read this passage learned that an advocacy group opposed the funding cuts on the grounds that society has "a moral obligation to protect public health"; this advocacy group urged people to listen to their "core values or convictions" and oppose the proposition. The data showed that of participants who did *not* read the moralizing passage, those who lacked a self-interested stake indicated less comfort publicly expressing their attitudes about the funding cuts than did participants who did have a self-interested stake, thus replicating results from Ratner and Miller (2001). For participants who had read the moralizing passage, however, no differences in comfort were found between vested and nonvested participants. The moralizing passage apparently increased nonvested participants' comfort speaking up.

Additional data from this study suggest that the moralizing passage liberated nonvested participants to act on their moral outrage rather than increasing their motivation to protest. First, measures of *private* attitudes did not respond to any of the manipulations; all participants in all conditions opposed the funding cuts. Persuasion produced by the moralizing passage thus cannot account for these results. Second, private attitudes predicted comfort protesting only when the manipulation granted them a self-interested stake, a moral stake, or both. Taken together these two findings suggest that the moralizing passage gave these participants the standing to feel comfortable acting on their privately held attitudes rather than increasing nonvested participants' felt obligation to speak up.

The finding that attitude change did not even accompany, let alone cause, the behavioral change found in the Effron and Miller (2008) study speaks to an important fact about the different ways persuasive arguments can affect behavior: They can either change people's attitudes, with their newly formed attitudes then leading people to change their behavior, or simply license people to act on the attitudes they already have. For those trying to exhort people to action, it is important to know whether the challenge they are facing is insufficient attitudinal support or insufficient comfort in acting on existing attitudes.

Arguments that may effectively change attitudes may not effectively license behavior, and vice versa. Kuran (1997), for example, talked about the powerful role played by the argument that slavery was un-Christian in the period leading up to the emancipation of slaves in America. As Kuran noted, this argument was not a decisive one when expressed earlier in the antislavery campaign, but it became increasingly decisive as abolition sentiment grew, suggesting that this rallying cry did not so much shift attitudes in an abolitionist direction as it did license already shifted attitudes. Thus, the argument functioned more to provide converts to the abolitionist cause with psychological standing—in this case, by giving them a moral stake in the slavery issue—than to produce more converts.

Psychological Standing From Links to Other Issues

Moralization, we argue, makes people feel comfortable protesting because it transforms an issue in which only certain people have a (material) stake into an issue in which everyone has a (moral) stake. In this way, granting nonvested individuals a moral stake links their outrage about an issue on which they lack the standing to protest (e.g., abortion) to an issue that everyone has the standing to protest (e.g., threats to the right to life or threats to the freedom to choose). This analysis suggests a more general strategy for granting people standing. If someone feels outraged over an issue but does not feel entitled to express his or her outrage, linking that issue to another one that the person does feel entitled to protest should give him or her the standing to speak up about the first issue (cf. McAdams, 1997).

The idea that linking issues together provides people with standing may explain a surprising asymmetry that has been observed in the relation of males' and females' attitudes toward social issues and their political participation in these issues. There are many social policy issues that men and women feel differently about but none more so than gun control. Women consistently report stronger pro-gun-control attitudes than men. Does this intensity gap, as it is called (for even men's attitudes are pro-gun control), mean that women are more likely than men to engage in political action in support of gun control? Historically, this has not been the case. Consistent with the general tendency of men to be more likely to engage in political action than women, the 1996 National Gun Policy survey, for instance, found that men were actually more likely to be involved in pro-gun-control activities. But by 1999 the participation gap between men and women had narrowed or reversed. Men were only marginally more likely to have joined or given money to a gun-control group, and women were actually more likely to have contacted a public official (Goss, 2003). So what happened in the late 1990s to lead women to finally convert their more intense pro-gun-control attitudes into action? According to Kristin Goss (2003), the answer is that during this time, the gun-control debate had been reframed in a way that empowered women to act. In particular, gun control, traditionally linked to crime prevention, now became a child safety issue, a change that allowed women to own their feelings. Because gender norms legitimate women's speaking out about child-related issues, linking women's outrage about gun control to child safety issues gave them more standing to protest lenient gun laws, thus allowing their political participation in the gun-control issue to be commensurate with their attitudes.

Psychological Standing From Personal Experience

We have argued that having a future stake, whether material or moral, in an issue grants people the standing to speak up about that issue. Having had past experience with an issue may also provide people with standing, even when this past experience does not implicate a future stake. The importance of standing of this type can sometimes be seen in the categories of members who are most and least likely to assume leadership roles in organizations.

As an example of how standing affects the leadership dynamics of organizations, consider the involvement of Whites in the NAACP (National Association for the Advancement of Colored People). The NAACP's goals remain the same as when it was founded almost 100 years ago:

> To promote equality of rights and eradicate caste or racial prejudice among the citizens of the United States; to advance the interest of colored citizens; to secure for them impartial suffrage; and to increase

> their opportunities for securing justice in the courts, education for their children, and complete equality before the law.

Whites have been represented in the membership of the organization since its inception, yet in the year 2007 only 1 of its over 400 branch presidents was White. Why would the percentage of Whites in leadership positions in the NAACP not mirror the percentage of Whites (over 10%) on the organization's membership rolls? Do White members tend to be less committed to the goals of the organization than Black members? Possibly, as the organization's explicit goal of promoting the welfare of African Americans seems to give Blacks a greater material stake in the organization than Whites. But there is another possibility, one that assumes that the Whites who join the NAACP are every bit as committed to the organization and its goals as are the Blacks who join it. This possibility points to a racial gap not in commitment but in standing. Whatever prevents members from running for, or being elected to, a leadership position in the NAACP, Whites have the additional hurdle, by virtue of their race, of not having the standing in the organization that Blacks do. Their entitlement, perhaps even to belong but certainly to lead an organization with the NAACP's mandate, is simply less than it is for Blacks.

What would give Whites the standing to hold a leadership position in the NAACP? News articles about the 2007 election of a White president to the Georgetown chapter of the NAACP provide a clue. These articles focused to a great extent on the past experience the new president had that gave her the standing to be president despite being White. One article described the individual as follows:

> A few highlights from the bio: Gunderson's from a predominantly black working-class suburb of Detroit; she aspires to be a civil rights lawyer, either at the Southern Poverty Law Center or the NAACP; she never planned to be the group's president this year, but someone nominated her, and she cared about the issues. She figured, Why not?
>
> One last thing: Gunderson is white. Listen to her speak, and you might never know. Her vocal inflection is unmistakably "urban."

The article went on to say,

> Now black students generally regard Gunderson as she regards herself: the sum of her parts. She has civil rights experiences—both participating in diversity-related groups in high school and in watching her many black friends endure prejudice. And, most important, she understands what it feels like to look around and only see people who don't look like you. (Samuelson, 2007, p. 37)

These passages imply that the new president's past—embodied in her "urban accent"—gives her a personal connection to the issues that the NAACP takes up. This personal connection gives her the standing to assume a leadership position, despite her race. Note that these experiences do not necessarily vest her with a material stake in promoting the NAACP's goals. It appears that past experiences can substitute for future stakes in providing people with standing. Not only did the president's past experiences make her feel comfortable running for office, they also apparently gave her legitimacy in voters' eyes as well.

That a lack of perceived standing can prevent a certain category of members from assuming leadership roles in organizations is also evident in the case of MADD (Mothers Against Drunk Driving). Despite admitting male members for the 25-year history of the organization, it was not until 2007 that it elected its first male president. Notably, a drunk driver had killed this male president's son ("MADD's First Male President to Honor Area Police," 2005). Formal eligibility requirements for the presidency of MADD require neither that you have lost a child to a drunk driver nor that you be a mother, but standing requirements favor candidates for whom both are true. As difficult as it is to imagine MADD members electing a male president, it is even more difficult to imagine them electing one who did not at least have the standing provided by personal experience with the issue.

Like leaders of cause organizations, advocates for causes frequently have a connection to the cause and readily publicize it (e.g., by identifying that they or someone close to them has been affected personally by it). Given our position that personal connections provide people with standing, publicizing these connections seem to be a reasonable strategy to establish one's entitlement to advocate.

Consider the case of James Brady, President Ronald Reagan's press secretary, who was shot while protecting Reagan from an assassination attempt. Brady and his wife, Sarah, have since campaigned vigorously and effectively for stricter gun-control laws. Why have the Bradys been effective advocates for their cause? One possibility is that their personal tragedy gives them credibility on the gun-control issue. This is possible, but it is difficult to see how their personal connection with the issue would give them expertise in the customary way one thinks about credibility. Indeed, one might argue that their experience biased them on the issue (cf. Kelley, 1972). But whether or not advocates have a meaningful connection to the cause they promote increases their credibility and hence their persuasiveness, it will increase their standing—which may ultimately be more critical to their effectiveness (Miller, Ratner, & Zhao, 2008). Advocates who emphasize their standing increase the likelihood that others will listen when they present their claims. Having standing helps advocates gain entry to the halls of power, access media outlets, and obtain positions in which they can more easily collect money, signatures, or other forms of support. Just as people would probably not elect a male president of MADD whose life had not been in some way touched by drunk driving, people are more likely to allow a victim of gun violence to speak publicly about his or her views on gun control.

Undermining Standing

Standing is necessary not only to protest a policy or promote a cause but also to protest particular actions of another person that one finds unjust or offensive. Suffering the consequences of such actions usually bestows psychological standing on the victim by virtue of his or her personal connection or material or moral stake. At least three factors, however, can undermine the standing derived from stakes and personal connection, thus making people uncomfortable expressing their outrage even when they directly experience the consequences of the injustice. These factors are the lack of an appropriate relationship to the perpetrator, the presence of a similar transgression in one's past, and the presence of others who have suffered a greater injustice than oneself but have declined to protest. We consider each factor in turn.

When Lack of Shared Group Membership With the Perpetrator Undermines Standing

Often people do not protest behavior that they find offensive because they feel that the relationship they have to the offending party does not give them the standing to do so. Consider the norms pertaining to the appropriate response to a publicly misbehaving young child. A quick perusal of parenting Web sites shows that the question of what norms guide the reaction to someone else's child are much discussed and debated and that issues of both obligation and entitlement arise. On the one hand, there is the question of whether anyone other than the parent is *obligated to* discipline the misbehaving child; on the other hand, there is the question of whether anyone outside of the child's family is *entitled to* discipline the misbehaving child. Reasons for not intervening thus sometimes reflect the denial of responsibility ("It's not my job to discipline someone else's child") but perhaps as often reflect the denial of entitlement ("It's not my place to discipline someone else's child"). Even though the offended person is directly affected by the child's misbehavior (and therefore has a personal stake in the situation), the absence of the requisite relationship with the perpetrator (i.e., the child) undermines the person's standing to speak up.

A similar dynamic arises when one hears derogatory comments made about an out-group by a member of the disparaged group. As an example, consider the circumstances of a male who hears misogynistic comments made by a female. However offended the male is by the comments, he will likely be less comfortable rebuking the female than he would another male. For one thing, he might construe the comment differently coming from a female than from a male; surely the comment is not, in fact, sexist if a woman would say it. A second possibility is that, personally offended though he may be, he may simply feel that it is not his place to rebuke the female speaker. As a woman, she has the standing to criticize her own group; as an out-group member, the man lacks the standing to chastise her.

Experimental evidence supports the claim that observers are sensitive to a critic's relation to the group that he or she criticizes. In one relevant line of work, a target person who made derogatory comments about Australians was perceived more negatively when he was presented as a non-Australian than when he was presented as an Australian (Hornsey, Trembath, & Gunthorpe, 2004). Importantly, both Australian and non-Australian participants displayed this effect, suggesting that negative perceptions of the non-Australian critic are driven by shared understandings of who has a right to criticize (Sutton, Douglas, Elder, & Tarrant, 2008; Sutton, Elder, & Douglas, 2006). Because critics lack the standing to criticize perpetrators with whom they do not share a social identity, they can expect more negative social consequences for their criticism relative to in-group critics.

The importance of a shared group membership with the offending party becomes especially salient when people's actions undermine this relationship. Consider the words of a Philadelphia prosecutor quoted in Buzz Bissenger's (1997) *A Prayer for the City* as he explains why he refuses to move to the suburbs despite the city's high taxes and deteriorating conditions: "I like being a Philadelphian. Once you leave Philadelphia you lose your standing to care and complain about it" (p. 14). In essence, the prosecutor is claiming that if you don't live and have a stake in a city, you are not entitled to complain. By leaving the city, you have changed your relationship with the offending party (in this case, Philadelphia itself); by moving, you have undermined the standing that your personal experience with the city granted you.

Research by Hornsey and Imani (2004) provides an empirical demonstration of how changing one's group membership relative to the offending party can undermine one's standing. These researchers presented Australian participants with anti-Australian comments attributed to a fellow Australian, a foreigner, or an Australian who had recently left the country and adopted British citizenship. Participants reacted just as negatively to the ex-Australian critic as they did to the foreign critic, which was significantly more negative than reactions to the Australian critic. For the same reason that Bissinger's prosecutor worried about the decision to move out of Philadelphia, people seem to feel that implicitly or explicitly renouncing their group membership revokes their license to criticize that group. If you have abandoned the requisite relationship with the perpetrator, you no longer have the standing to express your outrage.

When One's Past Behavior Undermines Standing

We have emphasized that having a personal connection to an issue in one's past gives one the standing to speak out about that issue in the present, as in the example of the White president of the NAACP. There is a particular kind of past experience, however, that undermines rather than grants standing. Having once committed a particular transgression undermines one's standing to protest a similar transgression in the future. For example, Israel frequently challenges European countries' criticism of Israel's treatment of Palestinians by referring to the Europeans' disregard for the fate of Jews during the 1930s and 1940s.

A lack of standing may also prevent members of groups who have committed particular transgressions from speaking up even when they themselves are victimized by those same transgressions. In a national phone survey, a full 50% of European American respondents who believed that they had been the target of racial discrimination chose not to speak up about their feelings of discrimination, whereas only 32% of African Americans who reported being the target of racial discrimination reported not speaking up (Dixon, Storen, & Van Horn, 2002; see Kaiser & Major, 2006). One interpretation of these observations is that European Americans felt they lacked the entitlement to protest reverse discrimination given their own racial group's historical and contemporary discrimination against African Americans. Similarly, in a survey of over 8,000 federal employees, only 23% of men who felt that they were victims of sexual harassment reported "asking or telling the [offender] to stop," whereas 41% of women reported confronting the harasser in this manner (U.S. Merit Systems Protection Board, 1995). Again, it is tempting to speculate that men, as members of a group that people likely view as the primary perpetrator of sexual harassment, feel that they lack standing to protest their own victimization by sexual harassment.

A similar dynamic occurs in families when parents find themselves inhibited from protesting some behavior in their children, such as smoking, that they themselves once did. It is not exactly hypocrisy to chastise your adolescent child for smoking if you smoked as a teenager (it would be hypocrisy if you continued to do it), but there is something illegitimate about it. At least the parent is unlikely to be seen by the child as having the right to protest his or her smoking. We have no standing to inveigh against the dangerous or immoral actions of others when we previously did the same thing ourselves.

When Others' Refusal to Protest Undermines Standing

A third factor that can undermine one's standing to protest a transgression is the behavior of others who have suffered more from the same transgression than oneself. It is widely recognized that the feeling of deprivation and the ensuing outrage that a perceived injustice produces are relative. Both depend on how the victim perceives his or her treatment to compare to that of other victims (Crosby, 1976; Davis, 1959; Gurr, 1970). Feelings of entitlement to be outraged are also relative. Victims who feel that others have more grounds to be outraged than they do, however outraged they may feel, will be inhibited from expressing it, if those worse off are not protesting their treatment. For example, the conciliatory tone taken by the Black South African leader Nelson Mandela toward the White government that had imprisoned him for 25 years likely had the calming effect it did at least partially because it diminished the standing to complain felt by other Black South Africans who had been less harshly treated than Mandela. In effect, their standing to resist a conciliatory stance toward a White minority, however strong their motivation to do so, was undermined by the grace Mandela showed under much greater pressure. It simply is difficult to feel that one is entitled to refuse to turn the other cheek when someone whose cheek has been slapped much harder does.

That the legitimacy of a complaint depends on the relative standing of the complainant is one reason people react so negatively to the often well-meaning attempts by others to console them about their misfortunes by pointing out that others are worse off. The claim that others have it worse off than you is a direct challenge to your standing to feel and express outrage. Being told that you are luckier than others may or may not diminish your sense of injustice, but it certainly will diminish the standing you think you have to express your sense of injustice. Your feelings of injustice may not be diminished by the words of others, but your sense of entitlement to those feelings will be.

Miller and Zak (2008) tested the hypothesis that people will be inhibited from protesting an injustice if they are confronted with another who fails to protest an even a worse injustice. The context they chose was the Ultimatum Game. This is a two-person game with the following rules. One person (the *proposer*) is provided with an amount of money (e.g., $5) and asked to decide how to distribute the money between himself or herself and the other player (the *responder*). The only power the responder has in the situation is to accept or reject the offer (ultimatum). If the responder accepts the proposer's offer, they each get the money specified in the proposal. If the responder rejects the proposer's offer, neither player gets any money. Because the game is played only once, and anonymously, reciprocation is not an issue.

The Ultimatum Game has generated more attention than any experimental game since the Prisoner's Dilemma Game and was introduced because of the irrational behavior that it generates. The game-theoretic solution to this game is for the proposer to offer the responder as little as possible; after all, the responder will be better off accepting whatever the proposer offers rather than rejecting it and getting nothing. This is not what typically happens, however. Participants in the proposer's position tend to offer at least 30% to 40% of the sum to the other player. Indeed, if they offer less than this, participants in the responder's position tend to reject it, thereby consigning both parties to nothing. This may represent a case of cutting off one's nose to spite one's face, but players in the responder's role apparently feel that receiving an insulting and unjust offer entitles them to deviate from the rational response. Their sense of justice is piqued, and they feel entitled to express their outrage by punishing the perpetrator.

Miller and Zak (2008) sought to determine how witnessing another person accept an even less generous offer would affect the willingness of Ultimatum Game players to accept an unfair offer they received themselves. They predicted that being confronted with the rational, level-headed response of another person in the face of an even more unequal split would increase the willingness of players to accept their offer. They reasoned that a player's perceived standing to reject an offer (i.e., be vindictive on principle) would be reduced if he or she witnessed another player accept a more insulting offer. That is, witnessing the lack of vindictive action from another player would undermine their intentions to "show" the proposer.

Participants played one round of the Ultimatum Game on a computer terminal with another (imaginary) person and alongside two other (imaginary) people playing an independent round of the game. Participants were unaware that the other actors in the game were imaginary. Participants were supposedly randomly assigned to the role of responder and told that their partner allocated to them $1.50 of the $5.00 pool of money and that in the other pair an even less fair amount was being offered to the other responder (i.e., $1). Participants' standing to protest was manipulated by the information they were given about the response of the other victim (who either accepted or rejected the offer). There was also a control condition, where participants were not told how the other victim responded to the $1 offer. It was assumed that participants who were informed that a more victimized other accepted the offer (i.e., turned the other cheek) would feel they had less standing to retaliate against their partner by rejecting their $1.50 offer. Consistent with prediction, only 28% of participants rejected the $1.50 offer when they were told that the other more deprived victim accepted a $1 offer as compared to 50% of participants who were not told how the other victim responded and 52% of participants who were told that the other victim rejected the $1 offer.

Support for the claim that the nonspiteful behavior of a worse-off victim would inhibit participants from acting on their feelings of spite rather than simply reduce their spitefulness came from measures that probed participants' feelings toward the other victim. Even though a fellow victim of an unfair split had more impact on participants' behavior when she accepted it than when she rejected it, participants liked the other victim more in the latter case than in the former case. In other words, participants may have been influenced more by the rational than irrational other, but they liked her less. This result is to be expected, of course, if the nature of the influence she wielded was that of undermining the participants' standing to act on their feelings of outrage rather than actually reducing their feeling of outrage.

In further support of this analysis, consider what happened when participants were asked whether they did or did not want to give the other victim, in another round of a similar game, the full control to decide on how to split up another pot of money. Consistent with the reasoning that participants resented the other victim who accepted an even worse offer for shaming them into accepting their offer, only 37% of participants in this condition chose to benefit their worse-off fellow victim, as compared to 70% when the other had rejected the worse offer. The fact that victims both derogated and punished other victims who deprived them of their standing suggests that standing, perhaps because of the freedom of action it provides, is something people do not like to lose.

Conclusions

This chapter addressed a powerful psychological barrier that can prevent outrage from being converted into social protest. Although there may be many barriers that hinder a person who is outraged by a situation from protesting it, such as the fear of retaliation, our focus has been on one particular psychological barrier: one's level of psychological standing. We argued that protesting an injustice in social spheres, as in legal spheres, can be regarded as a right or entitlement. (This claim does not deny that in some circumstances it might also be a responsibility as well.) Not everybody can legitimately protest an injustice no matter how outraged he or she is by it. A person's standing to protest is defined by a person's relation to the source of the outrage, with the particulars of standing varying considerably from context to context.

The concept of standing permits unique predictions as well as an alternative frame for viewing previous findings (e.g., the relation between self-interest and protest). As an example of the latter, consider an anomaly that exists in the intersection of procedural and distributive justice. On the one hand, research shows that people's satisfaction with an outcome often correlates more strongly with their perceptions of the fairness of the procedure that produced it than with its favorability. On the other hand, the likelihood of protesting an outcome is more strongly related to feelings of distributive injustice than procedural injustice (Tyler, Huo, & Lind, 1999). One possibility for why procedural justice sentiments are more predictive of satisfaction than of protest is that the feeling that it was an unjust procedure that produced an outcome, in the absence of the feeling that the actual outcome was unfair, leaves one with less standing to protest.

We began this chapter by illustrating how lacking a material stake in a perceived injustice deprives people of the psychological standing to protest, and we argued that people can obtain standing through other means as well. When people lack a material stake in an issue, they can nonetheless feel that they have the standing to protest if they observe other nonvested individuals protesting (Ratner & Miller, 2001) or if they perceive themselves as having a moral stake in the issue (Effron & Miller, 2008). Having a personal characteristic or history that explains to the world why one feels such outrage—as in the case of the White NAACP chapter president or James Brady—can also provide one with standing. At the same time, not just any connection to an issue will suffice. Having committed a particular transgression in the past or simply being a member of a group that has committed (or continues to commit) that transgression—as in the case of the Whites who are reluctant to report perceived racial discrimination—deprives one of the standing to protest that particular transgression. Finally, having a material stake in an issue's outcome is not always sufficient to license protest. Victims lack the standing to retaliate against a transgressor when others who have been more victimized by the transgression choose to turn the other cheek (Miller & Zak, 2008). This list of characteristics that give people standing is by no means exhaustive. By identifying some of the ways in which people can acquire and lose standing and by showing various ways in which standing shapes protest behavior, we hope to give readers a sense of the power and predictive utility of this psychological construct.

References

Bissenger, B. (1997). *A prayer for the city.* New York: Random House.

Crosby, F. (1976). A model of egoistical relative deprivation. *Psychological Review, 83*, 85–113.

Davis, J. A. (1959). A formal interpretation of the theory of relative deprivation. *Sociometry, 22*, 280–296.

Dixon, K. A., Storen, D., & Van Horn, C. E. (2002). *A workplace divided: How Americans view discrimination and race on the job*. New Brunswick: State University of New Jersey, Rutgers, John J. Heldrich Center for Workplace Development.
Effron, D. A., & Miller, D. T. (2008). Who is entitled to advocate: The effects of moralizing an issue on psychological standing. Unpublished manuscript.
French, J., & Raven, B. (1959). The bases of social power. In D. Cartwright (Ed.), *Studies of social power* (pp. 150–167). Ann Arbor, MI: Institute for Social Research.
Goss, K. A. (2003). *Disarmed*. Princeton, NJ: Princeton University Press.
Green, D. P., & Cowden, J. A. (1992). Who protests: Self-interest and White opposition to busing? *Journal of Politics, 54*, 471–496.
Gurr, T. R. (1970). *Why men rebel*. Princeton, NJ: Princeton University Press.
Hornsey, M. J., & Imani, A. (2004). Criticizing groups from the inside and the outside: An identity perspective on the intergroup sensitivity effect. *Personality and Social Psychology Bulletin, 30*, 365–383.
Hornsey, M. J., Trembath, M., & Gunthorpe, S. (2004). You can criticize because you care: Identity attachment, constructiveness, and the intergroup sensitivity effect. *European Journal of Social Psychology, 34*, 499–518.
Jackman, M. R. (1994). *The velvet glove: Paternalism and conflict in gender, class, and race relations*. Berkeley: University of California Press.
Kaiser, C. R., & Major, B. (2006). A social psychological perspective on perceiving and reporting discrimination. *Law and Social Inquiry, 31*, 801–830.
Kelley, H. H. (1972). Causal schemata and the attribution process. In E. E. Jones, D. E. Kanouse, H. H. Kelley, R. E. Nisbett, S. Valins, & B. Weiner (Eds.), *Attribution: Perceiving the causes of behavior* (pp. 79–94). Morristown, NJ: General Learning Press.
Kuran, T. (1997). *Private truths, public lies*. Cambridge, MA: Harvard University Press.
Lewin, K. (1951). *Field theory in social science* (D. Cartwright, Ed.). New York: Harper.
MADD's first male president to honor area police. (2005, October 17). *Oakland Tribune*. Retrieved May 15, 2008, from http://findarticles.com/p/articles/mi_qn4176/is_/ai_n15804170
McAdams, R. H. (1997). The origin, development, and regulation of norms. *Michigan Law Review, 96*(2), 338–433.
Mikula, G. (1993). On the experience of injustice. In W. Stroebe & M. Hewstone (Eds.), *European review of social psychology* (Vol. 4, pp. 223–244). Chichester, UK: Wiley.
Miller, D. T. (1999). The norm of self-interest. *American Psychologist, 54*, 1053–1060.
Miller, D. T. (2001). Disrespect and the experience of injustice. *Annual Review of Psychology, 52*, 527–553.
Miller, D. T., & Ratner, R. K. (1996). The power of the myth of self-interest. In L. Montada & M. J. Lerner (Eds.), *Current societal concerns about justice* (pp. 25–48). New York: Plenum.
Miller, D. T., & Ratner, R. K. (1998). The disparity between the actual and assumed power of self-interest. *Journal of Personality and Social Psychology, 74*, 53–62.
Miller, D. T., Ratner, R. K., & Zhao, M. (2008). *Vested interest and advocacy effectiveness: The benefits of psychological standing*. Unpublished manuscript.
Miller, D. T., & Zak, S. V. (2008). Disliking those who shame us into behaving well. Unpublished manuscript.

Olson, J. M., & Hafer, C. L. (2001). Tolerance of personal deprivation. In J. T. Jost & B. Major (Eds.), *The psychology of legitimacy* (pp. 157–175). New York: Cambridge University Press.
Randall, K. C. (1988). Universal jurisdiction under international law. *Texas Law Review*, *66*, 785–841.
Ratner, R. K., & Miller, D. T. (2001). The norm of self-interest and its effects on social action. *Journal of Personality and Social Psychology*, *81*, 5–16.
Regan, D. T., & Fazio, R. (1977). On the consistency between attitudes and behavior: Look to the method of attitude formation. *Journal of Experimental Social Psychology*, *13*, 28–45.
Samuelson, R. (2007, November 15–21). Under her skin. *Washington City Paper*, *27*(46).
Sears, D. O., Hensler, C. P., & Speer, L. K. (1979). Whites' oppositions to "busing": Self-interest or symbolic politics? *American Political Science Review*, *73*, 369–384.
Sivacek, J., & Crano, W. D. (1982). Vested interest as a moderator of attitude-behavior consistency. *Journal of Personality and Social Psychology*, *43*, 210–221.
Skitka, L. J., Bauman, C. W., & Sargis, E. G. (2005). Moral conviction: Another contributor to attitude strength or something more? *Journal of Personality and Social Psychology*, *88*, 895–917.
Solomon, R. C. (1990). *A passion for justice: Emotions and the origins of the social contract*. Reading, MA: Addison-Wesley.
Sutton, R. M., Douglas, K. M., Elder, T. J., & Tarrant, M. (2008). Social identity and social convention in responses to criticisms of groups. In Y. Kashima & K. Fiedler (Eds.), *Stereotype dynamics: Language-based approaches to the formation, maintenance and transformation of stereotypes*. Mahwah, NJ: Lawrence Erlbaum.
Sutton, R. M., Elder, T. J., & Douglas, K. M. (2006). Reactions to internal and external criticism of outgroups: Social convention in the intergroup sensitivity effect. *Personality and Social Psychology Bulletin*, *32*, 563–575.
Tyler, T. R., Huo, Y. J., & Lind, A. E. (1999). Two psychologies of conflict resolution: Differing antecedents of pre-experience choices and post-experience evaluations. *Group Processes and Intergroup Relations*, *2*, 99–118.
U.S. Merit Systems Protection Board. (1995). *Sexual harassment in the federal workplace: Trends, progress, continuing challenges*. Retrieved May 15, 2008, from http://www.mspb.gov/sites/mspb/pages/MSPB%20Studies.aspx
Zak, S. V., & Miller, D. T. (2008). Norms controlling reactions to norm violations. Unpublished manuscript.

CHAPTER 6

Deservingness, the Scope of Justice, and Actions Toward Others

JAMES M. OLSON, IRENE CHEUNG, and PAUL CONWAY
University of Western Ontario

CAROLYN L. HAFER
Brock University

Abstract: In this chapter, the authors report three studies that examined the psychological mechanisms underlying the effects of several factors previously shown to influence decisions to help or harm a target. In particular, the studies tested whether the perceived deservingness of a target for favorable (or unfavorable) treatment or, instead, the inclusion (or exclusion) of the target in the perceiver's "scope of justice" best accounted for the effects of variables that have been hypothesized by previous theorists to operate by means of the latter mechanism (inclusion–exclusion). In the first two studies, the authors manipulated the usefulness of the target, the degree of conflict between the target and the perceiver, and the similarity between the target and the perceiver. The results of these studies provided some support for a deservingness perspective but no support for a scope of justice perspective. In a third study, the authors tested whether exclusion from the scope of justice might indeed occur, but only in relatively extreme conditions. They conclude the chapter with suggestions for future research on this issue.

Scope of Justice

The "scope of justice" is a concept that has been used to explain harmful behavior (e.g., Opotow, 1990, 2001; Singer, 1996). Some theorists have suggested that harm-doing can result when targets are *excluded* from a perceiver's scope of justice, which occurs "when individuals or groups are perceived as outside the boundary in which moral values, rules, and considerations of fairness apply" (Opotow, 1990, p. 1). These theorists have argued that if targets are excluded from a perceiver's scope of justice, then actions toward the targets need not be influenced by fairness guidelines that usually affect behavior. For example, the targets may be treated negatively even though they did little to warrant such treatment according to justice principles (presumably, such treatment would result from nonjustice motives, such as self-interest). Exclusion from the scope of justice has been used to understand cases of extreme harm-doing, such as genocide or mass internment (e.g., Deutsch, 2000; Nagata, 1993; Opotow, 2005), but its typical application in empirical research has been to explain less severe responses, such as apathy about negative treatment or a willingness to recommend punishment (e.g., Brockner, 1990; Singer, 1996).

Researchers have identified several variables that are thought to increase the likelihood of a target being excluded from a perceiver's scope of justice (for a review, see Tyler, Boeckmann, Smith, & Huo, 1997, pp. 210–222). One hypothesized antecedent of exclusion relates to the perceived utility of the target, for instance, the extent to which the target is seen as harmful rather than beneficial (e.g., Leets, 2001; Opotow, 1993). Low perceived utility (e.g., high harmfulness) is hypothesized to increase the likelihood that a target will be excluded from the perceiver's scope of justice. A second proposed antecedent of exclusion is perceived conflict or threat, such as incompatibility between the target's and the perceiver's goals or competition between the target and the perceiver for scarce resources (e.g., Beaton & Tougas, 2001; Nagata, 1993). Greater conflict or threat is hypothesized to increase the likelihood of a target being excluded. A third proposed antecedent of exclusion is a lack of identification, such as perceived dissimilarity between the target and the perceiver (e.g., Singer, 1998; Wenzel, 2002). In general, less identification (e.g., greater dissimilarity between the target and the perceiver) is hypothesized to make it more likely that the target will be excluded from the perceiver's scope of justice.

Empirical investigations of the scope of justice concept have typically involved manipulations of one or more of the hypothesized antecedents of exclusion, followed by measures of participants' support for negative treatment of the target group. For example, Opotow (1993) asked participants whether they would support or oppose a construction project that would endanger a particular species of beetles. Participants supported the project more when the beetles were characterized as harmful for crops and the environment (low utility of beetles) than when the beetles were characterized as beneficial (high utility). Participants also supported the project more when it was characterized as important to humans (high conflict between humans and beetles) than when it was characterized as questionable in importance (low conflict). On the basis of a measure designed to assess the scope of justice, Opotow interpreted these findings as resulting from the effects of the manipulations on the relevance of fairness, such that low utility and high conflict narrowed participants' boundaries within which justice rules were applied, thereby resulting in the exclusion of the beetles (and greater willingness to harm them).

Deservingness as an Alternative Explanation

In a recent paper (Hafer & Olson, 2003), we offered a different perspective on research testing the scope of justice concept. We suggested that many findings in the scope of justice literature are open to alternative interpretations in terms of perceived *deservingness*, which refers to a state of compatibility in valence between a target's actions or traits and his or her outcomes (namely, good people receiving good outcomes and bad people receiving bad outcomes; see Feather, 1999; Lerner, 1981). In many past studies on the scope of justice, it is possible that the variables assumed to influence the perceived relevance of justice actually influenced the perceived deservingness of the target (whether the target deserved positive or negative treatment), and perceived deservingness then guided participants' responses. Many justice theorists have proposed that deservingness is a key component of fairness (e.g., Freudenthaler & Mikula, 1998; Heuer, Blumenthal, Douglas, & Weinblatt, 1999; Sunshine & Heuer, 2002). Thus, if participants in past studies treated a target person negatively because he or she deserved such treatment, they were not ignoring justice considerations (i.e., they were not excluding the target from their scope of justice). Instead, they believed they were treating the target fairly.

For example, Opotow's (1993) finding that participants supported the construction project more when the beetles were characterized as harmful rather than beneficial may have resulted from the harmful beetles being seen as deserving more negative treatment than the beneficial beetles. It is common for people to judge the worthiness of targets in terms of the positive or negative consequences of their actions (e.g., see Feather, 1999). Similarly, Opotow's finding that the beetles were treated more negatively when conflict with humans was high rather than low may have involved perceived deservingness. Interfering with an important project may have led to the inference that the beetles were more harmful or troublesome and did not deserve positive treatment, compared to the condition in which the project was of lower importance.

As mentioned earlier, theorists have also proposed that dissimilarity between a target and a perceiver increases the likelihood that the target will be excluded from the perceiver's scope of justice. Consistent with this reasoning, targets dissimilar to perceivers have been treated more harshly than similar targets in some studies (e.g., Singer, 1998, 1999). Once again, however, a plausible alternative explanation is that participants believed the dissimilar targets deserved more negative treatment than the similar targets. Humans tend to judge themselves very favorably (e.g., see Roese & Olson, 2007), so greater perceived similarity to the self might imply greater deservingness for positive outcomes.

The Present Research

The principal goal of the research reported in this chapter was to manipulate some of the variables that have been hypothesized to promote exclusion from the scope of justice—such as utility, conflict, and dissimilarity—and test whether exclusion or deservingness actually seemed to be the mechanism underlying the effects of the manipulations. In the first two studies, we measured participants' support for negative treatment of the target, as well as their perceptions of the target's deservingness for negative treatment and the relevance of fairness for the target (inclusion or exclusion of the target from the scope of justice). We predicted that the manipulations would affect participants' willingness to harm the targets and that perceived deservingness (but not relevance of fairness) would mediate at least some of these effects. In a third study, we tested whether exclusion from the scope of justice might indeed occur, but only in relatively extreme conditions. Specifically, we investigated whether fairness (e.g., deservingness) might have less impact on treatment decisions when identification with a target is either extremely close or extremely distant.

Beetles and Buildings

In one study, we used a scenario identical to the one developed by Opotow (1993) and replicated her manipulations. Participants read a story about a proposed construction project and its possible impact on a species of beetle, *Brachinus*, whose habitat would be destroyed by the project. The first information in the package provided background information about the species. Participants in the *dissimilar* condition learned that *Brachinus* is primitive, reflexive, and noncommunicative and that *Brachinus* parents play no role in the development of offspring. Participants in the *similar* condition learned that *Brachinus* is industrious, social, and communicative and that *Brachinus* parents are protective of their offspring.

The utility of the species was also manipulated through information about the beetles: Participants in the *detrimental* condition learned that *Brachinus* is a destructive species that destroys over 200 valuable plants, including human crops, costing the economy $1 billion each year. Participants in the *beneficial* condition learned that *Brachinus* is beneficial in several ways, including weed control, pollination of human crops, and pest control, benefitting the economy $1 billion each year.

Following the information about the beetles, participants were given details of the construction project. In the *high-conflict* condition, participants learned that the project was a reservoir for drinking water that was very much needed by the community and was supported by most citizens. In the *low-conflict* condition, participants learned that the project was an industrial complex that would have only questionable benefits and was considered to be unnecessary by most citizens.

After reading the scenario about the proposed construction project, participants indicated their support for the project on three items, such as "To what extent do you agree that this construction project should go ahead?" (from 1 = *strongly disagree* to 7 = *strongly agree*). A composite measure was formed by standardizing the three items and then averaging the *z*-scores; higher scores meant greater support for the project (i.e., more willingness to harm the beetles).

To measure the extent to which participants thought the beetles deserved protection, we asked participants to indicate their agreement (from 1 = *strongly disagree* to 7 = *strongly agree*) with six statements, such as "*Brachinus* deserves to have this land" and "In general, *Brachinus* deserves bad outcomes." A composite measure was formed by averaging the six responses (after reverse scoring some items); higher scores meant stronger perceptions that the beetles deserved favorable outcomes (e.g., deserved protection).

To measure the perceived relevance of fairness (or the extent to which the beetles were included in the scope of justice), we asked participants, "When making a final decision, to what extent should one consider how fairly *Brachinus* is treated?" (from 1 = *not at all* to 7 = *to a very great extent*) and "To what extent do you think fairness is relevant to this situation?" (from 1 = *not at all relevant* to 7 = *extremely relevant*). A composite measure was formed by averaging the two responses; higher scores meant greater perceived relevance of fairness (inclusion in the scope of justice).

Support for the Construction Project Preliminary analyses showed that sex of participant did not influence the results. A 2 (dissimilarity) × 2 (utility) × 2 (conflict) analysis of variance (ANOVA) on the composite measure of support for the construction project revealed that all three manipulations exerted significant main effects, but no interactions were significant.

As predicted, the main effect for utility, $F(1, 94) = 4.35, p < .05$, showed that participants supported the project more when *Brachinus* was detrimental than when *Brachinus* was beneficial ($M = 0.18$ vs. $M = -0.18$). Also as predicted, the main effect for conflict, $F(1, 94) = 5.58, p < .05$, indicated that participants supported the project more when it was important for humans (high conflict between humans and beetles) than when it was unimportant ($M = 0.20$ vs. $M = -0.21$). These two main effects paralleled findings in Opotow's (1993) study.

Contrary to predictions, the main effect for dissimilarity, $F(1, 94) = 4.66$, $p < .05$, indicated that participants supported the construction project (were willing to harm the beetles) more when the beetles were similar to humans than when they were dissimilar ($M = 0.18$ vs. $M = -0.19$). Interestingly, Opotow (1993) reported in her study that the pattern of means also indicated a negative effect of similarity (greater support for the construction project when the beetles were similar rather than dissimilar), although the difference was not reliable in the overall analysis. Perhaps insects that were relatively similar to humans were seen as bad (and, therefore, less deserving of protection) because they threatened participants' perceived superiority and distinctiveness from nonhuman animals (see Goldenberg et al., 2001; for parallel reasoning applied to intergroup relations, see Hornsey & Hogg, 1999).

Perceived Deservingness An ANOVA on the composite measure of perceived deservingness yielded two significant main effects, one for utility, $F(1, 93) = 10.14, p < .01$, and one for dissimilarity, $F(1, 93) = 3.80, p = .05$. The main effect for utility revealed that participants rated beneficial beetles as deserving more protection than detrimental beetles ($M = 5.54$ vs. $M = 4.72$), which paralleled a finding on the measure of support. The main effect for dissimilarity on deservingness showed that participants rated *Brachinus* as more deserving of protection when it was portrayed as dissimilar to humans than when it was portrayed as similar to humans ($M = 5.38$ vs. $M = 4.88$), which again paralleled a finding on the measure of support for the project (in this case, more favorable treatment of the dissimilar beetle than the similar beetle).

We examined whether perceived deservingness mediated the effects of the manipulations on support for the construction project with regression analyses, following the procedures recommended by Baron and Kenny (1986). Mediation would be suggested if (a) the manipulation predicted support for the project, (b) the manipulation predicted the proposed mediator (e.g., perceived deservingness), (c) the proposed mediator predicted support for the project even when the manipulation was controlled statistically, and (d) the manipulation no longer predicted support significantly

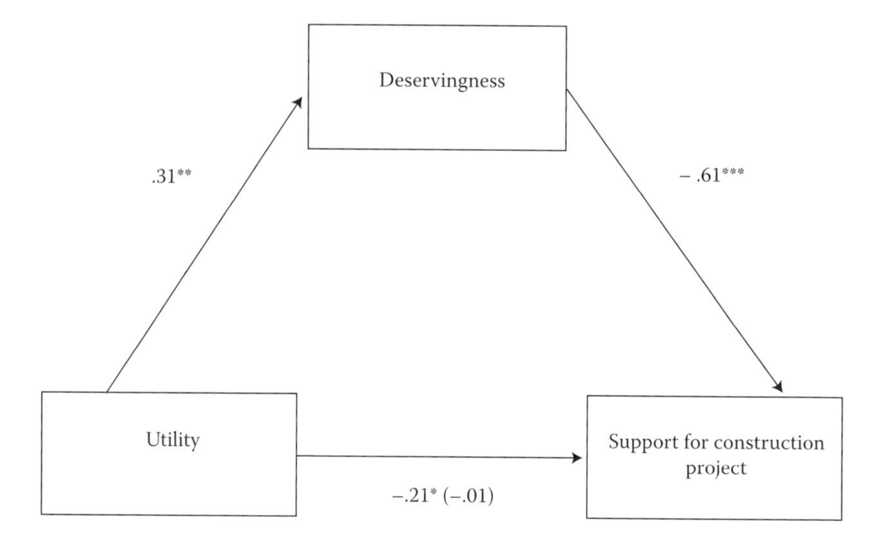

Figure 6.1 Mediation analyses for utility main effect, Study 1. *Note*. Standardized regression coefficients reported. Sobel test (using unstandardized coefficients), $z = 2.92$, $p < .01$. $^{***}p < .001$, $^{**}p < .01$, $^{*}p < .05$.

(or was significantly reduced in predictive strength) when the proposed mediator was controlled statistically.

We first examined mediation of the utility main effect. Figure 6.1 summarizes the results involving perceived deservingness. The regression analyses showed, as would be expected from the ANOVA results, that the manipulation of utility significantly predicted participants' support for the construction project ($\beta = -.21$, $p < .04$). The manipulation of utility also significantly predicted perceived deservingness ($\beta = .31$, $p < .01$). Finally, when the main effect and deservingness were entered into the equation simultaneously, deservingness predicted treatment over and above the manipulation ($\beta = -.61$, $p < .001$), whereas utility no longer predicted treatment significantly ($\beta = -.01$, *ns*). A Sobel test revealed that the reduction in the β for utility when deservingness was controlled was significant ($z = 2.92$, $p < .01$).

We next examined mediation of the dissimilarity main effect. Figure 6.2 summarizes the results. The regression analyses showed that the manipulation of dissimilarity significantly predicted participants' support for the construction project ($\beta = .21$, $p < .04$). The manipulation of dissimilarity also predicted perceived deservingness, but only marginally significantly ($\beta = -.18$, $p < .07$). When the main effect and deservingness were entered into the equation simultaneously, deservingness predicted treatment over and above the manipulation ($\beta = -.60$, $p < .001$), whereas dissimilarity no longer predicted treatment significantly ($\beta = .09$, *ns*). A Sobel test revealed

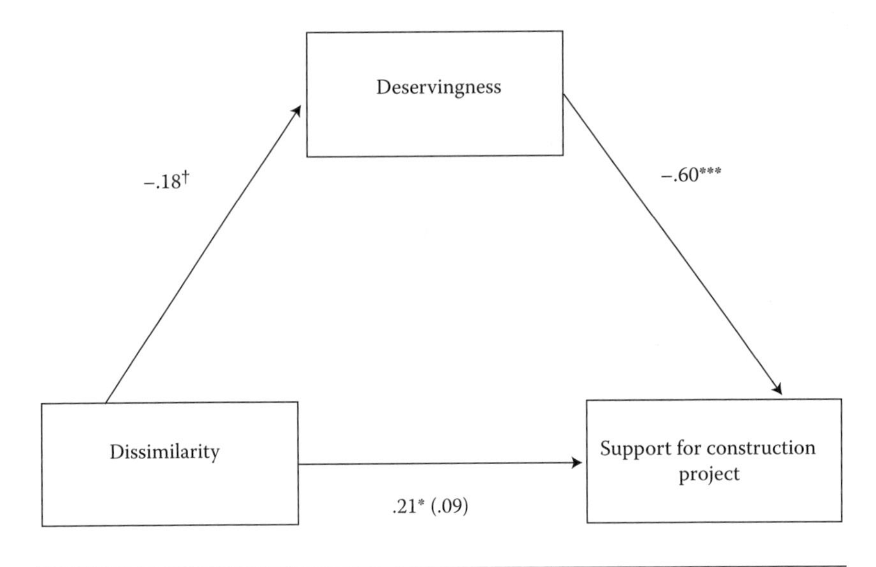

Figure 6.2 Mediation analyses for dissimilarity main effect, Study 1. *Note*. Standardized regression coefficients reported. Sobel test (using unstandardized coefficients), $z = 1.79$, $p < .08$. ***$p < .001$, *$p < .05$, †$p < .07$.

that the reduction in the β for dissimilarity when deservingness was controlled was marginally significant ($z = 1.79$, $p < .08$).

The regressions to examine possible mediation of the main effect of conflict showed that the manipulation did not predict perceived deservingness (β = −.08, *ns*). Therefore, deservingness could not have mediated the effect of conflict on support for the construction project.

Perceived Relevance of Fairness An ANOVA on the composite measure of the perceived relevance of fairness revealed no significant effects. The overall mean on the relevance measure was 4.03 on a 7-point scale, where 1 = *not at all relevant* and 7 = *extremely relevant*. Thus, participants thought that fairness was moderately relevant to the situation, and this perception did not vary across conditions.

Regression analyses to explore possible mediation showed that none of the manipulations predicted the perceived relevance of fairness (all $|\beta| < .12$, *ns*). Thus, perceived relevance could not have mediated the obtained effects of the manipulations on support for the construction project.

The results of Study 1 provided some support for a deservingness perspective on people's willingness to harm a target group but no support for a scope of justice perspective. The effects of utility and dissimilarity on support for the construction project were fully or marginally mediated by perceived deservingness, whereas the perceived relevance of fairness did not relate to support for the project. The utility manipulation affected support in the expected direction (more willingness to harm a detrimental species), but the dissimilarity manipulation yielded a surprising result (more willingness to harm a similar species). As noted earlier, this result actually paralleled the direction of difference in Opotow's (1993) study. In the next study, we turned to human targets to see whether similarity would exhibit the more intuitive relation to harm-doing.

The manipulation of conflict in Study 1 did not appear to be mediated by either deservingness or scope of justice. It is possible that high conflict altered how participants thought about fairness, perhaps changing their conceptions of justice (without rendering justice irrelevant). For example, under conditions of high conflict (the reservoir construction project, which was important for the community), the distributive rule of "survival of the fittest" may have been seen as an appropriate guide to the treatment of the beetles (see Lerner, 1977). In the current context, such a guide may have led participants to be willing to harm *Brachinus*.

Criminals and Cruelty

In the second study, we examined harm-doing in the form of punishing criminal behavior. Participants (students at the University of Western Ontario) read a story about an assault that occurred in a bar, in which one man used a broken beer bottle to slash the face of another man, causing serious facial injuries and permanent damage to the victim's vision in one eye. In the *dissimilar* condition, the perpetrator was said to be a 32-year-old mature student at the University of British Columbia, which meant that he was older than, and lived far away from, the participants. In the *similar* condition, the perpetrator was said to be a 20-year-old student at the University of Western Ontario (like the participants). We predicted that, unlike the results in Study 1, perceived dissimilarity would increase participants' willingness to punish (harm) the perpetrator, because the target was a human rather than a species of insect. We also predicted that this effect of dissimilarity would be mediated by perceived deservingness.

We also manipulated the utility of the perpetrator. In the *high-utility* condition, participants read that the perpetrator had connections to a major drug dealer whom police had been pursuing for several years, and if the perpetrator was not sent to jail, he could act as an informant to help police apprehend the dealer. In the *low-utility* condition, participants read that the perpetrator had connections to a major drug dealer whom police had been pursuing for several years, but no mention was made of the perpetrator potentially serving as an informant. We predicted that we would replicate the finding in the first study that greater utility would be associated with more favorable treatment. We predicted a different result from Study 1, however, in the mediation analyses for utility. In Study 1, the effect of utility was mediated by deservingness, presumably because beneficial beetles were seen as more meritorious than detrimental beetles. In Study 2, the utility manipulation was expected to disengage it from deservingness. Our manipulation of utility increased the target person's potential usefulness (if he was not sent to jail) without implicating any positive characteristics—indeed, his usefulness rested on his connections with a bad person. Therefore, just treatment of the high-utility perpetrator was expected to focus on the more *restorative* justice approach of having the perpetrator make amends for his transgression by helping the police apprehend a drug dealer rather than on giving the perpetrator his just deserts (in this case, just punishment; see Gromet & Darley, 2006). Thus, neither perceived deservingness nor the perceived relevance of justice was expected to mediate this finding.

After reading about the assault, participants answered two questions about treatment of the perpetrator: "Do you think that the perpetrator (the guy who cut the victim's face) should be sent to jail?" (from 1 = *no, definitely not sent to jail* to 9 = *yes, definitely sent to jail*) and "Should the prosecutor consider making a deal with the perpetrator, such as offering a reduced sentence in return for cooperating with the authorities?" (from 1 = *no, definitely not* to 9 = *yes, definitely*). Responses to these items were only weakly correlated ($r = -.23$) and yielded different results in the analyses, so they will be presented separately.

To measure the extent to which participants thought the perpetrator deserved punishment, they were asked, "Do you think the perpetrator deserves mild or severe punishment?" (from 1 = *very mild punishment* to 9 = *very severe punishment*). Two items measured the perceived relevance of fairness (or the extent to which the perpetrator was included in participants' scope of justice): "In answering the previous questions, to what extent was justice for the perpetrator important in your responses?" (from 1 = *not at all important* to 9 = *very important*) and "To what extent were your previous responses based on what was fair for the perpetrator?" (from 1 = *not at all* to 9 = *very much*). A composite measure was formed by averaging the two responses.

Punishment of the Perpetrator A 2 (dissimilarity) × 2 (utility) × 2 (sex of participant) ANOVA on participants' responses to the question asking whether the perpetrator should be sent to jail revealed only a main effect for utility, $F(1, 190) = 7.06$, $p < .01$. As expected, participants supported jailing the low-utility perpetrator more than the high-utility perpetrator ($M = 7.42$ vs. $M = 6.74$).

An ANOVA on participants' ratings of whether a deal should be made with the perpetrator revealed only an interaction between dissimilarity and sex of the participant, $F(1, 190) = 4.73$, $p < .04$. Inspection of the means showed that men supported a deal with the similar perpetrator more than with the dissimilar perpetrator ($M = 6.41$ vs. $M = 5.64$), whereas women's ratings were less discrepant for the similar versus dissimilar perpetrators ($M = 5.70$ vs. $M = 6.15$). These results indicate that the dissimilarity manipulation exerted its expected effect for men but not for women. In retrospect, this pattern makes sense, because the perpetrator was a man.

Perceived Deservingness An ANOVA on participants' ratings of whether the perpetrator deserved mild or severe treatment revealed only one effect: an interaction between dissimilarity and sex of the participant, $F(1, 190) = 6.05$, $p < .02$. Inspection of the means showed that men rated the similar perpetrator as deserving milder punishment than the dissimilar perpetrator ($M = 4.65$ vs. $M = 5.54$), whereas women's ratings did not differ for the similar versus dissimilar perpetrators ($M = 5.57$ vs. $M = 5.37$). Note that this pattern of means corresponds exactly to the pattern underlying the same interaction on participants' responses to whether a deal should be made with the perpetrator.

As in Study 1, we examined mediation issues by conducting regression analyses. The analyses of participants' ratings of whether the perpetrator should be sent to jail showed, consistent with the ANOVA results, that the utility manipulation significantly predicted treatment ($\beta = -.20$, $p < .01$). The utility manipulation, however, did not significantly predict perceived deservingness ($\beta = -.09$, *ns*), which means that deservingness could not have mediated the effect of utility on treatment. Interestingly, when utility and deservingness were entered into the regression equation simultaneously, deservingness predicted participants' ratings of whether the perpetrator should be sent to jail when utility was statistically controlled ($\beta = .31$, $p < .001$), and utility remained a significant predictor when deservingness was controlled ($\beta = -.18$, $p < .01$). Thus, both utility and deservingness predicted treatment of the perpetrator, and the two relations appeared to be independent of one another.

The regression analyses of participants' ratings of whether a deal should be made with the perpetrator are summarized in Figure 6.3. These analyses showed that the interaction of dissimilarity and sex significantly predicted support for a deal ($\beta = -.26$, $p < .05$; the two main effects were also entered into the equation). Also, the interaction term significantly predicted perceived deservingness ($\beta = .32$, $p < .02$). When the interaction term and deservingness were entered into the equation simultaneously, deservingness predicted treatment over and above the interaction term ($\beta = -.23$, $p < .01$), whereas the interaction term no longer predicted treatment significantly ($\beta = -.19$, *ns*). A Sobel test revealed that the reduction in the β for the interaction term when deservingness was controlled fell just short of conventional levels of significance ($z = -1.92$, $p = .055$).

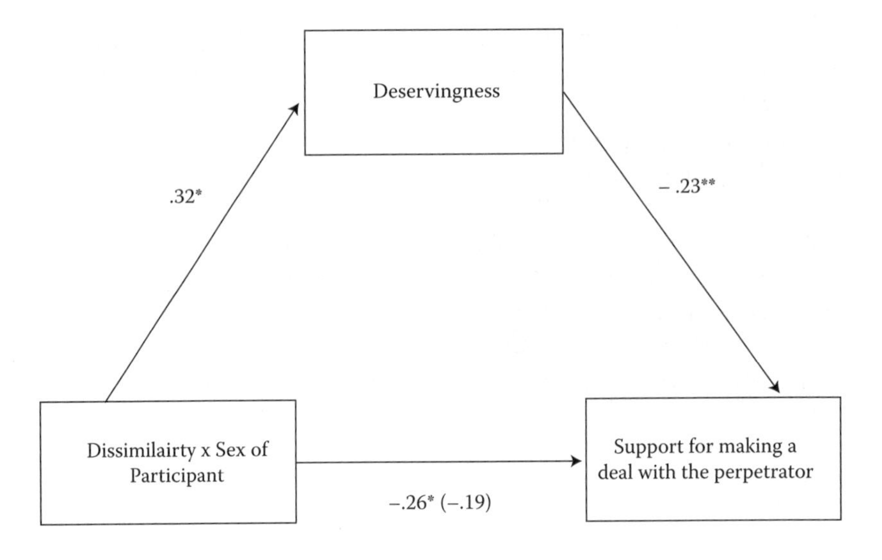

Figure 6.3 Mediation analyses for Dissimilarity × Sex interaction, Study 2. *Note*. Standardized regression coefficients reported. Sobel test (using unstandardized coefficients), $z = -1.92$, $p = .055$. $^{**}p < .01$, $^{*}p < .05$.

Thus, the regression analyses indicated that perceived deservingness mediated, at least marginally, the effect of dissimilarity on male participants' ratings of whether a deal should be made with the perpetrator. In contrast, deservingness did not seem to play any role in the effect of utility on participants' ratings of whether the perpetrator should be sent to jail. As noted in the introduction to this study, utility was not expected to affect deservingness in the scenario, because having connections to a drug dealer does not constitute a meritorious or admirable characteristic.

Perceived Relevance of Fairness An ANOVA on the composite measure of whether justice for the perpetrator was important revealed no significant main effects or interactions. The overall mean on the measure was 6.45, on a 9-point scale where 5 = *moderately important* and 9 = *very important*. Thus, participants generally considered justice to be important for, or relevant to, their responses to the treatment measures, and this perception did not differ across conditions.

Similarly, regression analyses yielded no evidence of mediation by the perceived relevance of fairness. With respect to the effect of utility on ratings of whether the perpetrator should be sent to jail, a regression showed that the utility manipulation did not significantly predict the perceived relevance of fairness ($\beta = -.02$, *ns*), which means that the latter variable could not have mediated the effect of utility on treatment. Likewise, with respect to the interaction of dissimilarity and sex on participants' ratings of whether a deal should be made with the perpetrator, a regression showed that the dissimilarity–sex interaction did not predict the perceived relevance of fairness ($\beta = .18$, *ns*). Thus, we found no evidence that the perceived relevance of fairness (inclusion in the participant's scope of justice) mediated the effects of the manipulations on treatment of the perpetrator.

In sum, the results of Study 2, like those of Study 1, provided some support for a deservingness perspective on people's willingness to harm a target but no support for a scope of justice perspective. The effect of dissimilarity on male participants' ratings of whether a deal should be made with the perpetrator seemed to be mediated at least marginally by perceived deservingness, whereas the perceived relevance of fairness did not appear to play a role in this finding. Although we did not predict sex differences in responses to the manipulation of dissimilarity, it is not surprising that female participants were insensitive to this manipulation, because the perpetrator was a man, and the crime was stereotypically masculine (assault causing bodily harm). Unlike Study 1 and Opotow's (1993) research, which focused on insects as targets, the direction of the dissimilarity effect on male participants was consistent with an intuitive analysis, in that the dissimilar man was treated more harshly than the similar man.

As predicted, the effect of utility on treatment of the perpetrator was not mediated by deservingness, because utility was manipulated by connections with a disreputable drug dealer. We think that the manipulation of utility affected recommended punishment for more restorative reasons. Participants were willing to make a deal with the useful perpetrator, but they did not admire him any more than the low-utility perpetrator. Thus, more favorable treatment did not reflect greater perceived deservingness or mean that justice was irrelevant.

Automobile Accidents and Accountability

The results from the two studies we have presented thus far suggest that exclusion from the scope of justice did *not* occur in conditions that previous researchers predicted *would* produce exclusion. Instead, harm-doing in these conditions was attributable to variations in the extent to which the target was seen to deserve positive or negative treatment.

Our studies thus raise the question: When, if ever, *are* justice considerations deemed irrelevant? (For other discussions of this issue, see Tyler et al. [1997], van den Bos & Lind [2002], and Wenzel [2002].) Although our findings have not thus far supported the scope of justice perspective, we continue to see some merit in Opotow's (1993) hypothesis that lack of identification with a target can lead to exclusion, but only at relatively extreme levels (e.g., hatred). Hate has been associated with a motivation to hurt others or to see them suffer (Royzman, McCauley, & Rozin, 2005), which might override justice concerns (see Opotow, 2005). But research and theorizing related to emotion and motivation have suggested an additional qualification of the identification hypothesis. Specifically, strong feelings of empathy and compassion have been associated with an altruistic motivation to increase the welfare of another individual (see Batson, 1991), which, paradoxically, might also override justice concerns. For example, Batson, Klein, Highberger, and Shaw (1995) found that individuals will forgo justice considerations to help those with whom they strongly empathize but for whom helping would violate justice principles.

Taken together, these perspectives lead to the intriguing prediction that identification may be related in a *curvilinear* fashion to exclusion: Relationships that are characterized *either* by strong empathy and emotional attachment *or*, at the other end of the continuum, by hate or emotional antagonism might *both* lead people to exclude targets from their scope of justice (for more details, see Hafer & Olson, 2003; Hafer, Olson, & Peterson, 2008). The following study tested this curvilinear prediction.

Participants read a story about a car accident in which a driver's brakes failed, and as a result, the driver hit and killed a child. Our first independent variable, which we labeled *identification*, ranged from extreme emotional attachment to extreme emotional antagonism, with moderate levels of identification in between. In the extreme attachment condition, participants were asked to imagine that the driver was their identical twin sibling, who was also their closest friend and confidant. In the extreme antagonism condition, participants were asked to imagine that the driver was their most hated enemy. Four more moderate degrees of identification were also created by varying the participants' liking of and geographical proximity to the driver. We predicted that fairness or justice would be seen as more relevant to punishment decisions in the four moderate conditions compared to the two extreme conditions.

We also reasoned that punishment of the driver in the moderate identification conditions (when justice was presumably more relevant) would be based primarily on what the driver was seen to deserve (see Darley & Pittman, 2003). Instead of *measuring* perceived deservingness and conducting mediation analyses, however, as in our first two studies, we *manipulated* a variable that is very closely connected to deservingness (see Spencer, Zanna, & Fong, 2005). Research and theorizing by Feather, Darley, and others (e.g., Darley & Pittman, 2003; Feather, 1999) have suggested that perceived deservingness for punishment mirrors how responsible a perpetrator is believed to be for a transgression. Thus, in the present study, we manipulated the responsibility of the driver in causing the accident, which was expected to influence strongly the perceived deservingness of the driver for punishment. In the low-responsibility condition, the driver had just brought the car in for a checkup, and all had appeared fine. In the high-responsibility condition, the driver knew that the car needed repairs and might be unsafe but chose not to have the car fixed. We predicted that participants' punishment decisions would be more influenced by the responsibility–deservingness manipulation—an indication that justice was more relevant—in the moderate identification conditions than in either the extreme attachment or the extreme antagonism conditions.[1]

After reading the story of the car accident, participants were asked three questions about the punishment they wanted the driver to receive: the charges they would lay (from 1 = *no charges* to 6 = *manslaughter*), the specific punishment they would want applied to the driver (from 1 = *no punishment* to 6 = *criminal conviction, jail sentence*), and, if the driver was sued, the damages they would want awarded to the family (from 1 = *none* to 5 = > *$500,000*). We standardized these items and calculated an average to create a composite measure of punishment. We measured the perceived relevance of fairness (or the extent to which the driver was included in participants' scope of justice) by asking participants to rate how much their decisions about punishment were influenced by what would be fair and just and also by what was morally right (from 1 = *not at all influenced* to 5 = *completely influenced*). The mean of these two items formed a composite measure of the perceived relevance of fairness. Finally, to measure the possible role of empathy (which was expected to be strongest in the extreme attachment condition), we asked participants the extent to which their decisions were influenced by their concern about the driver's welfare.

Influence of Responsibility on Punishment of the Perpetrator A 6 (identification) × 2 (responsibility) ANOVA on participants' punishment decisions yielded a main effect for identification and a main effect for responsibility, both of which were superseded by a significant two-way interaction, $F(5, 230) = 4.46$, $p < .002$. As predicted, responsibility—and, therefore, deservingness—had a relatively greater influence on punishment decisions in the moderate identification conditions than in the extreme conditions (see Figure 6.4). Within-cell correlations between the perceived responsibility of the driver and punishment decisions showed a similar conceptual pattern; although correlations between responsibility and punishment were always strong (at least .63), they were relatively weaker in the two extreme conditions (see Figure 6.5).

Perceived Relevance of Fairness An ANOVA on the composite measure of perceived relevance of fairness yielded a significant main effect for the identification manipulation, $F(5, 230) = 2.41$, $p < .04$. There was also a significant quadratic trend for identification, $p < .03$ (see Figure 6.6). We followed up these effects with a planned contrast that showed, as expected, that fairness was seen as less influential in the extreme emotional attachment and

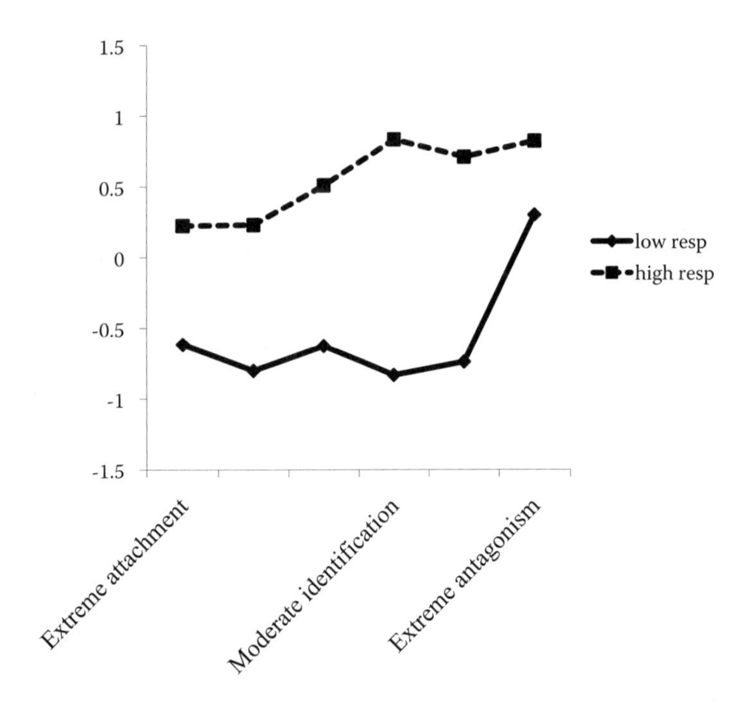

Figure 6.4 Mean (standardized) punishment decisions as a function of the identification and responsibility manipulations.

extreme emotional antagonism conditions compared to the four moderate identification groups ($M = 3.49$ vs. $M = 3.77$), $t(236) = -2.45$, $p < .02$.

Perceived Relevance of Concern for the Driver's Welfare An ANOVA on the extent to which participants reported that their punishment decisions were based on their concern for the driver's welfare yielded a significant main effect for identification, $F(5, 230) = 7.13$, $p < .001$. An a priori contrast revealed that participants in the extreme emotional attachment condition ($M = 3.05$) reported being more influenced by their concern for the driver's welfare than did participants in all other groups (*M*s = 2.68, 2.39, 2.50, 2.33, and 1.75 in the conditions ranging from moderate identification to extreme emotional antagonism; overall M in these five conditions = 2.33), $t(236) = 4.04$, $p < .001$.

In summary, the results of this study suggested that justice *is* sometimes seen as less relevant in people's judgments about and actions toward others, despite the fact that we found little evidence of variations in the relevance of justice in our first two investigations. For the present study, people were relatively less affected by the responsibility information (and, therefore, by the perceived deservingness of the target for punishment) when the

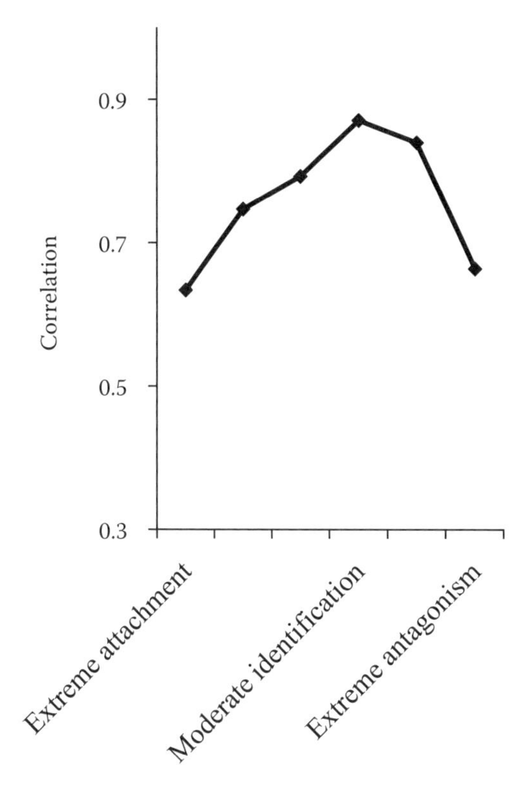

Figure 6.5 Within-cell correlations between perceived responsibility of driver and (standardized) punishment decisions as a function of the identification manipulation.

target was someone for whom they presumably felt extreme emotional attachment or extreme emotional antagonism. People also rated justice and morality as less influential in their decisions with regard to the target in these extreme cases. Finally, participants reported that their decisions were based on concern for the target's welfare to a greater extent in the extreme emotional attachment condition than in the other identification conditions. We assume that the relevant motivation in the extreme antagonism condition was a desire to hurt the target, regardless of justice considerations; future research, however, is needed to test this assumption.

Summary and Discussion of Studies

The main points shown by our research are as follows. First, some instances in which a behavior appears to result from exclusion from the perceiver's scope of justice may not be examples of exclusion—instead, the behavior may be the result of the perceiver's application of a justice rule such as

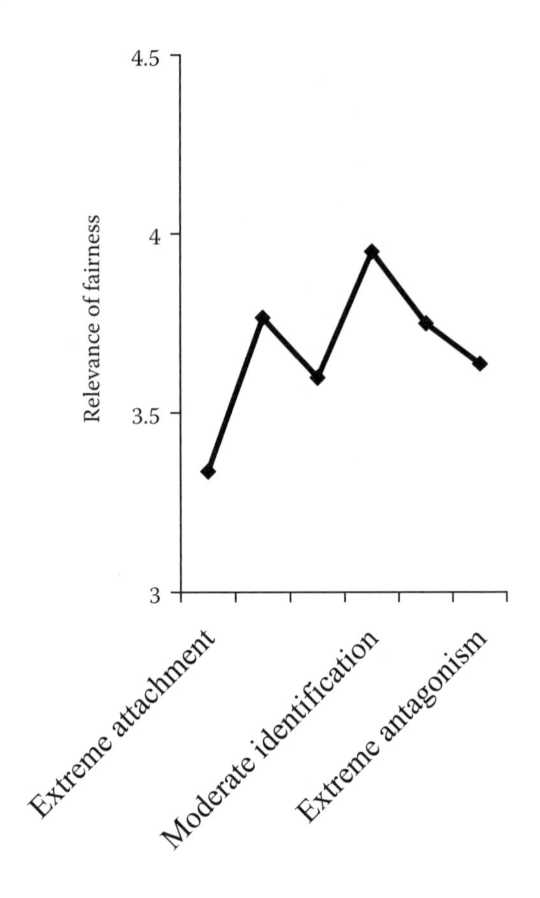

Figure 6.6 Mean ratings of the relevance of fairness as a function of the identification manipulation.

deservingness. For example, the beetles in Study 1 were given what they deserved as a function of the harm they inflicted on or the benefits they gave to human society, as well as their perceived similarity to humans. The criminal in Study 2 was given the punishment he deserved as a function of his similarity to the perceiver, at least by male participants.

Second, though behavior might often be a mixture of justice and other motives, or justice might be overridden by a coexisting motive, we propose that justice is rarely deemed irrelevant—perhaps only under certain extreme conditions, such as when a perceiver's relationship with a target is characterized by great emotional attachment or great emotional antagonism as in the twin and the hated enemy conditions in our third study. We should note, however, that even in these extreme conditions, the mean rating for the relevance of fairness was never below the midpoint of the scale, and our deservingness criterion always significantly correlated with treatment of the target, suggesting that whereas justice was seen as *less relevant* in certain cases, it was never perceived as *irrelevant* (for a discussion of what is meant by the relevance and irrelevance of justice, see Hafer & Olson, 2003). Perhaps the apparently consistent relevance of fairness in Study 3, as well as in Study 2, is not surprising, given that treatment of the target occurred in the context of the formal justice system. For Study 1, though, we think the high ratings for relevance of justice are striking, even more so because social desirability probably plays less of a role in participants' ratings of justice when the target is insects than when the target is human.[2]

Third, researchers invoking the concept of scope of justice have tended to overlook the possibility of exclusion for conditions with positive connotations, instead focusing on those with negative connotations, such as situations characterized by dissimilarity or conflict between the perceiver and the target. We suggest, however, that justice can be seen as less relevant in conditions with positive connotations as well, such as for those to whom we have strong emotional attachment or with whom we strongly empathize. This reasoning led us to predict a curvilinear association between the perceived relevance of fairness and identification, where the extremes of the identification variable were characterized by extreme attachment or extreme enmity. Curvilinear relations have rarely been considered in research on exclusion from justice and other moral considerations (for an exception, see Tyler & Lind, 1990).

Fourth, as we have argued elsewhere (Hafer & Olson, 2003), researchers have tended to confound negative treatment of targets with exclusion from a person's scope of justice and positive treatment with inclusion. Our reasoning behind the current set of studies, however, warns researchers to remain aware that *negative* treatment, such as when a perpetrator is punished, and positive treatment can be motivated by justice, and, similarly, *positive* treatment and negative treatment can be driven by motives other than justice, such as altruism.

Future Research on Scope of Justice

In this final section, we discuss several potential directions for future research on the concept of scope of justice. Specifically, we discuss the role of conflict, the size and malleability of individuals' boundaries for justice, and methodological issues.

Conflict and Exclusion from the Scope of Justice

In the first study presented in this chapter, we manipulated the degree of conflict between participants and targets. High conflict was associated with more negative treatment than low conflict, but mediation analyses showed that neither deservingness nor scope of justice accounted for the effects of conflict on treatment. We suspect that conflict influenced treatment by changing the definition of a "just" distribution to one in which competition and dominance were major themes. Lerner (1977), for example, argued that relationships characterized by conflicting interests can elicit justice rules akin to "survival of the fittest." Future research should investigate whether high conflict does indeed elicit different perceptions of justice rather than render justice concerns irrelevant.

The Size and Malleability of Individuals' Boundaries for Justice Another issue for future research is the investigation of factors that actually influence the "scope" of justice (e.g., its breadth). In the scope of justice literature, variables that might influence features of individuals' boundaries of justice (e.g., the heterogeneity of targets included in the scope) have, somewhat ironically, been ignored. Instead, researchers have focused on characteristics of a specific *target* (e.g., its utility) that affect whether it will be excluded from all participants' scopes of justice.

Several approaches could be used to explore variables influencing individuals' scopes of justice. First, one could experimentally manipulate variables that might influence perceivers' scopes of justice and measure how individuals' scopes change as a function of these manipulations. For example, external threats, such as the threat of terrorism, might shrink people's scopes. On the other hand, events that prime inclusiveness might expand scopes.

A second approach would be to investigate chronic individual differences that may be associated with the breadth of participants' scopes of justice. For example, the Universal Orientation Scale (Phillips & Ziller, 1997) measures individuals' tendencies to perceive similarities rather than differences between the self and others. *Social dominance orientation* (Pratto, Sidanius, Stallworth, & Malle, 1994) refers to individuals' support for intergroup status hierarchies. *Category width* (Pettigrew, 1958) refers to individuals' tendencies to see categories as broad rather than narrow. Perhaps people who are low in universal orientation, high in social dominance, and low in category width have narrower scopes (e.g., exclude more others from justice considerations) than people at the opposite ends of these dimensions. The results of the third study presented in this chapter suggest that one might also look to more emotion-based individual differences for variation in the breadth of people's scopes of justice, and those who possess high levels of a characteristic with positive connotations might sometimes have more narrow boundaries within which they see justice as relevant. For example, people very high in dispositional empathy (Davis, 1994) might actually have narrower scopes of justice than people lower in this trait, because the former strongly empathize with many targets and, thus, are motivated to extend positive outcomes and treatment to many targets regardless of what might be seen by others as relevant justice considerations.

Methodological Issues Another issue raised by our studies is how best to assess exclusion from the scope of justice. We assessed exclusion in the present research in two ways. First, in all three studies, we gathered explicit self-reports of the extent to which people believed that justice was a relevant consideration in a decision they made with respect to some target. Presumably, the more people reported that justice and fairness was relevant, important, or influential, the more the target was included in participants' scope of justice. Second, in Study 3, we manipulated a variable closely connected to deservingness (i.e., responsibility) and examined the extent to which this variable predicted or influenced participants' decisions about a target. The greater the influence of the deservingness–responsibility variable on participants' decisions about a target, the more the target is assumed to have been included in participants' scope of justice. On the basis of our results from these two strategies, we argued that deservingness is a justice rule that can account for what has often been labeled as *exclusion* in past research.

There are several limitations to these methods that should be addressed in future research. For example, there are limitations to self-reports (see Olson, Goffin, & Haynes, 2007). Social desirability demands may lead participants to say that they considered fairness even when they did not. Thus, asking people explicitly to report on the relevance of fairness, or even asking them to make decisions in light of what might obviously appear to be justice-related criteria (e.g., responsibility or other bases of perceived deservingness), may not always be an ideal method of gauging exclusion.

Two other problems with self-report measures, although perhaps less applicable to the influence of our responsibility manipulation in Study 3, could pertain to explicit ratings for the relevance of fairness. First, self-reports can suffer from participants' imperfect insight into the various motives and psychological processes that are driving their responses in a given situation (see Nisbett & Wilson, 1977). Thus, even in the absence of social desirability pressure, explicit measures of the extent to which justice is relevant may sometimes be inaccurate. Second, we asked people in the current studies to make a decision and then to report on the relevance of justice. These ratings might have reflected conscious or unconscious post hoc rationalizations of the decision they just made.

We tried to counteract shortcomings of any one study by using a variety of behavioral domains, by using different ways of phrasing the questions about the relevance of justice, and by both manipulating and measuring the concept of deservingness; future attempts to assess exclusion from the scope of justice using novel techniques are nonetheless warranted. As we have suggested elsewhere (Hafer & Olson, 2003), one such possibility is the use of subtle reaction time indicators of the extent to which the concept of justice is relevant or has been activated in a given situation (see Hafer, 2000).

One final methodological point is that the three studies reported here all rely on hypothetical scenarios. Future research is needed that tests whether our findings can be generalized to real behaviors in real-life situations. For example, surveys could ask people about their perceptions and behaviors with respect to either real-world issues, such as treatment of terrorists, or actual situations with which they have been involved, either as actors or as the recipients of others' treatment (e.g., Beaton & Tougas, 2001; Brockner, 1990). Problems with survey studies in this domain should be offset by also conducting high-impact experiments in which participants believe they are actually in a situation requiring them to make judgments about how to treat some target (see Lerner, 2003). The latter are more difficult, given potential ethical constraints, but not impossible. The research by Batson et al. (1995) noted earlier is one example.

In conclusion, we believe that the scope of justice is an important concept for research in social justice and related areas. The studies described in this chapter are an initial attempt to follow up our earlier analysis (Hafer & Olson, 2003; Hafer et al., in press) by empirically addressing some of the problems and ambiguities we see with previous research on the concept. We hope these results and discussion will motivate other scholars to further explore the interesting implications of the scope of justice.

Notes

1. Although we did not measure perceived deservingness in this study, a prior experiment that used exactly the same responsibility manipulation yielded a strong main effect of responsibility on a measure of deservingness, $F(1, 72) = 51.56, p < .001$, and participants' ratings of the perpetrator's responsibility for causing the accident correlated very highly with their ratings of the perpetrator's deservingness for punishment, $r(71) = .80, p < .001$ (Woznica, 2008).
2. Indeed, Opotow (1993) suggested that a strength of using beetles as the target in her original work on scope of justice was that social desirability was less an issue in the treatment of this animal than in the treatment of humans (or other more humanlike animals). Whereas our own beetles study might, therefore, have been less subject to social desirability concerns, our studies involving only humans as targets are perhaps more open to these influences.

References

Baron, R. M., & Kenny, D. A. (1986). The moderator–mediator variable distinction in social psychological research: Conceptual, strategic, and statistical considerations. *Journal of Personality and Social Psychology*, *51*, 1173–1182.

Batson, C. D. (1991). *The altruism question*. Hillsdale, NJ: Lawrence Erlbaum.

Batson, C. D., Klein, T. R., Highberger, L., & Shaw, L. L. (1995). Immorality from empathy-induced altruism: When compassion and justice conflict. *Journal of Personality and Social Psychology, 68*, 1042–1054.

Beaton, A. M., & Tougas, F. (2001). Reactions to affirmative action: Group membership and social justice. *Social Justice Research, 14*, 61–78.

Brockner, J. (1990). Scope of justice in the workplace: How survivors react to co-worker layoffs. *Journal of Social Issues, 46*(1), 95–106.

Darley, J. M., & Pittman, T. S. (2003). The psychology of compensatory and retributive justice. *Personality and Social Psychology Review, 7*, 324–336.

Davis, M. H. (1994). *Empathy: A social psychological approach*. Boulder, CO: Westview.

Deutsch, M. (2000). Justice and conflict. In M. Deutsch & P. T. Coleman (Eds.), *The handbook of conflict resolution* (pp. 41–64). San Francisco: Jossey-Bass/Pfeiffer.

Feather, N. T. (1999). *Values, achievement, and justice*. New York: Kluwer/Plenum.

Freudenthaler, H. H., & Mikula, G. (1998). From unfulfilled wants to the experience of injustice: Women's sense of injustice regarding the lopsided division of household labor. *Social Justice Research, 11*, 289–312.

Goldenberg, J. L., Pyszczynski, T., Greenberg, J., Solomon, S., Kluck, B., & Cornwell, R. (2001). I am *not* an animal: Mortality salience, disgust, and the denial of human creatureliness. *Journal of Experimental Psychology: General, 130*, 427–435.

Gromet, D. M., & Darley, J. M. (2006). Restoration and retribution: How including retributive components affects the acceptability of restorative justice procedures. *Social Justice Research, 19*, 395–432.

Hafer, C. L. (2000). Do innocent victims threaten the belief in a just world? Evidence from a modified Stroop task. *Journal of Personality and Social Psychology, 79*, 165–173.

Hafer, C. L., & Olson, J. M. (2003). An analysis of empirical research on the scope of justice. *Personality and Social Psychology Review, 7*, 311–323.

Hafer, C. L., Olson, J. M., & Peterson, A. A. (2008). Extreme harmdoing: A view from the social psychology of justice. In V. M. Esses & R. A. Vernon (Eds.), *Explaining the breakdown of ethnic relations: Why neighbors kill*. Oxford, UK: Blackwell.

Heuer, L., Blumenthal, E., Douglas, A., & Weinblatt, T. (1999). A deservingness approach to respect as a relationally based fairness judgment. *Personality and Social Psychology Bulletin, 25*, 1279–1292.

Hornsey, M. J., & Hogg, M. A. (1999). Subgroup differentiation as a response to an overly-inclusive group: A test of optimal distinctiveness theory. *European Journal of Social Psychology, 29*, 543–550.

Leets, L. (2001). Interrupting the cycle of moral exclusion: A communication contribution to social justice research. *Journal of Applied Social Psychology, 31*, 1859–1891.

Lerner, M. J. (1977). The justice motive: Some hypotheses as to its origins and forms. *Journal of Personality, 45*, 1–52.

Lerner, M. J. (1981). The justice motive in human relations: Some thoughts on what we know and need to know about justice. In M. J. Lerner & S. C. Lerner (Eds.), *The justice motive in social behavior* (pp. 11–35). New York: Plenum.

Lerner, M. J. (2003). The justice motive: Where psychologists found it, how they lost it, and why they may not find it again. *Personality and Social Psychology Review, 7*, 388–399.

Nagata, D. K. (1993). Moral exclusion and nonviolence: The Japanese American internment. In V. K. Kool (Ed.), *Nonviolence: Social and psychological issues* (pp. 85–93). Lanham, MD: University Press of America.

Nisbett, R. E., & Wilson, T. D. (1977). Telling more than we can know: Verbal reports on mental processes. *Psychological Review, 84*, 231–259.

Olson, J. M., Goffin, R. D., & Haynes, G. A. (2007). Relative versus absolute measures of explicit attitudes: Implications for predicting diverse attitude-relevant criteria. *Journal of Personality and Social Psychology, 93*, 907–926.

Opotow, S. (1990). Moral exclusion and injustice: An introduction. *Journal of Social Issues, 46*, 1–20.

Opotow, S. (1993). Animals and the scope of justice. *Journal of Social Issues, 49*, 71–85.

Opotow, S. (2001). Reconciliation in times of impunity: Challenges for social justice. *Social Justice Research, 14*, 149–170.

Opotow, S. (2005). Hate, conflict, and moral exclusion. In R. J. Sternberg (Ed.), *The psychology of hate* (pp. 121–153). Washington, DC: American Psychological Association.

Pettigrew, T. F. (1958). The measurement and correlates of category width as a cognitive variable. *Journal of Personality, 26*, 532–544.

Phillips, S. T., & Ziller, R. C. (1997). Toward a theory and measure of the nature of nonprejudice. *Journal of Personality and Social Psychology, 72*, 420–434.

Pratto, F., Sidanius, J., Stallworth, L. M., & Malle, B. F. (1994). Social dominance orientation: A personality variable predicting social and political attitudes. *Journal of Personality and Social Psychology, 67*, 741–763.

Roese, N. J., & Olson, J. M. (2007). Better, stronger, faster: Self-serving judgment, affect regulation, and the optimal vigilance hypothesis. *Perspectives on Psychological Science, 2*, 124–141.

Royzman, E. B., McCauley, C., & Rozin, P. (2005). From Plato to Putnam: Four ways to think about hate. In R. J. Sternberg (Ed.), *The psychology of hate* (pp. 3–35). Washington, DC: American Psychological Association.

Singer, M. S. (1996). Effects of scope of justice, informant ethnicity, and information frame on attitudes towards ethnicity-based selection. *International Journal of Psychology, 31*, 191–205.

Singer, M. S. (1998). The role of subjective concerns and characteristics of the moral issue in moral considerations. *British Journal of Psychology, 89*, 663–679.

Singer, M. S. (1999). The role of concern for others and moral intensity in adolescents' ethicality judgments. *Journal of Genetic Psychology, 160*, 155–166.

Spencer, S. J., Zanna, M. P., & Fong, G. T. (2005). Establishing a causal chain: Why experiments are often more effective than mediational analyses in examining psychological processes. *Journal of Personality and Social Psychology, 89*, 845–851.

Sunshine, J., & Heuer, L. (2002). Deservingness and perceptions of procedural justice in citizen encounters with the police. In M. Ross & D. T. Miller (Eds.), *The justice motive in everyday life* (pp. 397–415). New York: Cambridge University Press.

Tyler, T. R., Boeckmann, R. J., Smith, H. J., & Huo, Y. J. (1997). *Social justice in a diverse society.* Boulder, CO: Westview.

Tyler, T. R., & Lind, E. A. (1990). Intrinsic versus community-based justice models: When does group membership matter? *Journal of Social Issues*, *46*(1), 83–94.
Van den Bos, K., & Lind, E. A. (2002). Uncertainty management by means of fairness judgments. In M. P. Zanna (Ed.), *Advances in experimental social psychology* (Vol. 34, pp. 1–60). San Diego, CA: Academic Press.
Wenzel, M. (2002). What is social about justice? Inclusive identity and group values as the basis of the justice motive. *Journal of Experimental Social Psychology*, *38*, 205–218.
Woznica, A. (2008). *Explaining harmful actions: The breadth and malleability of university undergraduates' scopes of justice*. Unpublished senior honors thesis, University of Western Ontario, Canada.

CHAPTER 7

The Power of the Status Quo

Consequences for Maintaining and Perpetuating Inequality

DANIELLE GAUCHER, AARON C. KAY, and KRISTIN LAURIN

University of Waterloo

Abstract: In this chapter, the authors outline the evolution of system justification theory (SJT), with an emphasis on the consequences of the system justification (SJ) motive for perpetuating and maintaining inequality. They begin by reviewing evidence for out-group favoritism and anticipatory rationalizations—among the first effects cited as evidence for the existence of the SJ motive. They next review the motivational antecedents of SJ, namely, reduced perceptions of personal control, system threat, and perceived system inevitability. Finally, they discuss how the SJ motive encourages people to act in ways that perpetuate and maintain inequality. Specifically they show that the SJ motive (a) activates complementary stereotypes that portray a more balanced and fair social landscape; (b) promotes behavioral displays that are consistent with system-justifying stereotypes; (c) leads people to perceive group inequalities as due to essential, unchangeable differences between group members; and (d) encourages injunctification (i.e., the construal of what currently *is* as what *should be*) and the derogation of people who act counternormatively.

The signs and symptoms of group inequality in North America are prolific. For every dollar that men make, women make 70 cents (Canadian Labour Congress, 2008). Even controlling for important variables that could account for this difference, such as length of employment and education level, women still make less money than their male counterparts by simple virtue of their gender (Lips, 2003). The situation for visible minorities is similarly perturbing. Visible minorities continue to be underrepresented in important areas, such as politics and high-level management positions (Black & Hicks, 2006; Klie, 2007). Furthermore, current unemployment rates among visible minorities are double that of Caucasians, despite the fact that many members of visible minority groups, such as Southeast Asians, are more likely than Caucasians to hold university degrees (Canadian Census, 2006).

Given that Canadians strongly endorse values such as fairness and justice, why do these striking examples of inequality not cause a public outcry? Concern with issues of inequality tends to be left with activists and social workers, whereas the majority of citizens respond to evidence of inequality with inaction, denial, and even vehement justification. Indeed, Canadian history confirms that large-scale revolts are far more the exception than the rule. Why do people ignore and defend instances of injustice and the societal structures that maintain them?

System justification theory (SJT; Jost & Banaji, 1994) argues that people are motivated to justify the systems—the sets of rules and sociopolitical institutions—that have control over their lives, and they perceive them as more just and legitimate than they in fact are. People engage in a number of psychological strategies aimed at maintaining the belief in a fair and just system, even when the status quo is one of blatant inequality (e.g., Jost, 2001; Jost, Pelham, & Carvallo, 2002).

SJT has flourished in recent years as its insights into important social issues and utility for predicting human behavior have been recognized. In this chapter, we outline the evolution of SJT, with an emphasis on the consequences of the system justification (SJ) motive for perpetuating and maintaining inequality. We divided the research on SJ into three distinct generations: First-generation research primarily demonstrates that the SJ effect exists, second-generation research focuses on the psychological and motivational antecedents of the SJ motive, and third-generation research focuses on the downstream societal and personal consequences of the SJ motive.

First-Generation SJ Research: Demonstrations of SJ Effects

Out-Group Favoritism

Perhaps the most novel and controversial prediction to emerge from SJT is that of *out-group favoritism*, the phenomenon that members of disadvantaged groups tend to show a preference for the dominant social group over their own. Out-group favoritism demonstrates the strongest prediction of SJ theory: All people, regardless of their relative standing in the social hierarchy, justify their system. Although early out-group favoritism effects were often dismissed or even outright rejected as a general psychological phenomenon, as new experimental methodologies developed, evidence mounted in support of this type of intergroup perception. For example, a few studies have found that African Americans in the United States display an implicit preference for Whites over Blacks on a standard race Implicit Association Test (for a review, see Jost, Banaji, & Nosek, 2004; Nosek, Banaji, & Greenwald, 2002). An Internet study with over 13,000 African Americans supported this implicit preference for White out-group members (Nosek et al., 2002).

Out-group favoritism effects are not limited to the race domain or to implicit associations. The social psychological literature now features an abundance of evidence for out-group effects in the domains of gender (e.g., Blair & Banaji, 1996; Rudman & Kilianski, 2000), age (e.g., Levy & Banaji, 2002), and even less stable system affiliations such as university membership (e.g., Lane, Mitchell, & Banaji, 2005; for a review of out-group favoritism effects, see Dasgupta, 2004). Moreover, other types of unobtrusive measures such as the preference for out-group interaction partners (Jost et al., 2002, Study 2) have been taken as evidence of out-group favoritism. Admittedly, in the case of many unobtrusive measures of out-group favoritism, the precise reason for these phenomena remain somewhat contested. Also, the prevalence of out-group favoritism effects is still debated as ingroup favoritism—people's tendency to favor members of their ingroup—dominates the literature. SJT researchers, however, do not suggest that out-group favoritism is necessarily more common than ingroup favoritism. Rather, they contend that the fact that out-group favoritism exists at all is incredibly noteworthy, given the strong motivations to bolster perceptions of the self and one's group.

Anticipatory Rationalizations

SJT proposes that people will rationalize all aspects of their *current* social system. The *current* system structure and arrangements, however, are not the only aspects that may be rationalized. So that the legitimacy of a system is maintained, people may also come to rationalize imminent changes to the system. The more probable the new outcome is, the more likely it is that people will come to rationalize it. In this way, people come to anticipate the new status quo, and it is this anticipated status quo that gets justified. Anticipatory rationalizations—the tendency for people to justify imminent changes to one's system—was supported across a number of studies. During the George W. Bush–Al Gore presidential campaigns in the United

States, for instance, people rationalized (i.e., came to view as desirable and ideal) imminent changes to their sociopolitical system (Kay, Jimenez, & Jost, 2002). Regardless of people's initial political preference, they came to view the political leader that was most likely to win the election as the most desirable candidate. Republicans rated Bush as more desirable as the probability of his winning the presidency increased, whereas they rated him less favorably as the probability of his winning the presidency decreased. Democrats demonstrated a similar pattern of results, rating Bush as slightly more desirable as the probability of his winning the presidency increased.

Anticipatory rationalizations were not limited to the political domain. In another study—under conditions of high motivational involvement (i.e., when the tuition change was large and thus motivationally relevant to students)—the more probable the tuition changes were, the more likely students were to rationalize both increases and decreases in their university's tuition (Kay et al., 2002). Anticipatory rationalizations provided additional evidence for people's tendency to justify the system while highlighting the motive's breadth to influence people's judgments. When change to the system is imminent, people rationalize the change—regardless of their initial feelings or preferences.

Second-Generation SJ Research: Investigations of Why and When People Justify the Status Quo

Out of the work on out-group favoritism and anticipatory rationalizations rose questions about *why* and *when* people justify their systems: What psychological need does SJ fulfill? How do we know that SJ is motivated? And under what conditions are processes of rationalization most likely to occur? Recent research in our lab at the University of Waterloo has sought to elucidate the psychological and motivational antecedents of the SJ motive. Building on past research and theory (Jost, Banaji, & Nosek, 2004; Jost & Hunyady, 2005; Jost et al., 2007), we have accumulated experimental evidence suggesting that motivational antecedents of SJ phenomena include (but likely are not limited to) the need to view the world as orderly and under control, threats to the system, and the inevitability of one's system.

The Need to View the World as Orderly and Controllable

What psychological need does SJ fulfill? Jost and Hunyady (2005) theorized that people may justify the system as one way of satisfying their need for order and control. Beliefs in order and control help to ward off feelings of helplessness and the despair associated with viewing life as uncontrollable (Mikulincer, Gerber, & Weisenberg, 1990; Seligman, 1975, 1976). Typically, and especially in Western cultures, people rely on a sense of personal control in trying to avoid the perception of randomness and disorder. There are many instances, however, in which people's sense of order and personal control is threatened. Threats to one's sense of personal control can be rather mundane (i.e., catching a good or bad break) or quite striking (i.e., as in the case of getting a serious illness or being the innocent victim of a crime). Regardless of the severity of a given threat, we proposed that the discomfort it produces leads people to reaffirm their sense that things are under control, thereby dispelling negative affect. One way in which people might do this, we suggested, is by placing faith in external systems of control, such as religious institutions and the government—a process we coined *compensatory control.*

How do people reaffirm their belief that things are under control? Endorsing external systems, such as the government and religion, can help to reaffirm people's sense that things are under control. Religion, for instance, provides people with the view that all of the events in the world are under the control of an omnipotent supernatural power (e.g., God). People who believe in God can rest assured knowing that the world is not random, because God is purposely watching and controlling daily events. Governments likewise provide people with a sense of security that things are under control. Governments monitor and mandate the general workings of a system. They control the economy and directly influence people's daily life through legislation and development of laws. Thus we suggest that in the loss of personal control, one way people can reaffirm their sense that the world is under control is with compensatory means—by placing faith in external systems of control, such as religious institutions and the government. In this way, people can maintain their belief that things, in general, are not operating randomly or haphazardly.

Several studies have demonstrated the extent to which the psychological need to defend against randomness and chaos contributes to SJ. In three experimental studies, we tested the compensatory control hypothesis by manipulating people's perceptions of personal control and then assessing their endorsement of external agents of control, such as God and the government (Kay, Gaucher, Napier, Callan, & Laurin, 2008). To manipulate people's sense of controllability, we threatened or affirmed participants' sense of personal control by asking them to write about something positive that happened to them that they either did not have control over (e.g., winning the lottery) or did control (e.g., studied hard for an exam).

People whose sense of personal control had been threatened were more likely to engage in system-justifying tendencies (i.e., endorse the belief that God controls events in the world or defend the legitimacy of the current government) than were people whose feelings of personal control were affirmed. Importantly, compensatory control effects were found only when the external agent was perceived as both agentic and benevolent.

People's increased endorsement of God occurred when God was framed an omnipotent entity but not when God was framed as a creator who had no further control over the events in the world (see Figure 7.1). People's pre-existing beliefs about the government's benevolence determined their use of the external system as a compensatory means. Only those people who thought that the government had their best interests at heart increased their endorsement of the government following a loss of personal control (see Figure 7.2). This makes good sense. Relying on an external system of control that does not have one's best interests at heart would certainly not reduce the threat of chaos or randomness.

Compensatory control effects are not limited to Canadians or university students. Using data from the World Values Survey (2004), we found correlational evidence for the compensatory control effect across representative samples from 59 countries. In general, the less personal control people felt over their lives, the more they expressed a desire for governmental control, that is, the more faith they placed in their country's government. Again, this effect was moderated by benevolence, such that only people with governments whom they deemed as having their best interests at heart were likely to turn to their government when they experienced a loss of personal control (see Figure 7.3).

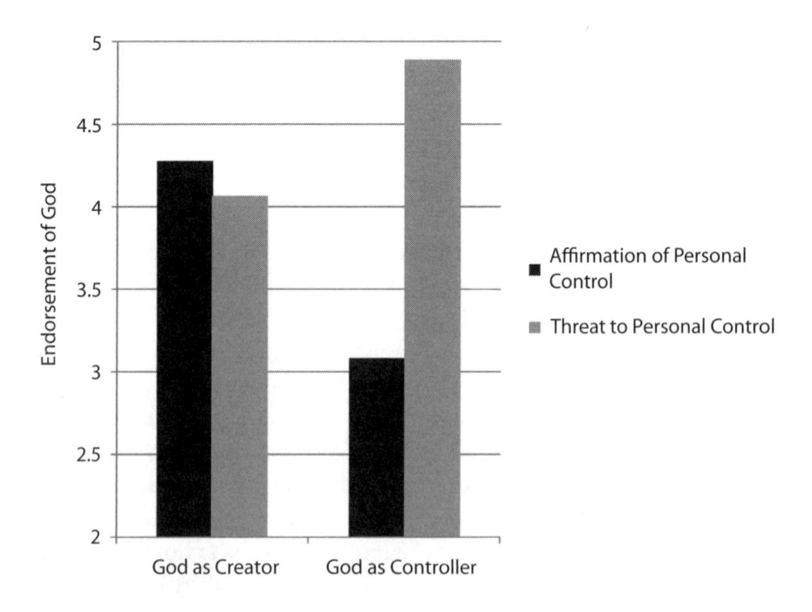

Figure 7.1 Endorsement of God as controller or creator expressed by participants whose sense of personal control was either threatened or affirmed.

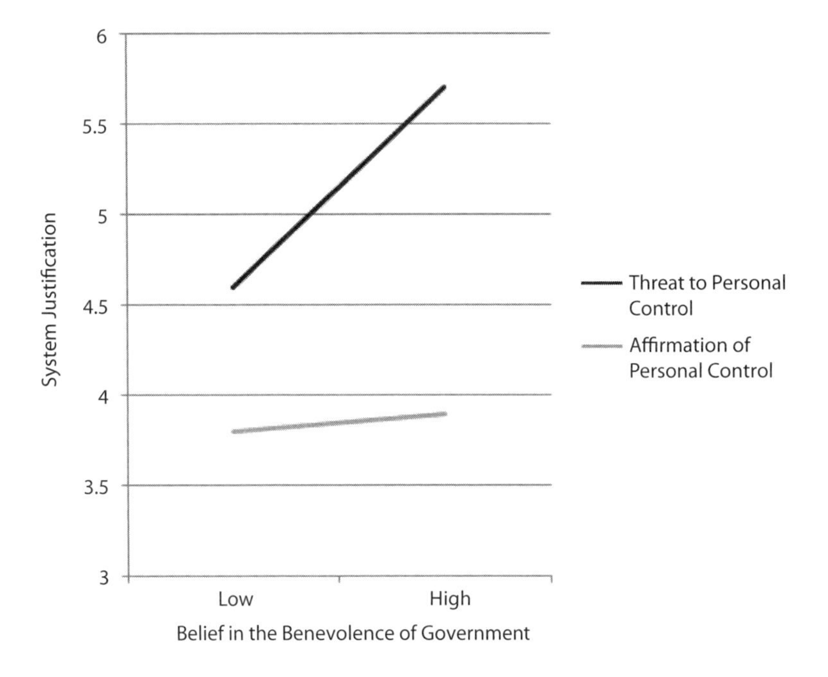

Figure 7.2 System justification of participants whose sense of personal control was either threatened or affirmed, depending on their belief in the benevolence of government.

System Threats

Is SJ a motivated response or simply a cognitive bias? To the extent that people have a given motivation, blocking that motivation causes it to become hyperactivated (Atkinson & Birch, 1970; Bargh, Gollwitzer, Lee-Chai, Barndollar, & Troetschel, 2001). This logic holds true for the SJ motive as well (see Jost & Hunyady, 2002). When people's SJ motive is blocked, they respond with a heightened desire to justify and rationalize. System threat—questioning the legitimacy of one's system—is one way to block people's SJ motive. The use of system threat to instigate rationalization processes is similar to past research demonstrating that self- and group-level threats instigate self and group defensive processes (e.g., Fein & Spencer, 1997; Sherman & Cohen, 2002).

Threats to the legitimacy of one's system can occur in the real world in the form of economic hardship (cf. Sales, 1972), disasters such as Hurricane Katrina (Napier, Mandisodza, Andersen, & Jost, 2006), and the terrorist attacks of 9/11 (Ullrich & Cohrs, 2007). Within the laboratory, vignettes of an outsider criticizing the legitimacy of one's current system are most often employed. The following is an example of a widely used Canadian system threat passage:

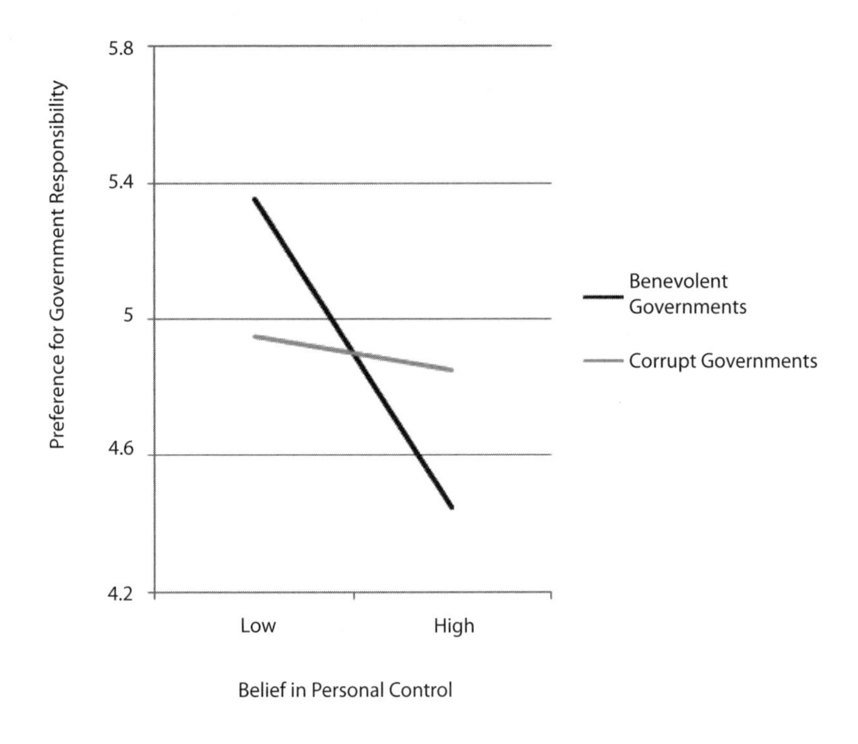

Figure 7.3 Preference for governmental responsibility as a function of beliefs about personal control and government benevolence.

> These days, many people in Canada feel disappointed with the nation's condition. Many citizens feel that the country has reached a low point in terms of social, economic, and political factors. People do not feel as safe and secure as they used to, and there is a sense of uncertainty regarding the country's future. It seems that many countries in the world are enjoying better social, economic, and political conditions than Canada. More and more Canadians express a willingness to leave Canada and emigrate to other nations.

This system threat manipulation has been shown to increase people's SJ tendencies while not affecting their general level of identification with the social system, self-esteem, or current mood (see Kay, Jost, & Young, 2005; Laurin, Kay, Gaucher, & Shepherd, 2008).

A number of studies have confirmed that system threats instigate rationalization processes. Specifically, system threat has been shown to increase people's (a) use of compensatory stereotypes (Jost, Kivetz, Rubini, Guermandi, & Mosso, 2005; Kay et al., 2005), (b) defense of political leaders (Banfield & Kay, 2008), (c) rationalization of system inequalities

(Kay, Gaucher, Peach, Zanna, & Spencer, 2008; Laurin et al., 2008), and (d) attraction to people who act in system-justifying ways (Lau, Kay, & Spencer, 2008).

Inevitability of the System

Researchers have reasoned that people justify their systems, at least in part, because they believe them to be inevitable (Laurin et al., 2008). An inevitable system is one that is both difficult to leave and unlikely to change—in other words, one that is likely to continue indefinitely to have its present influence over people's lives. The structure and composition of an inevitable system are predictable, unavoidable, unchanging, and persistent. Inevitability offers the promise that the system in its current form will continue to exert influence over people's lives for an indefinite period of time.

Evitable systems, however, offer the glimmer of hope that things may not always be as they are now. In this context, people should not be so motivated to perceive their systems *in their current forms* as just and fair—in the future they might leave, or the system might change. Indeed, Laurin et al. (2008) found that, for example, Canadian women who were reminded about how hard it was to leave Canada more strongly justified a current gender inequality compared to women who were reminded of the ease with which they could leave the country. Other manipulations of inevitability led to similar effects. In one study, participants were presented with information suggesting that the status quo in terms of women's representation in Canadian business either was going to persist for some time or was likely to change. Participants in the low-change (i.e., high-inevitability) condition subsequently displayed greater SJ tendencies; specifically, they showed greater resistance to redistribute social policies aimed at helping the disadvantaged as compared to participants in the high-change (i.e., low-inevitability) condition. Importantly, the *direction* of change did not affect our results, whether it was in the liberal or conservative direction; the mere presence of change—which presumably reflected a noninevitable system—led our participants to be less motivated to justify their system (Laurin et al., 2008).

Discussion

Justifying the system can be adaptive. It helps people cope with the existential and epistemic threats that arise from believing that one's system is illegitimate or operates unfairly. Also, it enables people to maintain a positive view of things that they cannot change and allows people to believe that things are under control. The intrapsychic benefits of SJ, however, stand in stark contrast to its more insidious consequences at the social and interpersonal levels.

Third-Generation SJ Research: Consequences of the SJ Motive

Social systems are often rife with inequality. Given that attempts to fix unfairness in the system require that people recognize the unfairness and desire to change it, it is easy to see how the SJ motive could interrupt the process of social change. Across a number of studies, we found that the SJ motive indeed carries with it the potential to maintain inequality. Specifically, the SJ motive (a) activates complementary stereotypes that portray a balanced and fair social landscape and promotes behavioral displays that are consistent with system-justifying stereotypes; (b) leads people to perceive group inequalities as due to essential, unchangeable differences between groups; and (c) encourages injunctification, as well as the derogation of people who act counter to the status quo.

SJ and the Activation of Complementary Stereotypes

In the face of inequality, one way to maintain an illusion of equality is to apply complementary stereotypes—to ascribe positive attributes or characteristics to stigmatized groups to compensate for their low status and/or to ascribe negative attributes or characteristics to valued groups to compensate for their high status. These are common in media representations: Consider people portrayed as unattractive but smart (nerds), attractive but dumb (models), athletic but oafish (jocks), poor but happy, or overweight but jolly, all of which abound in cultural messages. Although seemingly benign, these types of stereotypes may represent an especially potent source of support for societal inequalities. That is, complementary stereotypes allow people to feel like rewards and drawbacks in society are relatively balanced out and that the status quo is fair because no one group is left without a positive attribute to call its own—everyone has something.

Kay and Jost (2003, 2005; Kay, Czaplinksi, & Jost, 2009) have run several studies testing the role of complementary stereotyping in support for the status quo. In one study, under the guise of an impression formation task, participants were presented with one of four vignettes aiming to prime either complementarity or noncomplementarity. The vignettes varied the wealth and happiness of a fictional character, Mark, to demonstrate complementary stereotypes (rich and unhappy or poor and happy) or noncomplementary stereotypes (rich and happy or poor and unhappy). Next, as part of an ostensibly unrelated experiment, participants completed a scale measuring general satisfaction with the status quo. As predicted, participants who were exposed to the complementary stereotypes reported increased satisfaction with the status quo. This pattern of increased contentment with the status quo after exposure to complementary stereotyping has since emerged as a robust effect in other cultures (Kay, Czaplinski, et al., 2009), using different types of complementary stereotypes (Jost &

Kay, 2005) and with unobtrusive paradigms such as reaction time measures (Kay & Jost, 2003; for a review, see Kay et al., 2007).

Exposure to stereotypical *others* is not the only way in which complementary stereotypes can influence contentment with the status quo. Reminders of one's own complementary attributes have a similar effect. Kay, Laurin, and Shepherd (2008), for instance, found that after women were reminded of their communal nature (i.e., made to believe that they were especially cooperative, thoughtful, relationship oriented, and in tune with their emotions), they reported higher levels of contentment with the system of gender relations than women who were instead made to believe that they were particularly individualistic, competent, and achievement oriented. This finding is provocative because it suggests that complementary stereotyping of not only others but also the *self* can produce increased contentment with the status quo.

Of course, increased contentment with the status quo on its own does not necessarily imply the perpetuation of inequality. Direct links between contentment with the status quo and system-justifying behavior are necessary to make that claim, and such associations indeed exist. For example, Gaucher, Chua, and Kay (2008) found that people's premeasured contentment with the status quo predicted their level of activist behavior in response to an injustice in their current health care system. That is, the more content participants were with the status quo, the fewer protest postcards they completed when given the opportunity and the less likely they were to feel that Canadians "should be concerned" that people with lower socioeconomic status receive poorer quality health care than those with higher socioeconomic status. Similarly, Day, Yoshida, and Kay (2008) found that exposure to complementary stereotypes increased people's satisfaction with the status quo and decreased support for social programs aimed at helping the disadvantaged. To the extent that an increase in contentment with the status quo reduces social action and support for change, the relation between complementary stereotypes and the maintenance of inequality is hard to ignore.

Complementary stereotyping studies suggest that in people's day-to-day lives they may try to affirm their SJ motive by actively seeking out information, creating environments, and/or acting in ways that confirm complementary aspects of themselves and others. Indeed, Kay et al. (2008) found that following a system threat—when people's SJ tendencies were incited—women, but not men, portrayed themselves as more communal by rating themselves higher on a scale of communal traits. Similarly, recent findings show that the SJ motive causes men to be more attracted to women who act in system-justifying ways. Lau et al. (2008) had male participants read either a system threat or a system affirmation and then review the dating profiles of eight women, ostensibly taken from an online dating service.

Four of the eight profiles depicted women who fit the benevolently sexist stereotype of having vulnerability, purity, and the potential to make men feel competent. The other four profiles portrayed women who did not fit the stereotype associated with benevolent sexism—these women were described as possessing more agentic traits, such as ambition and adventurousness.

After reading each profile, male participants rated the extent to which they would like to meet the target woman, the extent to which they were interested in starting a relationship with her, and the extent to which they viewed her as an ideal romantic partner. As the results in Figure 7.4 indicate, after a system threat, men were more romantically interested in women who embodied benevolent sexist ideals (i.e., women who conformed to a complementary stereotype) than in women who contradicted this stereotype.

There are several implications of these findings. First, this research suggests that there may be social reinforcement of people who display characteristics that are consistent with complementary stereotypes. It is possible that on some level women are aware of men's preference for women who act in traditionally feminine ways and that this awareness encourages women to act accordingly. After all, people generally want to portray themselves as attractive potential mates on the interpersonal marketplace; for women, this may mean acting in a submissive and soft-spoken way. Indeed, past research carried out in nonromantic contexts, such as the workplace, has suggested that forces such as gender-role expectations and the desire to be accepted serve as encouragements for women to act in gender-stereotypic

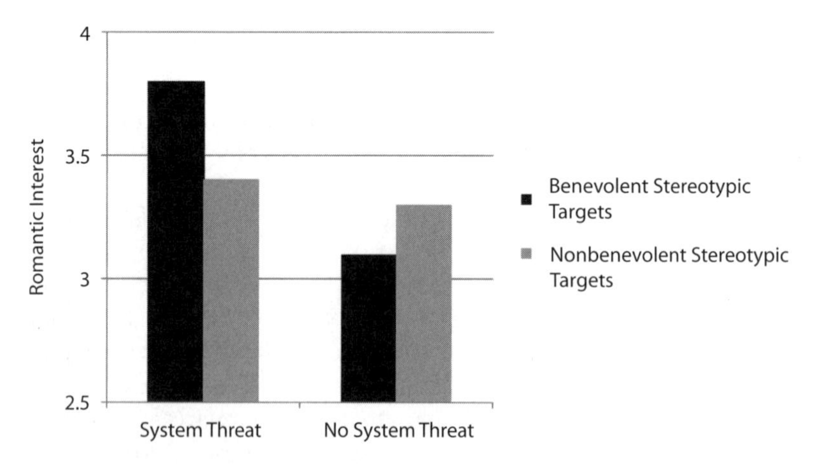

Figure 7.4 Romantic interest in benevolent-stereotypic and nonbenevolent-stereotypic women expressed by male participants either under system threat or not under system threat.

ways (Skrypnek & Snyder, 1982; Von Baeyer, Sherk, & Zanna, 1981; Zanna & Pack, 1975).

Unfortunately, acting according to traditional gender roles can have negative consequences for women. Although highly feminine traits such as compassion and empathy may be revered in some contexts, in most high-status positions, agentic qualities, such as self-confidence and assertiveness, are more socially prized. Thus, although women who embody more traditionally feminine traits may be liked more than men are, they will not necessary be respected or sought out for high-status positions (Eagly, Mladinic, & Otto, 1994).

Furthermore, men's preference for more submissive women, under conditions of system threat, serves to maintain traditional gender roles and power imbalances within romantic relationships. Moreover, this phenomenon may not serve all men particularly well either. Contrary to popular lay belief, partners in marriages involving one highly masculine and one highly feminine person report lower levels of satisfaction than partners in marriages involving any other personality combination. Indeed, partners in a marriage where both members are high in both feminine and masculine traits report the highest levels of satisfaction (Peterson, Baucom, Elliot, & Farr, 1989).

SJ Motive and Propensity to Perceive Group Inequalities as Essential and Unchangeable

When individuals are confronted with instances of injustice, their most commonly demonstrated defense against it is victim derogation, or the ascription of traits or characteristics that portray the victims of a putative injustice as deserving of their fate (Hafer & Bègue, 2005). At the level of group injustices, victim derogation is an effective means of SJ. Instead of blaming aspects of the system for inequality between groups, people can blame the inequality on essential, unchangeable differences between groups, thereby maintaining their belief that system is fair and legitimate. Across a number of studies, we found that people indeed justified the system by claiming that essential, unchangeable differences between groups, rather than flaws inherent to the system, were responsible for group inequalities. For example, as described earlier, women who believed the Canadian system to be inevitable more strongly endorsed the notion that essential and genuine differences between men and women, as opposed to unfairness in the Canadian system, were the cause of a gender inequality in financial outcomes (Laurin et al., 2008).

Essentialism has been associated with a broad host of behaviors and cognitions that maintain inequality, such as stereotype endorsement (Bastian & Haslam, 2006, 2007), prejudice (Haslam, Bastian, Bain, & Kashima, 2006), and the need for closure (Keller, 2005). Furthermore, essentialist beliefs have similar ideological underpinnings to biological determinism—the belief that group differences exist specifically because biology has destined things to be a particular way. With biological explanations for behavior comes the associated belief that biologically based behavior cannot be changed or at least is very resistant to attempts to change it (Keller, 2005).

Last, when people believe that inequalities in society are due to innate differences between groups, rather than unfairness in the system, their attention is drawn away from systematic problems with the existing social structures and instead drawn toward the blameworthy attributes of the disadvantaged group members themselves. This is problematic because the essentialist assumption itself is questionable and highly difficult to validate empirically: The essence or natural state of members of particular groups would prove a difficult concept to measure.

SJ and Injunctification: Perceiving That What "Is" Is What "Ought" to Be

Across several studies we have found that the SJ motive encourages people to engage in *injunctification*, that is, it leads people to show a general preference for the status quo, whatever the status quo may be.[1] When people's SJ motive was activated (e.g., under system threat), participants injunctified the composition of political structures, school funding policies, the gender distribution of academic faculties, and the degree of representation of women in areas traditionally dominated by men, such as business and politics (Kay, Gaucher, Peach, et al., 2008). In each case, the activation of the SJ motive led participants to perceive the status quo (e.g., there *are* few women in CEO positions) as ideal (e.g., there *should be* fewer women than men in CEO positions). Injunctification offers insight into the exact process of how people come to justify the status quo while highlighting the immense power of the status quo in determining people's ideals. Perhaps most interesting, however, is injunctification's downstream consequences on people's behavior and its implications for perpetuating inequality.

We reasoned that a natural consequence of believing that something is the way it *should be* is reducing people's motivation for system change. After all, believing that something is the way it *should be* implies that it is unnecessary to change the way that things currently are. While examining people's injunctification of current school funding practices, we found support for our hypothesis. In one study, university students read about current school funding norms for either a system that they were dependent on (one that had control over the outcomes in their lives) or systems on which they were not dependent on (did not have any control over the outcomes in their lives). In each condition, participants read that current funding policy operated by distributing funding to schools based on performance. Participants were then asked the extent to which they thought that the funding policy was the way that it should be (i.e., most desirable and ideal). As predicted, people rated the funding policy as more desirable for the system on which they were dependent than for systems on which they were not dependent. Furthermore, the more ideal the participants rated the funding norm, the less likely they were to endorse changes to current school funding policy. Outside the lab, the ramifications of this research are clear. It is difficult to modify antiquated public policy. Not only do proposals for new policy have to contend with questions about practicality, but they also have to overshadow the preexisting policy with its imbued sentiment that things are the way that they *should be.*

A similarly disturbing implication of injunctification is people's tendency to derogate others who act counter to the descriptive norm. In one study, Kay, Gaucher, Peach, et al. (2008, Study 4) presented Canadian-born women with information about the current status quo of women in business. Presented as part of a reputable Statistics Canada report on current gender issues in the workplace, women were made to believe that there were either very few or many female CEOs in the top Fortune 300 companies. Also, half of the participants were first given a system threat intended to heighten their propensity to engage in injunctification. Participants then completed our main dependent variables, but before leaving the lab the female experimenter told participants that she is a MBA student conducting this research study as part of her course work. Participants were then asked to rate the female business student's performance; this served as our index of derogation. As Figure 7.5 shows, under conditions of system threat, female participants who were told that there were few female CEOs not only subsequently believed there *should be* few female CEOs but also derogated a female business student by rating her as less likable,

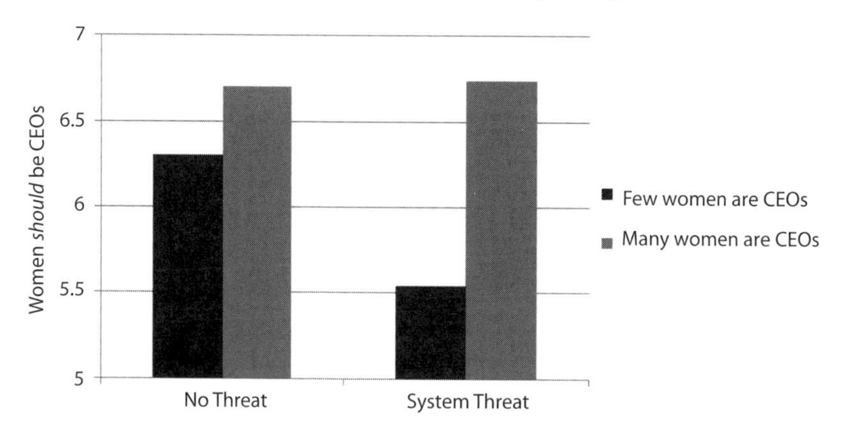

Figure 7.5 Belief that women should be CEOs expressed by female participants, either under system threat or not under system threat, who were told that there were either few or many female CEOs.

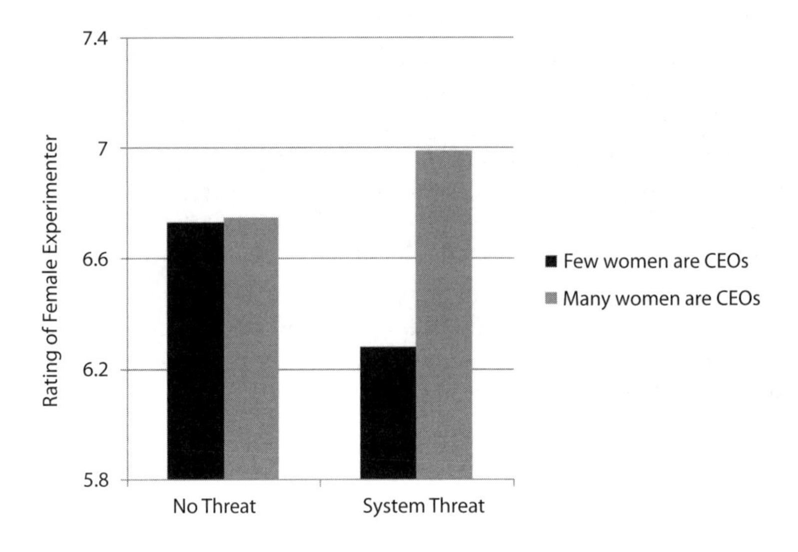

Figure 7.6 Average ratings of the likeability, competence, professionalism, and performance as experimenter of a female business student provided by female participants, either under system threat or not under system threat, who were told that there were either few or many female CEOs.

competent, and professional (see Figure 7.6). These findings are particularly noteworthy, as great lengths were taken to ensure that participants knew the high degree of impact that their ratings would have on the business student's grades and potential outcomes.

Conclusion

Regardless of how objectively unfair the status quo may be, people justify and rationalize it. First-generation research highlighted two of the first pieces of evidence for SJ effects: out-group favoritism and anticipatory rationalizations. Although the prevalence and exact mechanism through which out-group favoritism effects occur remain somewhat contested, there is a burgeoning set of data supporting their existence. Second-generation research has provided a deeper understanding of the motivational antecedents of justifying the system. People's need to view the world as orderly and under control contributes to their endorsement of social systems, and threats to the legitimacy of one's system, as well as conditions of high system inevitability, heighten people's desire to justify. Third-generation research has illuminated the consequences of SJ for the maintenance and perpetuation of inequality. The SJ motive activates complementary stereotypes that portray a more balanced and fair social landscape and promotes behavioral displays that are consistent with system-justifying stereotypes. It also leads people to perceive group inequalities as due to essential, unchangeable differences between group members and encourages injunctification, leading to a resistance to change and derogation of people who act counter to the descriptive norms.

The message of this chapter may be unsettling to people who are interested in social change. Insights into processes of SJ, however, not only highlight the difficulties of redressing social inequality but can also shed light on ways in which social inequality can be fought more effectively. For instance, piggybacking on people's natural tendency to justify and deem legitimate leaders of their system, we wondered whether framing change as system motivated (i.e., by a system insider versus a system outsider) could get people to support proenvironmental changes to their system. Preliminary evidence supported our hunch. After reading about proenvironmental changes on campus, students more strongly supported the proposed system changes when they were proposed by a system insider than by a system outsider (Gaucher, Peach, & Kay, 2008).

Similarly, in another study, we were able to affect women's motivation to enter the political domain by changing women's perception of the current status quo of women in politics. Women led to believe that there are many women in politics reported less fear of succeeding in the political domain than their counterparts who were led to believe that there were few female politicians (Gaucher, Friesen, & Kay, 2008). These findings suggest an interesting point of intervention for those interested in social change. When the status quo is one of equality rather than inequality, people will likely justify more equitable arrangements.

Note

1. More specifically, we believe that injunctification is the process by which people come to view current descriptive norms (what currently *is*) as injunctive (what *should be*). Typically, descriptive and injunctive norms are thought of as two separate types of social information that affect people's behavior. We propose, however, that because people are motivated to justify their social systems, descriptive norms (what currently *is*) directly influence people's beliefs about what *should be* (i.e., their injunctive norms). In this way, injunctification is not simply a cognitive bias that occurs in all situations but a motivated process that is incited by people's desire to justify their social systems.

References

Atkinson, J. W., & Birch, D. (1970). *The dynamics of action*. New York: Wiley.

Banfield, J. C., & Kay, A. C. (2008). [Justifying the powerful]. Unpublished raw data, University of Waterloo, Ontario, Canada.

Bargh, J. A., Gollwitzer, P. M., Lee-Chai, A. Y., Barndollar, K., & Troetschel, R. (2001). The automated will: Nonconscious activation and pursuit of behavioral goals. *Journal of Personality and Social Psychology, 81*, 1014–1027.

Bastian, B., & Haslam, N. (2006). Psychological essentialism and stereotype endorsement. *Journal of Experimental Social Psychology, 42*(2), 228–235.

Bastian, B., & Haslam, N. (2007). Psychological essentialism and attention allocation: Preferences for stereotype-consistent versus stereotype-inconsistent information. *Journal of Social Psychology, 147*(5), 531–541.

Black, J. H., & Hicks, B. M. (2006). Visible minority candidates in the 2004 federal election, *Canadian Parliamentary Review, 29*(2), 26–32.

Blair, I. V., & Banaji, M. R. (1996). Automatic and controlled processes in stereotype priming. *Journal of Personality and Social Psychology, 70*(6), 1142–1163.

Canadian Census. (2006, March). *Educational portrait of Canada, 2006 census: Highlights*. Retrieved August 29, 2008, from the Statistics Canada Web site: http://www12.statcan.ca/english/census06/analysis/education/highlights.cfm

Canadian Labour Congress. (2008, August). *Women's economic equality campaign teach-in workshop facilitator notes.* Retrieved August 29, 2008, from the Canadian Labour Congress Web site: http://canadianlabour.ca/en/Teach__Ins

Dasgupta, N. (2004). Implicit ingroup favoritism, outgroup favoritism, and their behavioral manifestations. *Social Justice Research, 17*(2), 143–169.

Day, M. V., Yoshida, E., & Kay, A. C. (2008). *Complementary forms of infrahumanization as a justification for inequality*. Manuscript in preparation.

Eagly, A. H., Mladinic, A., & Otto, S. (1994). Cognitive and affective bases of attitudes toward social groups and social policies. *Journal of Experimental Social Psychology, 30*(2), 113–137.

Fein, S., & Spencer, S. J. (1997). Prejudice as self-image maintenance: Affirming the self through derogating others. *Journal of Personality and Social Psychology, 73*, 31–44.

Gaucher, D., Chua, S., & Kay, A. C. (2008). [Contentment with the status quo and activist behaviour]. Unpublished raw data, University of Waterloo, Ontario, Canada.

Gaucher, D., Friesen, J. P., & Kay, A. C. (2008). [The bright side of system justification: Justifying equitable social arrangements]. Unpublished raw data, University of Waterloo, Ontario, Canada.

Gaucher, D., Peach, J. M., & Kay, A. C. (2008). [Justifying system motivated public policy]. Unpublished raw data, University of Waterloo, Ontario, Canada.

Hafer, C. L., & Bègue, L. (2005). Experimental research on just-world theory: Problems, developments, and future challenges. *Psychological Bulletin, 131*(1), 128–167.

Haslam, N., Bastian, B., Bain, P., & Kashima, Y. (2006). Psychological essentialism, implicit theories, and intergroup relations. *Group Processes and Intergroup Relations, 9*(1), 63–76.

Jost, J. T. (2001). *Outgroup favoritism and the theory of system justification: A paradigm for investigating the effects of socioeconomic success on stereotype content.* Mahwah, NJ: Lawrence Erlbaum.

Jost, J. T., & Banaji, M. R. (1994). The role of stereotyping in system-justification and the production of false consciousness [Special issue]. *British Journal of Social Psychology*, *33*(1), 1–27.

Jost, J. T., Banaji, M. R., & Nosek, B. A. (2004). A decade of system justification theory: Accumulated evidence of conscious and unconscious bolstering of the status quo. *Political Psychology*, *25*(6), 881–920.

Jost, J. T., & Hunyady, O. (2002). *The psychology of system justification and the palliative function of ideology*. Hove, UK: Psychology Press/Taylor & Francis.

Jost, J. T., & Hunyady, O. (2005). Antecedents and consequences of system-justifying ideologies. *Current Directions in Psychological Science*, *14*(5), 260–265.

Jost, J. T., & Kay, A. C. (2005). Exposure to benevolent sexism and complementary gender stereotypes: Consequences for specific and diffuse forms of system justification. *Journal of Personality and Social Psychology*, *88*, 498–509.

Jost, J. T., Kivetz, Y., Rubini, M., Guermandi, G., & Mosso, C. (2005). System-justifying functions of complementary regional and ethnic stereotypes: Cross-national evidence. *Social Justice Research, 18,* 305–333.

Jost, J. T., Napier, J. L., Thorisdottir, H., Gosling, S. D., Palfai, T. P., & Ostafin, B. (2007). Are needs to manage uncertainty and threat associated with political conservatism or ideological extremity? *Personality and Social Psychology Bulletin*, *33*(7), 989–1007.

Jost, J. T., Pelham, B. W., & Carvallo, M. R. (2002). Non-conscious forms of system justification: Implicit and behavioral preferences for higher status groups. *Journal of Experimental Social Psychology*, *38*(6), 586–602.

Kay, A. C., Czáplinski, S., & Jost, J. T. (2009). Left-right ideological differences in system justification following exposure to complementary versus noncomplementary stereotype exemplars. *European Journal of Social Psychology, 39*(2), 290–298.

Kay, A. C., Gaucher, D., Napier, J. L., Callan, M. J., & Laurin, K. (2008). God and the government: Testing a compensatory control mechanism for the support of external systems of control. *Journal of Personality and Social Psychology*, *95*(1), 18–35.

Kay, A. C., Gaucher, D., Peach, J. M., Zanna, M. P., & Spencer, S. J. (2008). *Towards an understanding of the naturalistic fallacy: System justification and the shift from is to ought*. Manuscript submitted for publication.

Kay, A. C., Jimenez, M. C., & Jost, J. T. (2002). Sour grapes, sweet lemons, and the anticipatory rationalization of the status quo. *Personality and Social Psychology Bulletin*, *28*(9), 1300–1312.

Kay, A. C., & Jost, J. T. (2003). Complementary justice: Effects of "poor but happy" and "poor but honest" stereotype exemplars on system justification and implicit activation of the justice motive. *Journal of Personality and Social Psychology*, *85*(5), 823–837.

Kay, A. C., Jost, J. T., Mandisodza, A. N., Sherman, S. J., Petrocelli, J. V., & Johnson, A. L. (2007). Panglossian ideology in the service of system justification: How complementary stereotypes help us to rationalize inequality. In M. P. Zanna (Ed.), *Advances in experimental social psychology* (Vol. 38, pp. 305–358). San Diego, CA: Academic Press.

Kay, A. C., Jost, J. T., & Young, S. (2005). Victim derogation and victim enhancement as alternate routes to system justification. *Psychological Science, 16*(3), 240–246.

Kay, A. C., Laurin, K., & Shepherd, S. (2008). [The system-justifying properties of compensatory self-stereotyping]. Unpublished raw data, University of Waterloo, Ontario, Canada.

Keller, J. (2005). In genes we trust: The biological component of psychological essentialism and its relationship to mechanisms of motivated social cognition. *Journal of Personality and Social Psychology, 88*(4), 686–702.

Klie, S. (2007). Workplaces not inclusive enough. *Canadian HR Reporter, 20*(13), 1.

Lane, K. A., Mitchell, J. P., & Banaji, M. R. (2005). Me and my group: Cultural status can disrupt cognitive consistency. *Social Cognition, 23*(4), 353–386.

Lau, G. P., Kay, A. C., & Spencer, S. J. (2008). Loving those who justify inequality: The effects of system threat on attraction to women who embody benevolent sexist ideals. *Psychological Science, 19*, 20–21.

Laurin, K., Kay, A. C., Gaucher, D., & Shepherd, S. (2008). *Perceptions of inevitability as a cause of system justification.* Manuscript submitted for publication.

Levy, B. R., & Banaji, M. R. (2002). *Implicit ageism*. Cambridge, MA: MIT Press.

Lips, H. M. (2003). The gender pay gap: Concrete indicator of women's progress toward equality. *Analyses of Social Issues and Public Policy, 3*(1), 87–109.

Mikulincer, M., Gerber, H., & Weisenberg, M. (1990). Judgment of control and depression: The role of self-esteem threat and self-focused attention. *Cognitive Therapy and Research, 14*, 589–608.

Napier, J., Mandisodza, A., Andersen, S. M., & Jost, J. T. (2006). System justification in responding to the poor and displaced in the aftermath of Hurricane Katrina. *Analyses of Social Issues and Public Policy, 6*, 57–73.

Nosek, B. A., Banaji, M., & Greenwald, A. G. (2002). Harvesting implicit group attitudes and beliefs from a demonstration Web site [Special issue]. *Group Dynamics: Theory, Research, and Practice, 6*(1), 101–115.

Peterson, C. D., Baucom, D. H., Elliot, M. J., & Farr, P. A. (1989). The relationship between sex role identity and marital adjustment. *Sex Roles, 21*, 775–787.

Rudman, L. A., & Kilianski, S. E. (2000). Implicit and explicit attitudes toward female authority. *Personality and Social Psychology Bulletin, 26*(11), 1315–1328.

Sales, S. M. (1972). Economic threat as a determinant of conversion rates in authoritarian and nonauthoritarian churches. *Journal of Personality and Social Psychology, 23*(3), 420–428.

Seligman, M. E. P. (1975). *Helplessness: On depression, development and death.* San Francisco: Freeman.

Seligman, M. E. P. (1976). *Learned helplessness and depression in animals and men.* Morristown, NJ: General Learning Press.

Sherman, D. K., & Cohen, G. L. (2002). Accepting threatening information: Self-affirmation and the reduction of defensive biases. *Psychological Science, 11*(4), 119–123.

Skrypnek, B. J., & Snyder, M. (1982). On the self-perpetuating nature of stereotypes about women and men. *Journal of Experimental Social Psychology, 18*(3), 277–291.

Ullrich, J., & Cohrs, J. (2007). Terrorism salience increases system justification: Experimental evidence. *Social Justice Research, 20*, 117–139.

Von Baeyer, C. L., Sherk, D. L., & Zanna, M. P. (1981). Impression management in the job interview: When the female applicant meets the male (chauvinist) interviewer. *Personality and Social Psychology Bulletin*, *7*(1), 45–51.
World Values Survey. (2004). *European and World Values Surveys four-wave integrated data file, 1981–2004, v.20060423, 2006. Surveys designed and executed by the European Values Study Group and World Values Survey Association. File Producers: ASEP/JDS, Madrid, Spain and Tilburg University, Tilburg, the Netherlands. File Distributors: ASEP/JDS and GESIS, Cologne, Germany.* Retrieved April 1, 2007, from http://www.worldvaluessurvey.org
Zanna, M. P., & Pack, S. J. (1975). On the self-fulfilling nature of apparent sex differences in behavior. *Journal of Experimental Social Psychology*, *11*(6), 583–591.

CHAPTER 8

System Justification

How Do We Know It's Motivated?

JOHN T. JOST, IDO LIVIATAN, JOJANNEKE VAN DER TOORN, ALISON LEDGERWOOD, and ANESU MANDISODZA
New York University

BRIAN A. NOSEK
University of Virginia

Abstract: According to system justification theory, people are *motivated* to defend and legitimize social systems that affect them. In this chapter, the authors review 15 years of theory and empirical research bearing on the motivational underpinnings of system justification processes. They begin by explaining why people are motivated to system justify (i.e., it serves certain social and psychological needs). They then describe five lines of evidence that corroborate the motivational claims of system justification theorists. Specifically, they find that (a) individual differences in self-deception and ideological motivation are linked to system justification, (b) system threat elicits defensive responses on behalf of the system, (c) people engage in biased information processing in favor of system-serving conclusions, (d) system justification processes exhibit properties of goal pursuit, and (e) the desire to legitimize the system inspires greater behavioral effort. The authors conclude by discussing the implications of a motivational approach for understanding conditions that foster resistance to versus support for social change.

Joanne Martin (1986, p. 217) raised a fundamental question in her Ontario Symposium chapter, "The Tolerance of Injustice": Why do "people who are clearly in a disadvantaged position—such as the poor, the underpaid, and victims of discrimination—often tolerate situations that seem unjust to an outside observer?" Variations of this question are at the core of theory and research by historians, social scientists, and particularly social psychologists, including Deutsch (1974), Lerner (1980), Crosby (1984), Tyler and McGraw (1986), Major (1994), Sidanius and Pratto (1999), and Olson and Hafer (2001). Our work on system justification as a motivated, goal-directed process follows very much in their footsteps.

History reveals a staggering number of instances of decent people (as well as indecent people) not merely passively accepting but sometimes even actively justifying and rationalizing social systems that are seen as extremely unjust by outsiders, often in retrospect. The caste system in India has survived largely intact for 3,000 years, with 150 million Indians to this day declared "Untouchables" (Ghose, 2003). The institution of slavery survived for more than 400 years in Europe and the Americas. Colonialism was also practiced for centuries and still is in some places (as is slavery), and the apartheid system in South Africa lasted for 46 years. These social systems were (or still are) bolstered by motivated social cognition through the use of stereotypes, rationalizations, ideologies, and legitimizing myths (e.g., Faust, 1981; Frederickson, 2002; Jackman, 1994; Jost & Hamilton, 2005; Kay et al., 2007; Sidanius & Pratto, 1999). That is, there are profound psychological factors that motivate individuals to accept, even support, the existing social system, even if that system entails substantial costs and relatively few benefits for them individually and for the community as a whole (Jost, Banaji, & Nosek, 2004).

Despite being the wealthiest society in history, the United States is a country in which 37 million citizens (approximately 12.6% of the population) are living in poverty ("Rising Economic Tide Fails to Lift Poor, Middle Class," 2006). Poverty rates for Blacks and Latinos in the United States are near to one in four (National Index of Violence and Harm, 2007). At the same time, the combined net worth of the 400 wealthiest Americans exceeds $1 trillion (Mishel, Bernstein, & Allegretta, 2005), with CEOs earning approximately 500 times the salary of their average employee, up from a factor of 85 just one decade ago (Stiglitz, 2004). Theories of motivation that stress self-interest, identity politics, and the thirst for justice would likely predict that these facts would elicit widespread protest, rebellion, and moral outrage on the part of the disadvantaged. For instance, Gurr (1970) summarized his prominent theory of relative deprivation this way: "Men are quick to aspire beyond their social means and quick to anger when those means prove inadequate, but slow to accept their limitations" (p. 58). More recently, Simon and Klandermans (2001) argued, "Feelings of illegitimate inequality or injustice typically result when social comparisons reveal that one's ingroup is worse off than relevant out-groups" (p. 324). "Quickness to anger" in economic and other spheres, however, occurs more rarely than one would expect, and the "sense of injustice" is surprisingly difficult to awaken (Deutsch, 1974). Despite the fact that most Americans explicitly espouse egalitarian ideals, public opinion polls show that a strong majority perceives the economic system to be fair and legitimate (Jost, Blount, Pfeffer, & Hunyady, 2003, pp. 55–57). In one particularly dramatic example, more than 80% of survey respondents belonging to the poorest economic classification endorsed the belief that "large income differences are necessary to get people to work hard" (Jost, Pelham, Sheldon, & Sullivan, 2003, p. 24).

Those few who do campaign for social and economic change are, generally speaking, not the ones who would benefit the most from it, and they are frequently subjected to some measure of resentment and disapproval for their efforts (e.g., Diekman & Goodfriend, 2007; for evidence of "backlash effects" against those who criticize or otherwise threaten the status quo, see also Frank, 2004; Kaiser, Dyrenforth, & Hagiwara, 2006; Rudman & Fairchild, 2004). Moral outrage, in other words, is often more easily directed at those who dare to *challenge* the system than at those who are responsible for its persistent failings. The poet W. H. Auden (1939/1977, p. 402) exercised considerable social psychological insight into this phenomenon when he wrote,

> There is a merciful mechanism in the human mind that prevents one from knowing how unhappy one is. One only realizes it if the unhappiness passes, and then one wonders how on earth one was ever able to stand it. If the factory workers once got out of factory life for six months, there would be a revolution such as the world has never seen.

This mechanism—like rationalization in general—is indeed merciful in certain psychological respects, because it helps people cope with and adapt to realities, including unwelcome realities (e.g., Gilbert, Pinel, Wilson, Blumberg, & Wheatley, 1998; Jost & Hunyady, 2002, 2005; Kay, Jimenez, & Jost, 2002; Lyubomisky & Ross, 1999; McGuire & McGuire, 1991; O'Brien & Major, 2005; Pyszczynski, 1982; Taylor & Brown, 1988). But it is also potentially costly at the societal level, insofar as it undermines the motivation to push for progress and social change (Wakslak, Jost, Tyler, & Chen, 2007). The goal of this chapter is to shed light on the motivational dynamics underlying system justification tendencies to better understand some of the psychological obstacles to social innovation, system change, and the attainment of justice-related goals.

Stalking the "Merciful Mechanism"

We have, in essence, been stalking Auden's "merciful mechanism" for 15 years now, and, if nothing else, we have given it a name: "system justification" (Jost & Banaji, 1994). Specifically, we have argued that in addition to having well-known motives for ego and group justification that are assumed to serve personal and collective self-esteem and interests, people are also motivated to defend, bolster, and rationalize the *social systems* that affect them—to see the status quo as good, fair, legitimate, and desirable (Jost et al., 2004; Jost, Burgess, & Mosso, 2001; Kay et al., 2007).[1] System justification theory does not suggest that people *always* perceive the status quo as completely fair and just; as with other motives (including ego and group justification motives), the strength of system justification motives is expected to vary considerably across individuals, groups, and situations. In short, we are merely suggesting that people are prone to emphasize their system's virtues, downplay its vices, and consequently see the societal status quo as better and more just than it *actually* is (Jost & Hunyady, 2005).[2]

For the purposes of indicating the breadth of situations in which we think that system justification processes can operate, we adopt Parsons's (1951) very general definition of a "social system" as a structured network of social relations, that is, a "system of processes of interaction between actors" (p. 25; see also Thorisdottir, Jost, & Kay, 2009). The property of "systematicity" implies that there exists some sustained differentiation or hierarchical clustering of relations among individuals and/or groups within the social order (Blasi & Jost, 2006), such as status, distributions of resources, and the division of social roles. Presumably, such systems can be relatively tangible, such as families, institutions, organizations, and even society as a whole, or they can be more abstract and intangible, such as the unwritten but clearly recognizable rules and norms that prescribe appropriate interpersonal and intergroup behavior. Indeed, research on system justification theory has shown that regardless of whether the system is operationalized as a nation, the government, the economic system, specific institutions, or even the network of social norms, it engenders the kind of psychological attachment that leads people to defend and bolster its legitimacy.

An important tenet of system justification theory is that for those who occupy a relatively advantaged position in the social system, the three motives of ego, group, and system justification are generally consonant, complementary, and mutually reinforcing. For those who are disadvantaged, however, these three motives are often in conflict or contradiction with one another, and different individuals may make different choices about how to resolve these conflicts (Jost & Burgess, 2000; Jost et al., 2001; O'Brien & Major, 2005). Accordingly, several studies have shown that the more African Americans subscribe to system-justifying beliefs, such as the belief that inequality in society is fair and necessary, the more they suffer in terms of self-esteem and neuroticism and the more ambivalent they feel toward fellow ingroup members (Jost & Thompson, 2000). These results were replicated and extended by O'Brien and Major (2005), who demonstrated that negative consequences arose only for those members of disadvantaged groups who were relatively highly identified with their own group; this is consistent with system justification theory given that it is only under these circumstances that a true conflict between group and system justification motives exists. Distancing from (or disidentification with) one's own group is another way of resolving the conflict between group and system justification motives for members of disadvantaged groups.

Because ego, group, and system justification motives are in contradiction for those who are disadvantaged by the status quo, such individuals are on average less likely than those who are advantaged to see the existing system as fair and legitimate (Jost et al., 2001; Sidanius & Pratto, 1999). It is ironic that under some circumstances, however—such as when the salience of individual or collective self-interest is very low—members of disadvantaged groups can be the most ardent supporters of the status quo (Jost, Pelham, et al., 2003). This phenomenon is difficult to explain from the standpoint of other prominent theories in social psychology, such as social identity theory and social dominance theory (see Jost et al., 2004). Evidence of enhanced system justification among the disadvantaged is somewhat more consistent with cognitive dissonance theory (Jost, Pelham, et al., 2003), but there is nothing in dissonanace theory to suggest that people, when faced with an incompatibility between the belief in the integrity of oneself (or one's group) and the belief in the integrity or legitimacy of the system, would ever opt for the system over the self or the ingroup (see also Blasi & Jost, 2006). In fact, prominent interpretations of cognitive dissonance theory that emphasize the need to maintain the integrity of the self-concept would lead to the expectation that people should resolve dissonance in a *self*-serving rather than a system-serving manner (e.g., Aronson, 1968, 1999).[3]

To take just one example, Jost, Pelham, et al. (2003) analyzed data from a survey study involving over 3,000 nationally representative respondents to the General Social Survey in the 1980s and 1990s who were asked whether they believed that "large differences in income were *legitimate and necessary*" either "to get people to work hard" or "as an incentive for individual effort." Results indicated that a majority of respondents accepted both justifications for economic inequality. Furthermore, these justifications were most enthusiastically endorsed by the very lowest income respondents, who did not show any of the self-serving or group-serving patterns of attribution that one might otherwise expect. These results (and others) suggest that nearly everyone holds at least some system-justifying attitudes and that, paradoxically, it is sometimes those who are the worst off who are the strongest defenders of the system.[4]

These findings are broadly consistent with the observations of political scientist Jennifer Hochschild (1981, pp. 1–2), who wrote,

> The American poor apparently do not support the downward distribution of wealth. The United States does not now have, and seldom ever has had, a political movement among the poor seeking greater economic equality. The fact that such a political movement could succeed constitutionally makes its absence even more startling. Since most of the population have less than an average amount of wealth—the median level of holdings is below the mean—more people would benefit than would lose from downward redistribution. And yet never has the poorer majority of the population, not to speak of the poorest minority, voted itself out of its economic disadvantage.

System justification theory, we propose, may help to explain why "the dog doesn't bark," as Hochschild (1981) put it in a well-known allusion to a Sherlock Holmes story.

Henry and Saul (2006) conducted an investigation of system justification tendencies among the disadvantaged in a study of the social and political attitudes of children in Bolivia, where 63% of the population lives below the poverty line and over a third of the population earns less than $2 per day. The poorest of the poor in Bolivia are the Indigenous peoples, who are direct descendants of the Incan and other native tribes living in Bolivia at the time of the Spanish conquest as long as 5 centuries ago. Today, the descendants of the Spanish conquerors compose 15% of the Bolivian population, but they still control most of the wealth in the country, as well as the governmental leadership positions, including (at the time of the study) the presidency.

In a particularly stringent test of the system justification hypothesis, Henry and Saul (2006) asked 10- to 15-year-old children for their opinions about the legitimacy of the Bolivian government. They sampled children from each of three groups—the severely disadvantaged Indigenous group, the relatively privileged Hispanic group, and a middle-status or "mixed" group of Mestizos (who have both Indigenous and Hispanic ancestors). Despite being much poorer and having parents who were far less likely to hold professional occupations (or even good jobs), Indigenous children were no less politically knowledgeable than children of other groups. They were, however, significantly more likely to approve of the Hispanic-run government, more likely to endorse the suppression of speeches against the government, and less likely to be cynical or distrusting of the government, compared to children who were better off.

A skeptical reader might wonder—as did Spears, Jetten, and Doosje (2001)—whether data such as these reflect the actual internalization of favorable attitudes toward the system (and the out-group) or whether they are the result of impression management processes, that is, insincere displays of deference (see also Scott, 1990). There is no way to answer this question with respect to the Bolivian children studied by Henry and Saul (2006), but other research using implicit measures that reduce opportunities for impression management, such as the Implicit Association Test (Nosek, Greenwald, & Banaji, 2007), suggest that favorability toward the social system and toward high-status out-groups is readily observable in implicit social cognition (see Dasgupta, 2004; Jost et al., 2004; Jost, Nosek, & Gosling, 2008). In these studies, substantial proportions of members of disadvantaged groups—including dark-skinned Morenos in Chile (Uhlmann, Dasgupta, Elgueta, Greenwald, & Swanson, 2002), poor people and the obese (Rudman, Feinberg, & Fairchild, 2002), Yale undergrads randomly assigned to low-status versus high-status residential colleges (Lane, Mitchell, & Banaji, 2005), gays and lesbians (Jost et al., 2004), Latinos and Asians (Jost, Pelham, & Carvallo, 2002), and African Americans (Ashburn-Nardo & Johnson, 2008; Ashburn-Nardo, Knowles, & Monteith, 2003; Jost et al., 2004; Nosek, Smyth, et al., 2007)—exhibit implicit biases in favor of more advantaged *out*-group members. Furthermore, the magnitude of implicit out-group favoritism among the disadvantaged is positively correlated with individuals' scores on measures of system justification (Ashburn-Nardo et al., 2003) and political conservatism (Jost et al., 2004).

The Palliative Function of System Justification

Although the findings described so far are reasonably conclusive they do not really answer the question of *why* people would justify the social system even at the expense of personal and group interests and esteem. In prior work we suggested that system justification serves certain psychological functions, without speculating about its evolutionary origins. Specifically, we proposed that system justification serves a set of relatively proximal epistemic, existential, and relational functions that help to manage uncertainty and threat and smooth out social relationships (Jost & Hunyady, 2005; Jost, Ledgerwood, & Hardin, 2008; Jost, Pietrzak, Liviatan, Mandisodza, & Napier, 2007; Kay, Gaucher, Napier, Callan, & Laurin, 2008). System justification is therefore reassuring because it enables people to cope with and feel better about the societal status quo and their place in it (see also Major, Quinton, & McCoy, 2002).

Along these lines, Jost and Hunyady (2002) suggested that system justification serves the *palliative function* of reducing negative affect associated with perceived injustice and increasing positive affect and therefore satisfaction with the status quo. This idea is reminiscent of Marx's notion that religion is the "opiate of the masses" or the "illusory happiness of the people." As Turner (1991) noted, "Presumably Marx thought that drugs were taken as a source of illusions and hallucinations and also as a palliative, a form of consolatory flight from the harshness of the real world" (p. 320). In several studies we find that giving people the opportunity to justify the system does indeed lead them to feel better and more satisfied and to report feeling more positive emotions and fewer negative emotions (e.g., Jost, Wakslak, & Tyler, 2008; Wakslak et al., 2007). Furthermore, chronically high system justifiers, such as political conservatives, are *happier* (as measured in terms of subjective well-being) than are chronically low system justifiers, such as liberals, leftists, and others who are more troubled by the degree of social and economic inequality in our society (Napier & Jost, 2008b).

The hedonic benefits of system justification, however, come with a *cost* in terms of decreased potential for social change and the remediation of inequality. Wakslak and colleagues (2007) demonstrated that system-justifying ideologies, whether measured or manipulated through a mind-set-priming technique, do indeed serve to reduce emotional distress—including negative affect in general and guilt in particular—but they also reduce "moral outrage." This last consequence is particularly important, because moral outrage motivates people to engage in helping behavior and to support social change (Carlson & Miller, 1987; Montada & Schneider, 1989). Thus, the reduction in moral outrage made people less inclined to help those who are disadvantaged, measured in terms of research participants' degree of support for and willingness to volunteer for or donate to a soup kitchen, a crisis hotline, and tutoring or job training programs for the underprivileged (see also Jost, Wakslak, & Tyler, 2008).

How Do We Know It's Motivated?

Many scholars and others are prepared to believe that attitudes and behaviors are commonly system justifying in their *consequences* but not necessarily that people are *motivated* to see the societal status quo as fair, legitimate, and desirable. Some skeptics have suggested that those who acquiesce are simply the *passive* recipients of ideology or are *compelled* by authorities to comply with the status quo, but they do not really *believe* in it (e.g., Scott, 1990; Spears et al., 2001). Others might accept that system-justifying attitudes are internalized because of social learning but deny that they have a motivational basis (e.g., Huddy, 2004; Mitchell & Tetlock, 2009; Reicher, 2004; Rubin & Hewstone, 2004). These theoretical alternatives provide an opportunity to clarify our own theoretical claims and to assess the empirical evidence for our specific propositions. In particular, we believe that these alternative interpretations underestimate the pervasiveness and goal-directed nature of system justification tendencies, that is, the ways in which people actively and purposively (but not necessarily consciously) rationalize existing social arrangements (see also Jost et al., 2007). They also overestimate the extent to which system-justifying beliefs will be responsive to reason and evidence (e.g., see Ledgerwood, Jost, Mandisodza, & Pohl, 2009) and are therefore unrealistically optimistic about the prospects for social change. We should point out that even if we are correct that system justification is a motivated process, this does not mean that people who engage in it are either irrational or malevolent (see also Jost, 2006). Rather, we have suggested that system justification serves a host of normal, typically adaptive epistemic, existential, and relational needs.

In the remainder of this chapter, we describe five lines of evidence that, especially in conjunction, lead to the conclusion that system justification is, as we have suggested, a motivated, goal-directed process. Specifically, we will show that (a) system justification is linked to individual differences in self-deception and ideological motivation; (b) situations of system threat tend to elicit defensive responses on behalf of the system; (c) system justification leads to selective, biased information processing in favor of system-serving conclusions; (d) system justification exhibits several other properties of goal pursuit, including the Lewinian properties of "equifinality" and "multifinality"; and (e) the desire to make the system look good and fair inspires behavioral efforts in terms of task persistence and performance. As we (and other researchers) make progress on each of these lines of evidence, the motivational case for system justification is strengthened.

Personality and Individual Differences: Self-Deception and Other Motives

Conceptualizing system justification tendencies as a goal-directed process suggests that their strength should be sensitive to individual differences in certain intrapsychic motives as well as the endorsement of ideological beliefs that are supportive of the status quo (see also Jost, Glaser, et al., 2003). Specifically, a system-justifying goal can be reached if one can distort perceptions of the status quo so as to avoid confronting the discrepancy between its actual state and personal or shared moral standards. One way of accomplishing this would be to engage in the process of self-deception. Thus, individuals' endorsement of system-justifying belief systems, such as political conservatism, should be correlated with their scores on self-deception, even in nonpolitical contexts.

To investigate this possibility, we conducted a study involving more than 8,500 online respondents. In this study, participants completed a single-item measure of political orientation along with Paulhus's (1984) measure of socially desirable responding, which includes individual subscales tapping motivational concerns related to self-deceptive enhancement and impression management. As can be seen in Figure 8.1, we observed a modest but consistent linear relationship between liberalism–conservatism and self-deception, $r(8629) = .12$, $p < .0001$, as well as a weaker but significant relationship between liberalism–conservatism and impression management, $r(8747) = .07$, $p < .0001$. Although the cross-sectional, correlational nature of these findings warns against drawing firm conclusions, this study does provide initial support for the notion that system justification tendencies are motivated.

Further evidence comes from research by Jost, Blount, et al. (2003) on "Fair Market Ideology," which is defined as the tendency to believe not merely that market-based procedures and outcomes are efficient (which many people believe) but also that they are inherently fair and just, which is an ethically normative position that no economist would seek to defend (with the possible exception of Milton Friedman, 1962). Endorsement of Fair Market Ideology appears to reflect a stable, individual difference variable that can be measured with items such as the following:

- "The free market system is a fair system."
- "Common or 'normal' business practices must be fair, or they would not survive."

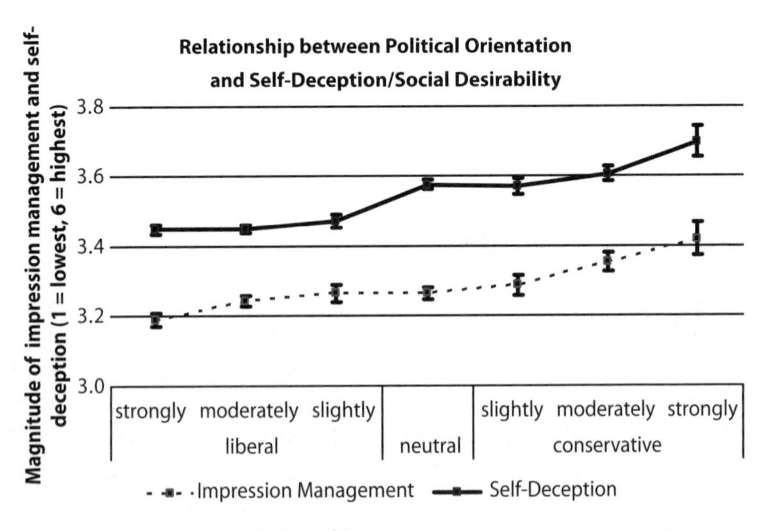

Figure 8.1 Relationship between political orientation and self-deception/social desirability. *Note:* Error bars indicate standard error of the mean.

- "Profitable businesses tend to be more morally responsible than unprofitable businesses."
- "The most fair economic system is a market system in which everyone is allowed to independently pursue their own economic interests."

In seven samples of MBA and non-MBA students at the University of Chicago, Stanford University, Boston University, and New York University, the tendency to subscribe to Fair Market Ideology was widespread and predicted by self-deceptive enhancement (Jost, Blount, et al., 2003). It was also positively correlated with the desire to believe in a just world in which people "get what they deserve and deserve what they get" (Lipkus, Dalbert, & Siegler, 1996).

Furthermore, self-reported political conservatism and individual's scores on Jost and Thompson's (2000) Economic System Justification scale both predicted scores on the Fair Market Ideology scale. Indicating that there are behavioral consequences of endorsing Fair Market Ideology, participants who scored more highly on the Fair Market Ideology scale tended to downplay or minimize the seriousness of high-profile corporate scandals and to recommend more lenient sentences for those involved. These findings suggest that system-justifying ideologies are linked to individual differences in self-deception and other forms of motivated social cognition and that self-deceptive tendencies are associated with *system*-serving biases as well as self- and group-serving biases (see also Elster, 1982; Jost, 1995; Jost & Banaji, 1994; Jost, Blount, et al., 2003; Turner, 1991).

Situational Effects: Defensive Responses to System Threat

The second category of evidence suggests that people respond defensively to threats directed at the societal status quo (Jost & Hunyady, 2002; Kay, Jost, & Young, 2005), much as they respond defensively to threats directed at their own self-esteem and threats to their group identity or status. It is well-known that conservative Republican president George W. Bush's approval rating shot up 40 percentage points immediately after 9/11, even before he had time to do anything about the attacks at all, and it stayed very high (around 70% or more) for about a year. One possibility is that these effects were due to the president's personal charisma or leadership style (Cohen, Ogilvie, Solomon, Greenberg, & Pyszczynski, 2005), but an alternative explanation is that system threat stimulates ideological defense of the social system and its representatives. The latter account is more plausible because according to the results of Gallup Poll research, Americans' opinions of nearly *every* system-level authority and agency became more favorable in the aftermath of 9/11, including Congress, the military, and the police (Jones, 2003). Likewise, trust in government to handle both domestic and international issues increased immediately after 9/11. Were Americans pursuing a cold calculus of evidence for these assessments, then the 9/11 attacks should have, if anything, indicated a failure of government to protect its citizens and decreased overall trust.

Because there are always many influences on public opinion that are difficult to disentangle, it is necessary to adopt an experimental approach to investigate cause and effect. We have developed several different paradigms for manipulating a sense of system threat in the laboratory. In one paradigm, participants are exposed to one of two passages, ostensibly written by a journalist, and they are instructed to try to remember the passage for a memory test later in the experiment (Kay et al., 2005; Jost, Kivetz, Rubini, Guermandi, & Mosso, 2005). An example of a "system-threatening" passage is as follows:

> These days, many people feel disappointed with the nation's condition. Many citizens feel that the country has reached a low point in terms of social, economic, and political factors. People do not feel as safe and secure as they used to, and there is a sense of uncertainty regarding the country's future. It seems that many countries in the world, such as the United States and Western European nations, are enjoying better social, economic, and political conditions than Israel. More and more Israelis express a willingness to leave Israel and emigrate to other nations.

The "system-affirming" passage reads as follows:

> These days, despite the difficulties the nation is facing, many people feel satisfied with the nation's condition. Many citizens feel that Israel has reached a stable point in terms of social, economic, and political factors. People feel safer and securer than they used to, and there is a sense of confidence and optimism regarding the country's future. It seems that compared with many countries in the world, the social, economic, and political conditions in Israel are relatively good. Fewer and fewer Israelis express a willingness to leave Israel and emigrate to other nations.

Exposure to the high-system-threat (versus the low-system-threat) passage does *not* significantly affect individual state self-esteem, measured with Heatherton and Polivy's (1991) scale, and it does not affect collective self-esteem, measured with Luhtanen and Crocker's (1992) scale or any of the individual subscales of those measures (Kay et al., 2005). That is, the manipulation does not threaten individual or collective self-esteem. It does, however, lead to a (presumably temporary) decrease in the perceived legitimacy of the status quo, and our motivational account therefore suggests that the threat should cause people to bolster the sagging legitimacy of the system (either directly or indirectly) when they have an opportunity to do so.

In accordance with this prediction, participants assigned to the high-system-threat condition rate powerful people as more intelligent and more independent and, conversely, the powerless as less intelligent and independent (Kay et al., 2005, Experiment 1a). System threat also leads people to rate the powerful as less happy (and the powerless as happier), consistent with work by Kay and Jost (2003; see also Jost & Kay, 2005; Kay et al., 2007; Oldmeadow & Fiske, 2007) on the system-justifying potential of complementary (or compensatory) stereotypes. Similarly, system threat increases judgments of obese people as lazier but more sociable, relative to normal weight people (Kay et al., 2005, Experiment 1b). In an example from recent American history, Napier, Mandisodza, Andersen, and Jost (2006) argued that the aftermath of Hurricane Katrina may have posed a threat to the perceived legitimacy of the governmental system, and this threat may have motivated journalists and ordinary citizens to engage in stereotyping and victim blaming in an effort to satisfy system justification motivation.

Ullrich and Cohrs (2007) conducted four experiments in which they exposed participants to a different kind of system threat—one in which the salience of terrorism as a threat to the social order was emphasized (see also Fischer, Greitemeyer, Kastenmüller, Frey, & Oßwald, 2007). This manipulation led participants to score significantly higher (compared to various control conditions) on a German translation of Kay and Jost's (2003) general or diffuse system justification scale, which contained items such as the following:

- "In general you find society to be fair."
- "Most policies serve the greater good."
- "In general the German political system operates as it should."

There is evidence, then, from multiple laboratories and several countries indicating that exposure to system threat induces people to respond defensively, showing stronger system justification on direct and indirect measures. Lau, Kay, and Spencer (2008) demonstrated that system threat can even motivate people to make different choices concerning dating partners, causing men to prefer women who confirm sexist, system-justifying stereotypes over those who do not.

As Blasi and Jost (2006) pointed out, findings of this kind do not lend themselves to a purely rational, "cold cognitive" explanation. Why should people become more prejudiced toward overweight people and, at the same time, more deferential to the powerful after reading a passage criticizing the United States? Why should reminding people about the terrorist threat increase their satisfaction with the political status quo? Why should thinking about the system's shortcomings alter the object of romantic desire? The apparent irrationality is not confined to North Americans and Germans. Exposure to a system threat passage led Israeli citizens to rely more heavily on stereotypes to rationalize social and economic inequalities between Ashkenazi and Mizrachi Jews (Jost et al., 2005). In sum, then, a wide range of system-justifying tendencies are increased in response to system-level threats, much as self-protective, ego-justifying motives become more pronounced when self-esteem is threatened.

Biased Judgment and the Desire for System-Serving Conclusions

Extensive research has shown that motivation can bias information processing, leading people to selectively attend to and process information that will allow them to reach desired conclusions (Kruglanski, 1996; Kunda, 1990; for a review, see Gollwitzer & Moskowitz, 1996). A third line of research on system justification theory has provided evidence suggesting that people engage in selective, biased information processing to reach system-justifying conclusions. Moreover, supporting the idea that such processes are goal directed, evidence has shown that these biases are sensitive to personal and situational factors that are linked to system justification motivation. For instance, an experiment by Haines and Jost (2000) revealed that people exhibited clear distortions in memory for the reasons given by the experimenter for creating power differences that favored the members of another group over the participants' own group. Specifically, they misremembered the reasons for the power differences as being more fair and legitimate than they actually were, recalling legitimate explanations when no explanation or even illegitimate explanations were given. This research suggests that the acceptance of pseudoexplanations allowed people to reconstrue inequality in legitimate terms, thereby satisfying their system justification motivation. Research by van der Toorn, Tyler, and Jost (2009) suggests further that being in a position of outcome dependence leads people to enhance the perceived legitimacy of authorities and institutions on which they depend (see also Kay & Zanna, 2009; Pepitone, 1950).

In a relatively direct demonstration of system-serving biases in social cognition, Ledgerwood et al. (2009) found that people see research evidence as stronger and more valid when it supports (versus challenges) the existence of the "American Dream" (see also Ho, Sanbonmatsu, & Akimoto, 2002; Mandisodza, Jost, & Unzueta, 2006; McCoy & Major, 2007). Using an experimental procedure developed by Pomerantz, Chaiken, and Tordesillas (1995), participants read and evaluated two studies: one concluding that hard work and determination lead to success (promeritocracy) and another concluding that there is no correlation (antimeritocracy). For each study, participants read an abstract, detailed methods and results sections, and three criticisms and three rebuttals. Across participants, a given study apparently supported either a promeritocracy or an antimeritocracy conclusion with no change in methods. For example, one study was a national telephone survey of over 800 adults in the workforce that, using a random-digit dialing procedure, "tested whether the success of American adults was more influenced by their parents' socioeconomic status, or by their own hard work and determination." Half of the participants read that this study's results supported the reality of the American Dream, namely, that a person's hard work and determination had a larger influence on a person's success than did his or her parental income and/or social status. The other half instead received an antimeritocracy conclusion that called into question the reality of the American Dream, suggesting that parental income and social status had a larger influence on a person's success than his or her own hard work and determination.

As hypothesized, Ledgerwood et al. (2009) found that participants judged the same study procedure as "more convincing" and "well conducted" when it supported the promeritocracy (versus antimeritocracy) conclusion. An internal analysis suggested that this was not just a case of people rationalizing their own personal, prior beliefs (cf. McCoy & Major, 2007). That is, the same pattern of results was evidence for those who explicitly disagreed (in a pretesting session held months earlier) that the United States is a meritocratic society in practice. Insofar as the belief that hard work leads to success can help rationalize existing inequalities in society (Jost, Pelham, et al., 2003; McCoy & Major, 2007; Sidanius & Pratto, 1999), this study suggests that system justification motivation leads people to engage in biased cognitive processing to maintain the apparent veracity of this belief. Reinforcing a motivational account, the results further revealed that the promeritocratic bias was exacerbated by system threat; under these circumstances the studies supporting the American Dream were seen as even more convincing than before, and the studies casting doubt on the American Dream were seen as even less convincing. This type of defensive reaction to system threat appears to reflect a motivated effort to restore legitimacy to the system.[5]

System Justification Exhibits Other Properties of Goal Pursuit

In addition to biasing information processing, system justification tendencies seem to follow properties of goal-directed behavior as well. One such property is the Lewinian property of equifinality. According to this property, satisfying the goal is the desired end-state, and there could be multiple, functionally interchangeable means of reaching the end-state. That is, there should be different ways of satisfying the system justification goal, including direct or indirect ways of legitimizing, for example, the economic system, the political system, or the system of gender relations in society or the family (Jost et al., 2007). Several studies have shown that people employ different strategies to restore legitimacy to the status quo, such as complementary stereotypic differentiation as well as more direct forms of system affirmation (Jost, Blount, et al., 2003; Jost & Hunyady, 2002; Kay et al., 2005), suggesting that these are two of the means that can be used, perhaps interchangeably, to justify and rationalize the status quo.

Another example comes from a previously unpublished experiment showing that system threat increases both economic and political routes to system justification (Liviatan & Jost, 2008). In this study, participants were first exposed either to a high-system-threat or a low-system-threat passage in a manipulation that was very similar to the one described earlier and used in research by Jost et al. (2005) and Kay et al. (2005). Next we gave some participants the opportunity to justify the system on political grounds (see items in Table 8.1) and other participants the opportunity to justify the system on economic grounds (see items in Table 8.2). Afterward, participants completed measures of positive and negative affect and were then given the opportunity to justify the system in the *other* domain.

Table 8.1 Items Used to Measure Justification of the Political System

1.	The American political system is the best system there is.
2.	The system of checks and balances ensures that no one branch of government can ever pursue unreasonable or illegal activities.
3.	It is part of the game of American politics to behave unethically. (R)
4.	Radical changes should be made in order to have a truly democratic political system in our country. (R)
5.	In general, the American political system operates as it should.
6.	There is no place in the world where civil liberties are better protected than right here at home.
7.	The political system lacks legitimacy because of the power of special interests. (R)
8.	The two-party electoral system is democracy at its best.
9.	Our governments have always tried to carry out diplomatic and military missions in the most humane way possible.
10.	Human rights and civil rights are constantly violated in the United States. (R)
11.	Our political actions in the international arena are guided entirely by selfish motives. (R)
12.	There are fundamental flaws in our political system, as clearly demonstrated in many previous elections. (R)
13.	America is a leader in the promotion of democracy around the world.
14.	The political system is unfair and cannot be trusted. (R)
15.	The main concern of our presidents has almost always been the public good.
16.	Some of our diplomatic and military interventions around the world can be classified as war crimes. (R)

Note: (R) = Items were reverse scored prior to analysis.

Table 8.2 Items Used to Measure Justification of the Economic System

1.	The way the free market system operates in the United States is fair.
2.	The American economic system is set up so that everyone is born with the same chance to succeed.
3.	The rules of our economic system only encourage greed and immorality. (R)
4.	Radical changes are needed to turn our economic system into a fair one. (R)
5.	Overall, Capitalism is the best economic system available.
6.	There is no country in the world where economic opportunities are better than in the United States.
7.	The American economic system unfairly increases the gap between rich and poor. (R)
8.	If incomes were more equal, nothing would motivate people to work hard.
9.	We should be embarrassed by the high rates of poverty in America. (R)
10.	No matter how much people try to stop it, there will always be widespread business corruption under Capitalism. (R)
11.	Economic markets do not reward people fairly. (R)
12.	Making incomes more equal means socialism, and that deprives people of individual freedoms.
13.	Under a free market system, people tend to get the outcomes they deserve.
14.	It is obvious that Capitalism is bad for most people in society. (R)
15.	Incomes cannot be made more equal because it's human nature to always want more than others have.
16.	Only a grand-scale economic revolution could create a better, more just distribution of resources in society. (R)

Note: (R) = Items were reverse scored prior to analysis.

We found, first, that system threat increased both economic and political routes to system justification, as can be seen in Figure 8.2. This is consistent with the principle of equifinality; there seem to be multiple ways of restoring legitimacy to the status quo following system threat (see also Kay et al., 2005). In addition, using a path model, we found that (adjusting for baseline levels of political orientation and system justification as measured weeks or months before) being assigned to the high-system-threat (versus low-system-threat) condition led to an increase in whichever type of system justification (economic or political) participants had the opportunity to endorse first (see Figure 8.3). The degree of system justification on that first measure was associated with a significant decrease in negative affect and a slight increase in positive affect,[6] which may indicate another property of goal pursuit, namely, that there is relief associated with fulfilling a goal, consistent with the hypothesized palliative function of system justification (Jost & Hunyady, 2002). Furthermore, the left over or residual negative affect following the first system justification opportunity significantly predicted the degree of system justification in the second opportunity. So,

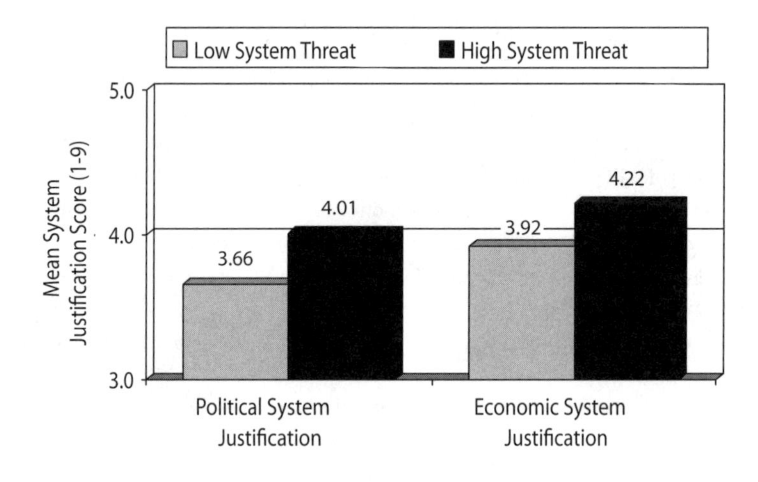

Figure 8.2 System threat increases multiple routes to system justification.

this study provides some evidence that system justification reduces negative affect and that negative affect motivates further efforts to engage in system justification (see also Jost, Wakslak, & Tyler, 2008).

System justification processes exhibit the property of multifinality as well. That is, attaining the system justification goal satisfies multiple needs, making it a potentially powerful motivational force (Jost &

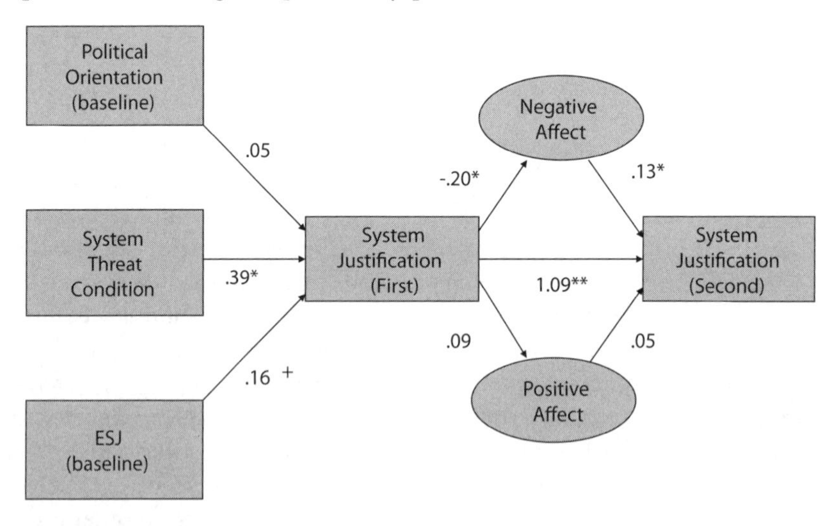

Figure 8.3 Path model illustrating the effects of system threat on system justification and positive and negative affect.

Hunyady, 2005). We theorize that system justification satisfies at least three important types of psychological needs, including:

- epistemic needs to reduce uncertainty and create a stable, predictable worldview;
- existential needs to manage threat and to perceive a safe, reassuring environment; and
- relational needs to achieve shared reality with important others, including friends and family members who have system justification needs of their own.

The possibility that there may be relational reasons, in addition to epistemic and existential reasons, to profess one's support for the status quo and to refrain from "upsetting the apple cart" is a relatively new addition to system justification theory (see Jost, Ledgerwood, & Hardin, 2008). We are suggesting that people engage in system justification at least in part because it facilitates social interaction with others. This idea is consistent with other research indicating that it is socially normative for people to derogate those who are perceived as complaining about discrimination and injustice in the system as well as those who seek to challenge or reform the status quo (e.g., Diekman & Goodfriend, 2007; Kaiser et al., 2006; Rudman & Fairchild, 2004). That is, it may be easier in general to establish common ground (or shared reality) concerning system-justifying (rather than system-challenging) beliefs.

System Justification Inspires Behavioral Effort

Our fifth line of evidence for the motivational basis of system justification is that the desire to make the system look good can inspire task persistence and performance, both of which are classic features of goal pursuit (Bargh, Gollwitzer, Lee-Chai, Barndollar, & Trotschel, 2001; Lewin, 1935; Wicklund & Gollwitzer, 1982). That is, Ledgerwood and colleagues (2009) sought to determine whether system justification motivation has significant *behavioral* ramifications, such as leading people to work harder in the service of the social system. They hypothesized that people would work harder when their behavior was seen as diagnostic of the American system and when successful performance could be seen as contributing to the legitimacy of that system.

To measure behavioral effort, Ledgerwood and colleagues (2009) asked a sample of college students to work on an anagram task in which they would unscramble a number of letter strings to create as many correct English words as possible. They manipulated whether the instructions before the task attributed success on the task to luck or to effort by suggesting that past research revealed anagram task performance to be mainly due to either luck or hard work. This factor was crossed with a manipulation of whether the task was framed as irrelevant or relevant to evaluating the legitimacy of American society by saying that the study was about "the relationship between effort and doing well in scrambled word tasks" (system-irrelevant condition) or "the relationship between effort and doing well in American society" (system-relevant condition).

In the system-irrelevant condition, Ledgerwood and colleagues hypothesized that people would (quite rationally) exert more effort and perform more successfully, as measured by the number of anagrams they solved correctly, when they believed that success on this task was due to effort rather than luck. In fact, this is what prior research on goal commitment, expectations, and performance motivation has shown (e.g., Atkinson, 1964; Bandura, 1997; Carver & Scheier, 1998; Heckhausen, 1977; Vroom, 1964). From a system justification perspective, however, attributing success purely to luck could threaten the perceived link between effort and success and therefore motivate participants to defend this link. If so, then participants might (paradoxically) work harder after learning that success is due to luck (versus effort) in an attempt to prove that hard work really does lead to success. Furthermore, insofar as this defensive tendency to work harder at a task when success is attributed to luck (versus effort) serves a system-justifying function, it should emerge only when the link between hard work and success is framed as relevant (versus irrelevant) to the social system.

This is indeed what the study showed. When the task was seen as largely irrelevant to American society as a whole, the pattern of results was consistent with prior research; namely, task performance was better when people believed that effort led to success than when they believed that luck was the key. When the task was seen as diagnostic of American society, however, participants worked harder and were more accurate in solving anagrams in an apparent effort to affirm that our system is in fact a meritocratic and therefore highly legitimate one. Ledgerwood and colleagues conducted a successful replication of this study with preselected participants who, explicitly at least, rejected the notion that American society is highly meritocratic in practice. Their persistence was measured on a set of impossible anagrams after telling them that success on the task was due to luck or effort and (for half the participants) that the task was diagnostic of American society. Results indicated that even people who self-consciously rejected the notion that society is meritocratic put behavioral effort into defending the system against antimeritocratic insinuations, but only when the task was seen as diagnostic of American society as a whole. Thus, it appears that the motivation to defend, bolster, and justify the societal status quo can inspire people to expend more effort than they otherwise would to restore perceived legitimacy to the system.

Concluding Remarks: System Justification as Motivated Goal Pursuit

Philip Mason (1971) once wrote, "That so many people for so much of history have accepted treatment manifestly unfair must always be puzzling to an observer from an individualist society, particularly in an age of revolt against privilege and inequality" (p. 13). Research on system justification is meant to solve just this puzzle. In this chapter we focused on recent evidence suggesting that people are motivated, at least to some degree, to defend, bolster, and justify the social systems that affect them (see also Jost et al., 2004; Kay et al., 2007).

For a number of reasons that should now be clear, we think that it is useful to adopt a goal systems framework in recognition of the motivational force of system justification tendencies. Such a framework, we think, helps to explain why system justification is so prevalent, insofar as it suggests that there are multiple means (e.g., social, economic, and political forms of system justification) of satisfying multiple needs (i.e., epistemic, existential, and relational needs). The preferred means presumably depend on both situational and individual differences (e.g., Kay et al., 2005; Kay, Czaplinski, & Jost, 2008). In addition, goals are often pursued nonconsciously (e.g., Bargh et al., 2001; Ferguson, Hassin & Bargh, 2007), and we are indeed finding evidence of implicit or nonconscious motivation to evaluate the system favorably in ongoing research.

The goal of system justification may be pursued nonconsciously for several reasons (see also Jost et al., 2004). First, because system justification may conflict with other goals and norms such as egalitarianism, people may be more likely to resist system-justifying conclusions when they are made explicit. But without awareness of the goal or its implications, system-justifying tendencies are less likely to be resisted. Second, acts of patriotism and other system-justifying efforts may be so frequent that they become overlearned and automatic, thereby becoming relatively effortless to use and effortful to avoid. And third, continuous conscious pursuit of system justification in multiple social domains would be exhausting, so it would be highly functional to develop a routinized capacity to maintain the subjective sense that existing social arrangements are fair and legitimate and to automatically defend the status quo against threat.

A motivational approach to system justification may ultimately help us to answer a question asked by Morton Deutsch (1974) many years ago, namely, "How do we awaken the sense of injustice?" Because a goal systems framework allows for the operation of competing goals—such as ego justification or group justification (Jost, Burgess, & Mosso, 2001), goals for novelty or accuracy (Kruglanski, 2004; Kunda, 1990), or the desire for retribution and other justice-related motives (Darley & Pittman, 2003; Lerner, 2003)—it can help to clarify the circumstances under which people will challenge or criticize the system. Such an approach will enable us to better understand the processes that give rise to clear and widespread defection from the motivational clutches of system justification (e.g., Jost, Ledgerwood, et al., 2008; Reicher, 2004).

The system justification goal will finally be abandoned when justifying the system no longer satisfies epistemic, existential, or relational needs. This may occur when the status quo itself offers no stability or certainty or may even be regarded as a source of threat rather than reassurance or when it has become counternormative to stick with an old regime when a new one is gaining in popularity. Under circumstances such as these, the motivational impetus of system justification tendencies would be low, and people might even work to change the status quo. Arguably, this is what happened in Eastern Europe in the late 1980s and early 1990s, when several decades of Soviet Communist rule in Hungary, Poland, Romania, the Czech Republic, and elsewhere came to an abrupt end. Once a new system or regime acquires an aura of inevitability, system justification motives should lead people to engage in rationalization processes that will bolster the new system and bury the old one (see also Blasi & Jost, 2006; Kay et al., 2002).

Kurt Lewin (1947) once wrote, "The study of the conditions for change begins appropriately with an analysis of the conditions for 'no change,' that is for the state of equilibrium" (p. 208). For this reason, our personal and professional interests in social change have led us to try to understand, to the best of our abilities, the social and psychological *obstacles* to change, that is, to analyze the psychological power of the status quo. Our success in this theoretical and practical endeavor will not only enable human beings to overcome the "merciful mechanism" that prevents recognition of unfairness and inequality in society but also help us to better promote a world in which justice principles such as equity, equality, and need are not merely palliative fictions but pillars of reality.

Authors' Note

Some of the ideas described in this chapter were first presented at the 2007 Morton Deutsch Award Ceremony at Columbia University's Teachers College and a conference held in honor of Joanne Martin's retirement from Stanford University's Graduate School of Business in June 2007. Thus, we are particularly grateful to Morton Deutsch and Joanne Martin for providing both distal and proximal sources of inspiration for this work. Subsequent presentations of this material took place at the Eleventh Ontario Symposium: The Psychology of Justice and Legitimacy in Waterloo, Canada, as well as the Summer Institute in Social Psychology in Austin, Texas, the International Society of Political Psychology conference in Portland, Oregon, the Harvard Law School and the Department of Psychology at Yale University. This research was supported by National Science Foundation Award No. BCS-0617558 to John T. Jost and by the Center for Catastrophe Preparedness and Response at New York University. We thank Aaron Kay and James Olson for their extremely helpful comments on an earlier draft.

Notes

1. System justification theory shares some similarities with just-world theory (Lerner, 1980), which posits that people are motivated to believe that we live in an orderly, predictable, and just world in which people get what they deserve. Although many consequences of system justification and the belief in a just world are the same, the underlying motives are theorized to be somewhat different. Whereas just-world theory concerns the desire for actual justice (i.e., the *justice motive*), system justification theory concerns the desire for the perception (or appearance) of justice (i.e., the *justification motive*). According to Lerner's (1980) theory, people will attempt to justify unjust outcomes only when they are unable to engage in behaviors that would restore it directly. In theory, then, people who strongly endorse the belief in a just world should, whenever possible, choose *more* just alternatives to the status quo. By contrast, we hypothesize that people who strongly endorse system-justifying belief systems are likely to support even an unjust status quo and to derogate potential alternatives.
2. It may be difficult to measure or otherwise assess the objective fairness of a given social system or situation (as well as its actual rather than perceived costs and benefits), but we regard this as a disciplinary challenge that must be confronted by social scientists. That is, social scientists should not merely concern themselves with descriptive questions about what people perceive as fair but also address, to the best of their abilities and on the basis of reason and evidence, thorny normative questions about justice (e.g., see Jost & Kay, in press; Tyler & Jost, 2007).
3. Another difference between system justification and cognitive dissonance theories is that the latter assumes people experience dissonance *only* when something that they choose is incongruent with other values, beliefs, and actions, leading them to engage in a rationalization *of their choice*. System justification theory goes further in predicting the post hoc rationalization of occurrences that are not of one's own choosing, such as unintended outcomes of one's own or others' behavior, as well as *anticipated* social and political events (e.g., see Kay et al., 2002).
4. Findings such as these—especially when taken in conjunction with the notion that political conservatism is a system-justifying ideology—may be confusing to those who assume that more advantaged members of society are generally the most politically conservative. The validity of this "self-interest" assumption, however, has been called into question repeatedly (e.g., Frank, 2004; Jost, Glaser, et al., 2003). In a recent cross-national investigation involving

respondents from 19 democratic countries, Napier and Jost (2008a) observed virtually no correlation between socioeconomic status (SES) and political orientation, in part because there is a weak but positive correlation between income and conservatism and a weak but negative correlation between education and conservatism. Results from this investigation also indicated that high-SES respondents may be drawn to right-wing ideology because of economic conservatism, whereas low-SES respondents may be drawn to it because of moral and ethnic intolerance.

5. The findings from this research were supportive of a motivational account in other ways as well. Specifically, Ledgerwood et al. (2009) found that the promeritocracy bias was enhanced for those who were chronically high (versus low) on the Economic System Justification scale and for women, who might be especially motivated to justify the system to rationalize their lower social status (see also Henry & Saul, 2006; Jost, Pelham, et al., 2003). Furthermore, motivated information processing, namely, selective cognitive elaboration, was found to statistically mediate the bias in favor of system-serving conclusions.
6. The degree of system justification on the first measure was also positively associated with the degree of system justification on the second measure, which suggests that adopting one means of attaining the system justification goal does not necessarily decrease the likelihood of adopting other means. Thus, although we did find evidence consistent with the notion of equifinality (i.e., that system threat leads people to show increased system justification in either economic or political domains), we did not obtain evidence that system justification in the first instance entirely satisfied the goal. The failure of one means to satiate the desire to pursue other means could suggest high *commitment* to attaining the system justification goal, insofar as commitment leads people to persist in goal pursuit even after "progress" has occurred (see Fishbach, Dhar, & Zhang, 2006).

References

Aronson, E. (1968). Dissonance theory: Progress and problems. In R. P. Abelson, E. Aronson, W. J. McGuire, T. M. Newcomb, M. J. Rosenberg, & P. H. Tannebaum (Eds.), *Theories of cognitive consistency: A sourcebook*. Chicago: Rand McNally.

Aronson, E. (1999). Dissonance, hypocrisy, and the self-concept. In E. Harmon-Jones & J. Mills (Eds.), *Cognitive dissonance: Progress on a pivotal theory in social psychology*. Washington, DC: American Psychological Association.

Ashburn-Nardo, L., & Johnson, N. (2008). Implicit outgroup favoritism and intergroup judgment: The moderating role of stereotypic context. *Social Justice Research, 21*, 490–508.

Ashburn-Nardo, L., Knowles, M. L., & Monteith, M. J. (2003). Black Americans' implicit racial associations and their implications for intergroup judgment. *Social Cognition, 21*, 61–87.

Atkinson, J. W. (1964). *An introduction to motivation*. New York: Van Nostrand.

Auden, W. H. (1977). *The English Auden: Poems, essays, and dramatic writings 1927–1939* (E. Mendelson, Ed.). London: Faber & Faber. (Original work published 1939)

Bandura, A. (1997). *Self-efficacy: The exercise of control.* New York: Freeman.

Bargh, J. A., Gollwitzer, P. M., Lee-Chai, A., Barndollar, K., & Trotschel, R. (2001). The automated will: Nonconscious activation and pursuit of behavioral goals. *Journal of Personality and Social Psychology, 81*, 1014–1027.

Blasi, G., & Jost, J. T. (2006). System justification theory and research: Implications for law, legal advocacy, and social justice. *California Law Review, 94*, 1119–1168.

Carlson, M., & Miller, N. (1987). Explanation of the relation between negative mood and helping. *Psychological Bulletin, 102*, 91–108.

Carver, C. S., & Scheier, M. F. (1998). *On the self-regulation of behavior.* New York: Cambridge University Press.

Cohen, F., Ogilvie, D. M., Solomon, S., Greenberg, J., & Pyszczynski, T. (2005). American roulette: The effect of reminders of death on support for George W. Bush in the 2004 presidential election. *Analyses of Social Issues and Public Policy, 5*, 177–187.

Crosby, F. J. (1984). The denial of personal discrimination. *American Behavioral Scientist, 27*, 371–386.

Darley, J. M., & Pittman, T. S. (2003). The psychology of compensatory and redistributive justice. *Personality and Social Psychology Review, 7*, 324–336.

Dasgupta, N. (2004). Implicit ingroup favoritism, outgroup favoritism, and their behavioral manifestations. *Social Justice Research, 17*, 143–169.

Deutsch, M. (1974). Awakening the sense of injustice. In M. Lerner & M. Ross (Eds.), *The quest for justice: Myth, reality, ideal.* Montreal, Canada: Holt, Rinehart & Winston.

Diekman, A. B., & Goodfriend, W. (2007). The good and bad of social change: Ambivalence towards activist groups. *Social Justice Research, 20*, 401–417.

Elster, J. (1982). Belief, bias, and ideology. In M. Hollis & S. Lukes (Eds.), *Rationality and relativism* (pp. 123–148). Oxford, UK: Blackwell.

Faust, D. G. (Ed.). (1981). *The ideology of slavery: Proslavery thought in the antebellum South, 1830–1860.* Baton Rouge: Louisiana State University Press.

Ferguson, M., Hassin, R., & Bargh, J. A. (2007). Implicit motivation: Past, present, and future. In J. Shah & W. Gardner (Eds.), *Handbook of motivation science.* New York: Guilford.

Fischer, P., Greitemeyer, T., Kastenmüller, A., Frey, D., & Oßwald, S. (2007). Terror salience and punishment: Does terror salience induce threat to social order? *Journal of Experimental Social Psychology, 43*, 964–971.

Fishbach, A., Dhar, R., & Zhang, Y. (2006). Subgoals as substitutes or complements: The role of goal accessibility. *Journal of Personality and Social Psychology, 91*, 232–242.

Frank, T. (2004). *What's the matter with Kansas?* New York: Metropolitan Books.

Frederickson, G. M. (2002). *Racism: A short history.* Princeton, NJ: Princeton University Press.

Friedman, M. (1962). *Capitalism and freedom.* Chicago: University of Chicago Press.

Ghose, S. (2003). The Dalit in India. *Social Research, 70*, 83–109.

Gilbert, D. T., Pinel, E. C., Wilson, T. D., Blumberg, S. J., & Wheatley, T. (1998). Immune neglect: A source of durability bias in affective forecasting. *Journal of Personality and Social Psychology, 75*, 617–638.

Gollwitzer, P. M., & Moskowitz, G. B. (1996). Goal effects on action and cognition. In E. T. Higgins & A. W. Kruglanski (Eds.), *Social psychology: Handbook of basic principles* (pp. 361–399). New York: Guilford.

Gurr, T. R. (1970). *Why men rebel.* Princeton, NJ: Princeton University Press.

Haines, E., & Jost, J. T. (2000). Placating the powerless: Effects of legitimate and illegitimate explanation on affect, memory, and stereotyping. *Social Justice Research, 13*, 219–236.

Heatherton, T. F., & Polivy, J. (1991). Development and validation of a scale for measuring state self-esteem. *Journal of Personality and Social Psychology, 60*, 895–910.

Heckhausen, H. (1977). Achievement motivation and its constructs: A cognitive model. *Motivation and Emotion, 1*, 283–329.

Henry, P. J., & Saul, A. (2006). The development of system justification in the developing world. *Social Justice Research, 19*, 365–378.

Ho, E. A., Sanbonmatsu, D., & Akimoto, S. A. (2002). The effects of comparative status on social stereotypes: How the perceived success of some persons affects the stereotypes of others. *Social Cognition, 20*, 36–57.

Hochschild, J. (1981). *What's fair?* Princeton, NJ: Princeton University Press.

Huddy, L. (2004). Contrasting theoretical approaches to intergroup relations. *Political Psychology, 25*, 947–967.

Jackman, M. R. (1994). *The velvet glove: Paternalism and conflict in gender, class, and race relations*. Berkeley: University of California Press.

Jones, J. M. (2003, September 9). Sept. 11 effects, though largely faded, persist. *Gallup Poll.* Retrieved September 27, 2004, from www.gallup.com/poll.content/?ci=9208

Jost, J. T. (1995). Negative illusions: Conceptual clarification and psychological evidence concerning false consciousness. *Political Psychology, 16*, 397–424.

Jost, J. T. (2006). The end of the end of ideology. *American Psychologist, 61*, 651–670.

Jost, J. T., & Banaji, M. R. (1994). The role of stereotyping in system-justification and the production of false consciousness. *British Journal of Social Psychology, 33*, 1–27.

Jost, J. T., Banaji, M. R., & Nosek, B. A. (2004). A decade of system justification theory: Accumulated evidence of conscious and unconscious bolstering of the status quo. *Political Psychology, 25*, 881–919.

Jost, J. T., Blount, S., Pfeffer, J., & Hunyady, G. (2003). Fair Market Ideology: Its cognitive-motivational underpinnings. *Research in Organizational Behavior, 25*, 53–91.

Jost, J. T., & Burgess, D. (2000). Attitudinal ambivalence and the conflict between group and system justification motives in low status groups. *Personality and Social Psychology Bulletin, 26*, 293–305.

Jost, J. T., Burgess, D., & Mosso, C. (2001). Conflicts of legitimation among self, group, and system: The integrative potential of system justification theory. In J. T. Jost & B. Major (Eds.), *The psychology of legitimacy: Emerging perspectives on ideology, justice, and intergroup relations* (pp. 363–388). New York: Cambridge University Press.

Jost, J. T., Glaser, J., Kruglanski, A. W., & Sulloway, F. (2003). Political conservatism as motivated social cognition. *Psychological Bulletin, 129*, 339–375.

Jost, J. T., & Hamilton, D. L. (2005). Stereotypes in our culture. In J. Dovidio, P. Glick, & L. Rudman (Eds.), *On the nature of prejudice: Fifty years after Allport* (pp. 208–224). Oxford, UK: Blackwell.

Jost, J. T., & Hunyady, O. (2002). The psychology of system justification and the palliative function of ideology. *European Review of Social Psychology, 13*, 111–153.

Jost, J. T., & Hunyady, O. (2005). Antecedents and consequences of system-justifying ideologies. *Current Directions in Psychological Science, 14*, 260–265.

Jost, J. T., & Kay, A. C. (2005). Exposure to benevolent sexism and complementary gender stereotypes: Consequences for specific and diffuse forms of system justification. *Journal of Personality and Social Psychology, 88*, 498–509.

Jost, J. T., & Kay, A. C. (in press). Social justice: History, theory, & research. In S.T. Fiske, D. Gilbert, & G. Lindsay (Eds.), *Handbook of social psychology* (5th edition) New York: McGraw Hill.

Jost, J. T., Kivetz, Y., Rubini, M., Guermandi, G., & Mosso, C. (2005). System-justifying functions of complementary regional and ethnic stereotypes: Cross-national evidence. *Social Justice Research, 18*, 305–333.

Jost, J. T., Ledgerwood, A., & Hardin, C. D. (2008). Shared reality, system justification, and the relational basis of ideological beliefs. *Social and Personality Psychology Compass, 2*, 171–186.

Jost, J. T., Nosek, B. A., & Gosling, S. D. (2008). Ideology: Its resurgence in social, personality, and political psychology. *Perspectives on Psychological Science, 3*, 126–136.

Jost, J. T., Pelham, B. W., & Carvallo, M. (2002). Non-conscious forms of system justification: Cognitive, affective, and behavioral preferences for higher status groups. *Journal of Experimental Social Psychology, 38*, 586–602.

Jost, J. T., Pelham, B. W., Sheldon, O., & Sullivan, B. N. (2003). Social inequality and the reduction of ideological dissonance on behalf of the system: Evidence of enhanced system justification among the disadvantaged. *European Journal of Social Psychology, 33*, 13–36.

Jost, J. T., Pietrzak, J., Liviatan, I., Mandisodza, A., & Napier, J. (2007). System justification as conscious and nonconscious goal pursuit. In J. Shah & W. Gardner (Eds.), *Handbook of motivation science*. New York: Guilford.

Jost, J. T., & Thompson, E. P. (2000). Group-based dominance and opposition to equality as independent predictors of self-esteem, ethnocentrism, and social policy attitudes among African Americans and European Americans. *Journal of Experimental Social Psychology, 36*, 209–232.

Jost, J. T., Wakslak, C., & Tyler, T. R. (2008). System justification theory and the alleviation of emotional distress: Palliative effects of ideology in an arbitrary social hierarchy and in society. In K. Hegtvedt & J. Clay-Warner (Eds.), *Advances in group processes*. London: Emerald/JAI.

Kaiser, C. R., Dyrenforth, P. S., & Hagiwara, N. (2006). Why are attributions to discrimination interpersonally costly? A test of system- and group-justifying motivations. *Personality and Social Psychology Bulletin, 32*, 1423–1536.

Kay, A. C., Czaplinski, S., & Jost, J. T. (2008). Left–right ideological differences in system justification following exposure to complementary versus noncomplementary stereotype exemplars. *European Journal of Social Psychology, 39,* 290–298.

Kay, A. C., Gaucher, D., Napier, J. L., Callan, M. J., & Laurin, K. (2008). God and the government: Testing a compensatory control mechanism for the support of external systems. *Journal of Personality and Social Psychology, 95,* 18–35.

Kay, A., Jimenez, M. C., & Jost, J. T. (2002). Sour grapes, sweet lemons, and the anticipatory rationalization of the status quo. *Personality and Social Psychology Bulletin, 28,* 1300–1312.

Kay, A. C., & Jost, J. T. (2003). Complementary justice: Effects of "poor but happy" and "poor but honest" stereotype exemplars on system justification and implicit activation of the justice motive. *Journal of Personality and Social Psychology, 85,* 823–837.

Kay, A. C., Jost, J. T., Mandisodza, A. N., Sherman, S. J., Petrocelli, J. V., & Johnson, A. L. (2007). Panglossian ideology in the service of system justification: How complementary stereotypes help us to rationalize inequality. *Advances in Experimental Social Psychology, 39,* 305–358.

Kay, A. C., Jost, J. T., & Young, S. (2005). Victim-derogation and victim-enhancement as alternate routes to system-justification. *Psychological Science, 16,* 240–246.

Kay, A. C., & Zanna, M. (2009). A contextual analysis of the system justification motive and its societal consequences. In J. T. Jost, A. C. Kay, & H. Thorisdottir (Eds.), *Social and psychological bases of ideology and system justification.* New York: Oxford University Press.

Kruglanski, A. W. (1996). Motivated social cognition: Principles of the interface. In E. T. Higgins, & A. W. Kruglanski (Eds.), *Social psychology: Handbook of basic principles* (pp. 492–520). New York: Guilford.

Kruglanski, A. W. (2004). *The psychology of closed mindedness.* New York: Psychology Press.

Kunda, Z. (1990). The case for motivated reasoning. *Psychological Bulletin, 108,* 480–498.

Lane, K. A., Mitchell, J. P., & Banaji, M. R. (2005). Me and my group: Cultural status can disrupt cognitive consistency. *Social Cognition, 23,* 353–386.

Lau, G. P., Kay, A. C., & Spencer, S. J. (2008). Loving those who justify inequality: The effects of system threat on attraction to women who embody benevolent sexist ideals. *Psychological Science, 19,* 20–21.

Ledgerwood, A., Jost, J. T., Mandisodza, A., & Pohl, M. (2009). Working for the system: Motivated defense of the American Dream. Manuscript submitted for publication.

Lerner, M. J. (1980). *The belief in a just world: A fundamental delusion.* New York: Plenum.

Lerner, M. J. (2003). The justice motive: Where social psychologists found it, how they lost it, and why they may not find it again. *Personality and Social Psychology Review, 7,* 388–399.

Lewin, K. (1935). *A dynamic theory of personality.* New York: McGraw-Hill.

Lewin, K. (1947). Group decision and social change. In T. M. Newcomb & E. L. Hartley (Eds.), *Readings in social psychology* (pp. 330–344). New York: Holt.

Lipkus, I. M., Dalbert, C., & Siegler, I. C. (1996). The importance of distinguishing the belief in a just world for self versus for others: Implications for psychological well-being. *Personality and Social Psychology Bulletin, 22*, 666–677.

Liviatan, I., & Jost, J. T. (2008). [The palliative function of system justification processes]. Unpublished raw data.

Luhtanen, R. K., & Crocker, J. (1992). A collective self-esteem scale: Self-evaluation of one's social identity. *Personality and Social Psychology Bulletin, 18*, 302–318.

Lyubomisky, S., & Ross, L. (1999). Changes in attractiveness of elected, rejected, and precluded alternatives: A comparison of happy and unhappy individuals. *Journal of Personality and Social Psychology, 76*, 988–1007.

Major, B. (1994). From social inequality to personal entitlement: The role of social comparisons, legitimacy appraisals, and group memberships. *Advances in Experimental Social Psychology, 26*, 293–355.

Major, B., Quinton, W., & McCoy, S. K. (2002). Antecedents and consequences of attributions to discrimination: Theoretical and empirical advances. *Advances in Experimental Social Psychology, 34*, 251–330.

Mandisodza, A. N., Jost, J. T., & Unzueta, M. (2006). "Tall poppies" and "American dreams": Reactions to rich and poor in Australia and the U.S.A. *Journal of Cross-Cultural Psychology, 37*, 659–668.

Martin, J. (1986). The tolerance of injustice. In J. M. Olson, C. P. Herman, & M. P. Zanna (Eds.), *Relative deprivation and social comparison: The Ontario Symposium* (Vol. 4, pp. 217–242). Hillsdale, NJ: Lawrence Erlbaum.

Mason, P. (1971). *Patterns of dominance*. London: Oxford University Press.

McCoy, S. K., & Major, B. (2007). Priming meritocracy and the psychological justification of inequality. *Journal of Experimental Social Psychology, 43*(3), 341–351.

McGuire, W. J., & McGuire, C. V. (1991). The content, structure, and operation of thought systems. In R. S. Wyer Jr. & T. K. Srull (Eds.), *Advances in social cognition* (Vol. IV, pp. 1–78). Hillsdale, NJ: Lawrence Erlbaum.

Mishel, L., Bernstein, J., & Allegretta, S. (2005). *The state of working America 2004/2005*. Ithaca, NY: Cornell University Press.

Mitchell, G., & Tetlock, P. E. (2009). Disentangling reasons and rationalizations: Exploring perceived fairness in hypothetical societies. In J. T. Jost, A. C. Kay, & H. Thorisdottir (Eds.), *Social and psychological bases of ideology and system justification*. New York: Oxford University Press.

Montada, L., & Schneider, A. (1989). Justice and emotional reactions to the disadvantaged. *Social Justice Research, 3*, 313–344.

Napier, J. L., & Jost, J. T. (2008a). The "anti-democratic personality" revisited: A cross-national investigation of working class authoritarianism. *Journal of Social Issues, 64*, 595–617.

Napier, J. L., & Jost, J. T. (2008b). Why are conservatives happier than liberals? *Psychological Science, 19*, 565–572.

Napier, J. L., Mandisodza, A. N., Andersen, S. M., & Jost, J. T. (2006). System justification in responding to the poor and displaced in the aftermath of Hurricane Katrina. *Analyses of Social Issues and Public Policy, 6*, 57–73.

National Index of Violence and Harm. (2007, November 17). *Last decade sees closing poverty gap between minorities and Whites, young and old, women and men. Large income gap between rich and poor persists.* Retrieved March 30, 2008, from, http://www.manchester.edu/links/violenceindex

Nosek, B. A., Greenwald, A. G., & Banaji, M. R. (2007). The Implicit Association Test at age 7: A methodological and conceptual review. In J. A. Bargh (Ed.), *Automatic processes in social thinking and behavior* (pp. 265–292). Philadelphia: Psychology Press.

Nosek, B. A., Smyth, F. L., Hansen, J. J., Devos, T., Lindner, N. M., Ranganath, K. A., Smith, C. T., Olson, K. R., Chugh, D., Greenwald, A. G., & Banaji, M. R. (2007). Pervasiveness and correlates of implicit attitudes and stereotypes. *European Review of Social Psychology, 18*, 36–88.

O'Brien, L. T., & Major, B. (2005). System justifying beliefs and psychological well-being: The roles of group status and identity. *Personality and Social Psychology Bulletin, 31*, 1718–1729.

Oldmeadow, J., & Fiske, S. T. (2007). System-justifying ideologies moderate status = competence stereotypes: Roles for belief in a just world and social dominance orientation. *European Journal of Social Psychology, 37*, 1135–1148.

Olson, J. A., & Hafer, C. L. (2001). Tolerance of personal deprivation. In J. T. Jost & B. Major (Eds.), *The psychology of legitimacy: Emerging perspectives on ideology, justice, and intergroup relations* (pp. 157–175). New York: Cambridge University Press.

Parsons, T. (1951). *The social system*. New York: Free Press.

Paulhus, D. L. (1984). Two-component models of socially desirable responding. *Journal of Personality and Social Psychology, 46*, 598–609.

Pepitone, A. (1950). Motivational effects in social perception. *Human Relations, 3*, 57–76.

Pomerantz, E. M., Chaiken, S., & Tordesillas, R. S. (1995). Attitude strength and resistance processes. *Journal of Personality and Social Psychology, 69*, 408–419.

Pyszczynski, T. (1982). Cognitive strategies for coping with uncertain outcomes. *Journal of Research in Personality, 16*, 386–399.

Reicher, S. (2004). The context of social identity: Dominance, resistance, and change. *Political Psychology, 25*, 921–945.

Rising economic tide fails to lift poor, middle class. (2006, August 29). *USA Today*. Retrieved March 30, 2008, from, http://www.usatoday.com/news/opinion/2006-08-29-economy_x.htm

Rubin, M., & Hewstone, M. (2004). Social identity, system justification, and social dominance: Commentary on Reicher, Jost et al., and Sidanius et al. *Political Psychology, 25*, 823–844.

Rudman, L. A., & Fairchild, K. (2004). Reactions to counterstereotypic behavior: The role of backlash in cultural stereotype maintenance. *Journal of Personality and Social Psychology, 87*, 157–176.

Rudman, L. A., Feinberg, J., & Fairchild, K. (2002). Minority members' implicit attitudes: Automatic ingroup bias as a function of group status. *Social Cognition, 20*, 294–320.
Scott, J. (1990). *Domination and the arts of resistance.* New Haven, CT: Yale University.
Sidanius, J., & Pratto, F. (1999). *Social dominance: An intergroup theory of social hierarchy and oppression.* New York: Cambridge University Press.
Simon, B., & Klandermans, B. (2001). Politicized collective identity: A social psychological analysis. *American Psychologist, 56*, 319–331.
Spears, R., Jetten, J., & Doosje, B. (2001). The (il)legitimacy of ingroup bias: From social reality to social resistance. In J. T. Jost & B. Major (Eds.), *The psychology of legitimacy: Emerging perspectives on ideology, justice, and intergroup relations* (pp. 332–362). New York: Cambridge University Press.
Stiglitz, J. E. (2004). *The roaring nineties.* New York: W. W. Norton.
Taylor, S. E., & Brown, J. D. (1988). Illusion and well-being: A social psychological perspective on mental health. *Psychological Bulletin, 103*, 193–210.
Thorisdottir, H., Jost, J. T., & Kay, A. C. (2009). On the social and psychological bases of ideology and system justification. In J. T. Jost, A. C. Kay, & H. Thorisdottir (Eds.), *Social and psychological bases of ideology and system justification.* New York: Oxford University Press.
Turner, D. (1991). Religion: Illusions and liberation. In T. Carver (Ed.), *The Cambridge companion to Marx* (pp. 320–337). Cambridge, UK: Cambridge University Press.
Tyler, T. R., & Jost, J. T. (2007). Psychology and the law: Reconciling normative and descriptive accounts of social justice and system legitimacy. In A. W. Kruglanski, & E. T., Higgins (Eds.), *Social psychology: Handbook of basic principles* (2nd ed., pp. 807–825). New York: Guilford.
Tyler, T. R., & McGraw, K. M. (1986). Ideology and the interpretation of personal experience: Procedural justice and political quiescence. *Journal of Social Issues, 42*, 115–128.
Uhlmann, E., Dasgupta, N., Elgueta, A., Greenwald, A. G., & Swanson, J. E. (2002). Subgroup prejudice based on skin color among Hispanics in the United States and Latin America. *Social Cognition, 20*, 198–225.
Ullrich, J., & Cohrs, J. C. (2007). Terrorism salience increases system justification: Experimental evidence. *Social Justice Research, 20*, 117–139.
Van der Toorn, J. M., Tyler, T. R., & Jost, J. T. (2009). *Justice or justification? The effect of outcome dependence on legitimization of authority.* Unpublished manuscript. New York University.
Vroom, V. H. (1964). *Work and motivation.* New York: Wiley.
Wakslak, C. J., Jost, J. T., Tyler, T. R., & Chen, E. S. (2007). Moral outrage mediates the dampening effect of system justification on support for redistributive social policies. *Psychological Science, 18*, 267–274.
Wicklund, R. A., & Gollwitzer, P. M. (1982). *Symbolic self-completion.* Hillsdale, NJ: Lawrence Erlbaum.

CHAPTER 9

Self-Regulation, Homeostasis, and Behavioral Disinhibition in Normative Judgments

KEES VAN DEN BOS
Utrecht University

Abstract: In this chapter, the author proposes two general messages: One message is that when the human organism is in homeostasis (such as when the human being is not busy with coping with personal uncertainty or not recovering from other salient self-threatening information), it can be expected that people will react in a relatively calm way to fair and unfair events and show composed reactions to other events that bolster or violate their social-cultural norms and values. Thus, the regulated and homeostatic self leads to calmer, better composed human beings. In the second part of this chapter, however, the author puts forward the hypothesis that the well-regulated self may also lead to less desirable reactions. That is, he proposes that a disinhibited self often is needed to overcome people's inhibition to intervening in moral dilemmas and that behavioral disinhibition therefore may have positive effects on how people respond to these dilemmas. Thus, the take-home message of this chapter is twofold: Behavioral disinhibition may be bad, except when it is not. At least sometimes, behavioral disinhibition can be conducive for the greater good.

Keywords: self-regulation, homeostasis, behavioral disinhibition, justice, normative judgments

In this chapter, I will focus on the role of self-regulation, homeostasis, and behavioral disinhibition in social psychology in general and normative judgments in particular. More specifically, I will first briefly discuss recent insights pertaining to the important role of self-regulatory and homeostatic processes in the way people form justice judgments. I will argue that when the human organism is in homeostasis (such as when the human being is not busy with coping with personal uncertainty or not recovering from other salient self-threatening information), we then can expect people to react in a relatively calm way to fair and unfair events and, more generally, to show composed reactions to other events that bolster or violate their social-cultural norms and values. On the other hand, under conditions of homeostatic imbalance, people may react very strongly, especially toward unfair events, and be agitated by other transgressions of their cultural worldviews.

Thus, I will start this chapter with the suggestion that the regulated and homeostatic self leads to calmer, better composed human beings. Related to this, conventional wisdom holds that behavioral disinhibition has negative effects on what humans do (e.g., Kant, 1785/1959; Le Bon, 1896/2001). In a subsequent part of this chapter, however, I will put forward the hypothesis that the regulated self may also lead to less desirable reactions. In particular, I will propose that at least some behavioral disinhibition is typically needed to overcome people's inhibition to intervening in moral dilemmas and that behavioral disinhibition therefore may have positive effects on how people respond to these dilemmas. Results of several experiments to be reviewed briefly here will support this disinhibition hypothesis: People with salient disinhibited behaviors (such as aggressive or powerful behaviors) or people with stronger predispositions toward behavioral disinhibition make decisions toward saving a greater number of lives in moral dilemmas, are more likely to behaviorally intervene in these dilemmas, and show a greater dislike of a desirable product they would receive by unethical means. Thus, in contrast with what various theories and worldviews dictate, behavioral disinhibition can be conducive for the greater good. I will end this chapter by listing some conclusions that can be drawn from the work on self-regulation and homeostasis in the justice judgment process and the role of behavioral disinhibition in moral decisions.

Self-Regulation in the Justice Judgment Process

In my work on the social psychology of normative judgments, I have derived huge inspiration from two of the earlier volumes in the Ontario Symposium series, namely, the legendary first volume on social cognition (Higgins, Herman, & Zanna, 1981) and the important fourth volume on relative deprivation (Olson, Herman, & Zanna, 1986). Taken together, these two volumes provide a very nice combination of (a) basic social psychology using a social cognition perspective (broadly defined) and (b) a focus on issues that really matter in people's lives and that are truly important for our understanding of what humans are and what is driving them. In my opinion, this best of the two worlds of more basic and more applied social psychology is still among the most exciting and the best of what social psychology has to offer, and this has been and still is a huge inspiration for me and a major force that is driving my work. So in some sense, one could say that with the present chapter I want to pay my respects to these two groundbreaking volumes that have had such an impact on social psychology in general and the social psychology of justice and legitimacy in particular.

One of the things that my colleagues and I studied following this combination of social cognition basics and justice research that matters is the issue of self-regulatory processes in the social psychology of justice judgments (see also Van Prooijen, Karremans, & Van Beest, 2006) and, more particularly, the experience of personal uncertainty, which is the result of people's being uncertain about themselves (see, e.g., Van den Bos, 2001; Van den Bos, Poortvliet, Maas, Miedema, & Van den Ham, 2005; see also Hogg, 2005; McGregor, Zanna, Holmes, & Spencer, 2001; Sedikides, De Cremer, Hart, & Brebels, in press). Uncertainty management models propose that people want to protect themselves from being in or thinking of situations in which they are uncertain about themselves (e.g., Van den Bos & Lind, 2002, in press). One way in which people can do this is by adhering to their cultural norms and values (Van den Bos et al., 2005). Experiences that are supportive of people's cultural worldviews lead people to be less uncertain about themselves or to be able to better tolerate the uncertainty (Van den Bos, Heuven, Burger, & Fernández Van Veldhuizen, 2006). As a result, it can be predicted that people who are uncertain about themselves or who have been reminded about their personal uncertainties will react very positively toward worldview-supportive experiences (Van den Bos, 2001). In contrast, experiences that threaten or impinge on people's worldviews do not help people at all to cope with their uncertainties, and hence people will respond very negatively toward these worldview-threatening experiences (Van den Bos et al., 2005). In this way, uncertainty models hypothesize that under conditions of uncertainty, people will react especially positively toward the occurrence of events or persons that uphold their cultural norms and values and particularly negatively toward transgressions of these concepts (e.g., Van den Bos & Lind, 2002; Van den Bos et al., 2005).

Although an elaborate overview of the empirical work on uncertainty management is beyond the scope of this chapter (for recent reviews, see, e.g., Hogg, 2007; Marigold, McGregor, & Zanna, in press; Sedikides et al., in press; Van den Bos & Lind, in press), results are in accordance with uncertainty management predictions. For example, research by Van den Bos (2001) was founded on the observation that in most, if not all, societies, being treated in a fair manner is in accordance with cultural norms and values, whereas being treated in an unfair way is a violation of these norms and values (e.g., Folger, 1984; Folger & Cropanzano, 1998; Tyler & Smith, 1998). Integrating this observation with the previously reviewed uncertainty management models, Van den Bos (2001) argued that this should imply that asking (as opposed to not asking) people to think about their uncertainties should lead them to react more positively toward fair events and more negatively toward unfair events. Findings reported by Van den Bos (2001) were supportive of this hypothesis.

Furthermore, salience of personal uncertainty may also moderate reactions to other experiences that bolster or violate people's cultural norms and values. For example, Van den Bos et al. (2005) built their third and fourth experiments on the observation that because of social identity concerns (e.g., Tajfel & Turner, 1979, 1986) and belongingness needs (e.g., Baumeister & Leary, 1995), praise of students' own university constitutes a bolstering of their cultural worldviews whereas criticism of the university represents a violation of participants' worldviews (Dechesne, Janssen, & Van Knippenberg, 2000). Following this line of research, Van den Bos et al. hypothesized and showed that when university students have been reminded about their personal uncertainties, they will react more positively toward information that is favorable about their own university and more negatively toward information that is unfavorable about their university (Van den Bos et al., 2005, Experiments 3 and 4). Moreover, the five experiments presented in the Van den Bos et al. article all suggest that, at least sometimes, models of uncertainty management (e.g., Van den Bos & Lind, 2002; see also Martin, 1999; McGregor et al., 2001) may better explain people's reactions to cultural-worldview-defense reactions than a viable alternative account (i.e., terror management theory; see Van den Bos, in press; Van den Bos & Lind, in press; Van den Bos et al., 2005; but see Landau et al., 2004; Routledge, Arndt, & Goldenberg, 2004). Recently, Yavuz and Van den Bos (2008) replicated these findings in Turkey.

Other studies also provide supportive evidence for predictions by related uncertainty management models. For instance, Hofstede (2001) showed that people high in uncertainty avoidance, compared to people low in uncertainty avoidance, are more conservative, are less tolerant of diversity, are less open to new experiences and alternative lifestyles, want immigrants to be sent back to their countries of origin, and reject people from other races as their neighbors. McGregor, Gailliot, Vasquez, and Nash (2007; McGregor & Marigold, 2003; McGregor et al., 2001) showed that people who are made uncertain about themselves react more defensively toward events that threaten their cultural worldviews and that people do so because in this way they want to restore their sense of self (namely, by being persons who can be certain about themselves; see also Martin, 1999). Related to this, Hogg (2000, 2005) showed that extreme self-uncertainty motivates people to endorse ideological belief systems related to orthodoxy, hierarchy, and extremism to a greater extent (see also Towler, 1984).

Taken together, these different research findings suggest that the uncertain self is an organism that is heavily involved in processes of self-regulation. As a result of these processes, the organism engages in courses of action pertaining to cultural-worldview defense and hence will react particularly positively toward fair events and other experiences that bolster cultural norms and values and will respond especially negatively toward unfair events and other issues that violate cultural worldviews. Personal uncertainty and other self-threatening events (see, e.g., Loseman, Miedema, Van den Bos, & Vermunt, 2009; Miedema, Van den Bos, & Vermunt, 2006) will activate psychological systems that people use to detect and handle alarming situations (Van den Bos, Ham, et al., 2008). In particular, findings suggest that after the alarm system has been activated, people may react especially positively to the experience of fair events (Van den Bos, Ham, et al., 2008), perhaps because the relational quality and inclusion message conveyed by fair treatment and other fair experiences (cf. Lind, 2001; Leung, Tong, & Ho, 2004; Tyler & Lind, 1992) may signal the soothing message that threats are manageable or that there is less reason for alarm than initially felt. In other words, experiences of fair events and other cultural-worldview-supportive experiences may switch an activated alarm system from "Code Red" back to "Code Orange" or "Code Green."

What I have discussed thus far suggests that personal uncertainty and other experiences that activate the human alarm system (Eisenberger & Lieberman, 2004; Eisenberger, Lieberman, & Williams, 2003; Lieberman & Eisenberger, 2004; Murray, Holmes, & Collins, 2005) disrupt the homeostasis of the human organism. I will now briefly discuss some propositions that may follow from explicitly adopting a homeostasis perspective on justice judgments and cultural worldviews.

The Homeostasis Hypothesis

Homeostasis is the property of, for instance, a living organism that regulates its internal environment so as to maintain a stable, constant condition. I think that adopting the homeostatic perspective more explicitly in social psychology in general and in the justice literature in particular may yield some interesting research hypotheses not fully tested in empirical research.

For example, the concept of homeostasis explicitly highlights that people can be conformers or regulators. Regulators try to maintain life system parameters at a constant level over possibly wide ambient environmental variations. On the other hand, conformers allow the environment to determine the parameter. To the best of my knowledge, the possibility of individual differences between people whether they tend to be active regulators or more passive conformers is something that has been not been studied in the modern social psychology of self-regulation (for an overview, see Baumeister & Vohs, 2004). The difference between regulators and conformers may have important implications for how we should conceive of how people manage self-threatening situations. For instance, some human beings may actively want to cope with personal uncertainty, whereas others may more passively experience this aversive feeling. This is not to say that conformers do not have behavioral adaptations allowing them to exert some control over a given parameter, but it does suggest that they react in more passive ways and hence with more passive emotions (e.g., sadness instead of anger) to events that disrupt the organism's homeostatic condition. These more conformist reactions to experiences of personal uncertainty have not received much attention in the uncertainty management literature. More generally, studying differential effects that regulators versus conformers may show in processes of self-regulation may be one of the research hypotheses that the homeostatic hypothesis can yield.

An advantage of homeostatic regulation is that it allows an organism to function effectively in a broad range of environmental conditions. One research hypothesis that is in accordance with the homeostasis hypothesis is that trying to restore stability comes at a price, because even an automatic regulation system requires additional energy (DeWall, Baumeister, Stillman, & Gailliot, 2007). Furthermore, most homeostatic regulation is controlled by the release of hormones into the bloodstream. Other regulatory processes, however, rely on simple diffusion to maintain a balance.

The psychology of homeostasis highlights that humans will use control mechanisms to restore the balance in the human body (Gailliot & Baumeister, 2007). All homeostatic control mechanisms have at least three interdependent components for the variable being regulated; the receptor is the sensing component that monitors and responds to changes in the environment. When the receptor senses a stimulus, it sends information to a control center, the component that sets the range at which a variable is maintained. The control center determines an appropriate response to the stimulus. The result of that response feeds to the effector, either enhancing it with positive feedback or depressing it with negative feedback.

Negative feedback mechanisms reduce or suppress the original stimulus, given the effector's output. Most homeostatic control mechanisms require a negative feedback loop to keep conditions from exceeding tolerable limits. The purpose is to prevent sudden severe changes within a complex organism.

Positive feedback mechanisms are designed to accelerate or enhance the output created by a stimulus that has already been activated. Unlike negative feedback mechanisms that initiate to maintain or regulate physiological or psychological functions within a set and narrow range, the positive feedback mechanisms are designed to push levels out of normal ranges. To achieve this purpose, a series of events initiates a cascading process that builds to increase the effect of the stimulus. This process can be beneficial but is rarely used by the human organism because of risks of the acceleration becoming uncontrollable.

Much physiological and psychological uproar results from the disturbance of homeostasis, a condition known as homeostatic imbalance. Many social psychological conditions can cause homeostatic imbalance. Perhaps one of the most important conditions, and hence one of the most pivotal antecedents of self-regulatory processes, is the threat of death (Gailliot, Schmeichel, & Baumeister, 2006). According to terror management theory (e.g., Greenberg, Solomon, & Pyszczynski, 1997; Pyszczynski, Greenberg, & Solomon, 1999; Solomon, Greenberg, & Pyszczynski, 1991), the fear of death is rooted in an instinct for self-preservation. Although human beings share this instinct with other species, only humans are aware that death is inevitable. This combination of an instinctive drive for self-preservation with an awareness of the inevitability of death creates the potential for paralyzing terror. Furthermore, the theory posits that this potential for terror is managed by a cultural anxiety buffer, a social psychological structure consisting of people's cultural worldviews and their self-esteem. To the extent that this buffer provides protection against death concerns, reminding individuals of their death should increase their need for that buffer. Thus, reminders of death should increase the need for the protection provided by the buffer and therefore lead to strong negative evaluations of events that impinge on the cultural worldview and lead to strong positive evaluations of things that uphold or provide an opportunity to reconstruct the worldview.

One proposition that follows from the homeostasis hypothesis is that the many intriguing findings obtained following the terror management framework (see, e.g., Greenberg et al., 1997; Pyszczynski et al., 1999; Solomon et al., 1991) could be viewed as an example of the fact that conditions (such as mortality salience) that lead to homeostatic imbalance trigger psychological (e.g., Van den Bos et al., 2005; Yavuz & Van den Bos, 2008) and physiological (e.g., Gailliot & Baumeister, 2007) mechanisms. Perhaps this can yield more refined insights into the psychological processes that mortality salience (and other concepts examined in ingenious terror management studies; e.g., Mikulincer, Florian, Birnbaum, &

Malishkevich, 2002; Taubman-Ben-Ari, Florian, & Mikulincer, 1999) yield in causing worldview-defense reactions. Besides the already mentioned possible difference between active self-regulators and more passive situation-conformers, and the important work by Baumeister, DeWall, Gailliot, and colleagues on self-regulatory processes following mortality salience (e.g., DeWall & Baumeister, 2007; Gailliot & Baumeister, 2007; Gailliot et al., 2006; Gailliot, Schmeichel, & Maner, 2007), a possibly new prediction that the homeostatic hypothesis would propose is that once the human organism has restored the imbalance in the human body caused by mortality salience or other self-threatening conditions, the human being will stop or lose interest in being engaged in processes of worldview defense and would like to return to more open-minded reactions to people with other cultural worldviews.

Future research may examine the sequential effects of mortality salience manipulations predicted by the homeostatic hypothesis put forward here. Research studying the possible limits of the homeostatic hypothesis may also be important. For example, perhaps future research will find evidence for the alternative hypothesis that because people may often be confronted with mortality salience in their lives (e.g., watching the news makes mortality salient almost every time), worldview-defense reactions may have become automatized, and hence homeostatic processes may not be sufficient to instigate more open-minded reactions to alternative worldviews.

In ideal circumstances, the previously mentioned homeostatic control mechanisms should prevent homeostatic imbalance from occurring, and when the imbalance does occur, the mechanisms should help to restore the imbalance or at least get the imbalance at a more tolerable level (Van den Bos & Lind, in press). In less ideal circumstances, the psychological mechanisms that are instigated following homeostatic imbalance may not restore the imbalance but in fact may make things worse. For example, an angry organism may start acting aggressively, which in effect may lead to even more negative moods within the individual, not more positive or neutral moods. Thus, both functional and dysfunctional mechanisms may be instigated by conditions leading to homeostatic imbalance. This may also be the reason why not many research findings have obtained evidence for the uncertainty management prediction that people following uncertainty salience or real personal uncertainty may engage in thoughts and behaviors that may lessen the uncertainty or make it more tolerable (Van den Bos et al., 2006; Van den Bos & Lind, in press).

Behavioral Disinhibition in Moral Dilemmas

In the previous two sections, I discussed the notion that personal uncertainty and other self-threatening conditions may activate the human alarm system, a psychological system that people use to detect and handle alarming situations and that prompts people to process more alertly what is going on in the situations in which they find themselves. Furthermore, I noted some predictions that may be derived from explicitly incorporating a homeostasis hypothesis in the literature on justice judgments and cultural worldviews. Taken together, these sections suggest that the regulated and homeostatic self leads to more calm, better composed human beings. For example, Kant (1785/1959) proposed in his rationalistic theory of ethics that people who stay calm in every situation and who hence think carefully about what is going on in the situation before they start acting in the situation may be better people and who may well do what is better for society at large. Thus, Kant was arguing that it would be good if people would act with somewhat more inhibition than they normally tend to do and that it certainly would be conducive for the greater good if people would engage in more inhibitory behaviors than disinhibited people, such as those in large crowds (see Le Bon, 1896/2001), aggressive persons (see Berkowitz, 1983), or those in powerful situations (see Kipnis, 1972). Related to this argument is the observation that a first effect of uncertainty is that it causes behavioral inhibition (Peterson & Flanders, 2002). In the present section, however, I will argue that sometimes it is good to purposefully disrupt the homeostasis of the human organism. More specifically, I will propose that behavioral *dis*inhibition may have positive consequences on how people respond to moral dilemmas.

Ever since the days of ancient Greeks such as Aristotle and Aristippus, people have been intrigued by the issue of what to do when confronted with dilemmas in which they have to choose between different actions, with each action having good reasons for doing that action (e.g., Beauchamp, 2001; Greene, Sommerville, Nystrom, Darley, & Cohen, 2001). One of the reasons that moral dilemmas are fascinating is that they involve difficult decisions about what to do. Common knowledge holds that people should think carefully about the pros and cons of their decisions when responding to difficult issues (e.g., Descartes, 1644; Locke, 1689/1997; Simon, 1955). Furthermore, it has been suggested that people should think carefully about how to respond to moral issues (e.g., Kant, 1785/1959). In a recent line of research, we took a somewhat different perspective and proposed that precisely because moral dilemmas involve difficult decisions about what to do, people may feel inhibited concerning how to act in these dilemmas (Van den Bos, Müller, et al., 2008). Building on this line of reasoning, we tried to show that disinhibited behavioral states indeed may help to overcome intervention inertia in moral dilemmas. In studying these issues, we defined behavioral disinhibition as a state in which people do not or only weakly care about what others think of their actions. We labeled the inhibitory tendency to intervene in moral dilemmas as *intervention inertia.*

Many different moral dilemmas have been studied (see, e.g., Beauchamp, 2001; Greene et al., 2001; Hauser, 2006; Thomson, 1976). In our studies, we examined three different moral dilemmas. That is, we studied how people overcome inhibitory tendencies to intervene in the footbridge dilemma, the trolley dilemma, and dilemmas regarding how to respond toward obtaining desirable products that are the result of another person behaving in unethical ways.

Consider an important moral dilemma familiar to contemporary moral scholars, the footbridge dilemma (Greene et al., 2001). In this dilemma, a runaway trolley is headed for five people who will not be able to leave the railway track in time before the trolley will overrun and kill them. You are standing next to another person on a footbridge that spans the track. The only way for you to save the five people is to push the one person off the footbridge onto the track below. The person will die if you do this, but the person's body will stop the trolley from reaching the others. What will you do? People typically indicate that they are not willing to sacrifice the one person to save the five people on the track. Many explanations for this finding have been suggested (see, e.g., Greene et al., 2001; Hauser, 2006). We proposed that a largely overlooked psychological reaction that helps explain the finding is that people feel inhibited to intervene in the dilemma to save the greater number of lives. Thus, we reasoned that precisely because the footbridge dilemma involves a difficult decision regarding what to do in the dilemma, people may feel inhibited concerning how to act in the dilemma. Furthermore, we argued that if this line of reasoning has merit, then it should be the case that reminding people of disinhibited behaviors (such as aggression or behavior in powerful positions) should reduce inhibition to intervene in the dilemma.

Now consider a related dilemma, the trolley dilemma (Thomson, 1976, 1986). As before, a trolley threatens to kill five people. The only way to save them is to hit a switch that will turn the trolley onto an alternate set of tracks, where it will kill one person instead of five. What will you do now? In this dilemma, people usually respond that they are willing to sacrifice the one person to save the five people. Our analysis suggests (and the data of the third experiment by Van den Bos, Müller, et al., 2008, show) that people in the trolley dilemma, compared to the footbridge dilemma, feel less inhibited to intervene in the dilemma to save the greater number of lives. Because intervention inhibition is expected to be stronger in the footbridge dilemma than in the trolley dilemma, we predicted that disinhibited behavioral states would help to reduce intervention inhibition more in the footbridge dilemma.

In the first five experiments of the Van den Bos, Müller, et al. (2008) paper, we tested various components of the disinhibition hypothesis in the footbridge and trolley dilemmas, revealing evidence for the hypothesis on various important dependent variables: The experiments revealed that making aggression salient to people increases their willingness to save the greater number of lives in the footbridge dilemma (Experiments 1 and 2), that making power salient leads to less inhibition to intervene in the footbridge dilemma (Experiment 3; see also Kumagai & Van den Bos, 2008), that directly making behaving in disinhibited ways salient reduces intervention inhibition in the trolley dilemma (Experiment 4), and that low levels of predisposition toward behavioral inhibition increases behavioral decisions to intervene in the footbridge dilemma (Experiment 5). Thus, the findings show the disinhibition effect in both the trolley and the footbridge dilemmas, but we also reveal that the effect is particularly powerful in the footbridge dilemma in which overcoming intervention inhibition is more of an issue than in the trolley dilemma.

In the sixth experiment, Van den Bos, Müller, et al. (2008) showed the disinhibition effect in another dilemma on a dependent variable that is important in consumer behavior: product satisfaction. Specifically, participants of Experiment 6 worked together with another participant. Suddenly, the other participant enacted unethical behavior that increased the chances of both participants winning desirable products. Participants thus faced a dilemma in how to respond toward having a very good chance of obtaining a desirable product but obtaining this desirable good by means of unethical behavior. Now we know that most personal norms disapprove of unethical and unfair behaviors by others (Folger, 1984). Relatively few people have the resources needed, however, to openly resist or protest strong behavioral demands (e.g., Milgram, 1974). Therefore, we expected that when disinhibited behavior was not salient, participants would be inhibited to intervene against the other person's behavior and hence would indicate that they would be satisfied with the desirable product. States of behavioral disinhibition, however, by definition lower people's concerns about what others think of their reactions. Therefore, when disinhibited behavior is salient, participants should find it easier to resist the other person's unethical behavior and hence to indicate that they are not satisfied with the product obtained by unethical means. We thus examined whether product satisfaction indeed was reduced when disinhibited behavior was (as opposed to was not) made salient, thereby revealing that disinhibited states may lead people to dislike desirable goods that result from unethical behavior.

Specifically, participants in the majority of our experiments completed tasks that made aggression (Experiments 1 and 2), power (Experiment 3), or disinhibited behavior salient (Experiments 4 and 6). With these tasks we activated associations of information that served to remind our participants of behavior they performed in the past. Thus, the aggression manipulation reminded people about aggressive behavior they performed in the past. The power manipulation reminded participants about their behavior when occupying powerful positions. The disinhibited behavior manipulation reminded participants about their behaving with fewer behavioral constraints than they normally experience. Thus, for example, in the condition in which behavioral disinhibition was salient, participants completed three open-ended questions that asked them about their thoughts and feelings about their behaving with fewer constraints than they normally experience. This condition was contrasted with the disinhibition nonsalient condition in which participants were asked to respond to questions that asked them about their thoughts and feelings about their behaving in a normal way during a regular day.[1]

An important element of our line of reasoning is that reminding people of different conditions such as being aggressive, holding powerful positions, or acting in disinhibited ways may all lead to less behavioral inhibition. To ground this assumption we conducted pretests to show that our manipulations that made aggression, power, and disinhibited behavior salient indeed all weakened behavioral inhibition. In the pretests, aggression, power, or disinhibited behavior were made salient after which a state version of a widely used scale to measure behavioral inhibition was assessed (Carver & White, 1994). Example items asked to what extent participants agreed with statements such as "At this moment, I worry about making mistakes" and "At this moment, I would feel pretty worried or upset when I think or know somebody is angry at me."

The results of the aggression pretest showed the anticipated pattern that when aggression was salient, participants reacted with less state behavioral inhibition than when aggression was not salient. Power, being a concept about which different cultures may hold different beliefs (Hofstede, 2001), was pretested among Dutch and Japanese participants. Results revealed a main effect of the salience manipulation, indicating that when power was salient, participants reacted with less behavioral inhibition than when power was not salient, and a main effect of country, showing that Japanese participants were more behaviorally inhibited than Dutch participants. The country effect is in accordance with cultural differences between Japan and the Netherlands (Hofstede, 2001), and the nonsignificant interaction effect between country and the salience manipulation indicated that power salience equally lowered state inhibition in both groups of participants.

Finally, as intended, the pretest of our manipulation that made disinhibited behavior salient (versus nonsalient) showed that when disinhibited behavior was salient, it led to less behavioral inhibition than when disinhibited behavior was not salient. We also assessed whether the manipulation of disinhibition salience, being a new and key independent variable for the present line of research, influenced a state version of Carver and White's (1994) Behavioral Activation Scale, which it did not, suggesting that the manipulation of behavioral disinhibition salience weakened behavioral inhibition and did not affect behavioral activation.

We also assessed whether our salience manipulations influenced participants' affective feelings. To this end, participants completed the Positive and Negative Affect Schedule (Watson, Clark, & Tellegen, 1988) after completing the salience manipulations and before responding to the moral dilemmas. Findings indicated no significant effects on affective states or specific affective reactions. These findings suggest that the brief tasks that we used in our salience manipulations did not influence affect and that affect cannot explain the effects of our salience manipulations.

Our first experiment was presented to the participants as two separate studies. In the first study, participants completed tasks that made aggression either salient or not salient. In the second study, participants responded to the footbridge dilemma. As predicted, we found that participants were more willing to sacrifice one person to save the lives of five others when aggression was salient than when aggression was not salient.

Thus, the findings of our first experiment show that when aggression is salient, it may lead people to save the greater number of persons in the footbridge dilemma. In our second experiment, we tried to replicate this effect, and in addition we examined whether a similar effect would be found in the trolley dilemma. As in the first experiment, participants in the footbridge dilemma were more inclined to save the five people when aggression was salient than when aggression was not salient. Within the trolley condition, there was no significant effect of the aggression salience manipulation. To put it differently, when aggression was not salient, the effect commonly observed when people are responding to trolley or footbridge dilemmas was replicated: Participants in the aggression nonsalient condition were more willing to save the lives of the five persons in the trolley dilemma than in the footbridge dilemma. As a result of aggression being salient, however, there was no significant difference between the two dilemmas in the condition in which aggression was salient.

The results of our first two experiments show that when a state of behavioral disinhibition, such as aggression, is salient, it may lead people to save the greater number of persons in the footbridge dilemma, in which people are assumed to normally experience strong inhibitory tendencies to intervene in the dilemma. In our third experiment, our dependent variable directly assessed inhibition to intervene in dilemmas, and we examined whether power, another social condition that we found to weaken behavioral inhibition (see also Leroy & Brauer, 2007), may lead people to feel less inhibited in how to respond to moral dilemmas. In our third experiment, therefore, power was or was not made salient, after which we assessed the extent to which participants felt inhibited to intervene in either the footbridge dilemma or the trolley dilemma. As predicted, participants were less inhibited to intervene in the footbridge dilemma when power was salient than when power was not salient. Within the trolley condition, we found no significant effect of the power salience manipulation. In other words, in the power nonsalient condition, intervention inhibition was stronger in the footbridge dilemma than in the trolley dilemma, whereas in the power salient condition, the two dilemmas did not differ in experienced intervention inhibition.

The findings of our third experiment thus revealed that power can lead people to reduce their conventional inhibitory tendencies to intervene in footbridge dilemmas. Power did not affect intervention inhibition in the trolley dilemma where inhibition to intervene typically is less strong. Similar effects were recently obtained among Japanese students, using an implicit power-priming manipulation (Kumagai & Van den Bos, 2008). In the fourth experiment of the Van den Bos, Müller, et al. (2008) paper, we examined among a nonstudent population whether a manipulation that directly makes behavioral disinhibition salient (rather than indirectly through power or aggression) would lower intervention inhibition in the trolley dilemma. In the experiment, Dutch citizens walking in the city center of Utrecht participated and were randomly assigned to one of the two salience conditions, asking participants to imagine either their behaving with fewer constraints than they normally experience (behavioral disinhibition salient) or their behaving in a normal way during a regular day (behavioral disinhibition nonsalient). Participants' ages ranged between 14 and 62 years, and the mean age was 29.45 years. Of the participants, 28% had completed a university degree, 42% had completed a lower form of advanced education, 28% had completed high school only, and 2% had not completed any level of education. Results indeed showed that participants were less inhibited to intervene in the trolley dilemma when their behaving in a disinhibited way was salient as opposed to not salient.

Thus, the findings of our fourth experiment reveal that making behavioral disinhibition directly salient can lower intervention inhibition in the trolley dilemma. In our fifth experiment, we examined the influence of stable individual differences in the level of behavioral inhibition on behavioral decisions of whether to intervene in the footbridge or trolley dilemmas. Participants in this experiment were German students whose dispositional levels of behavioral inhibition were measured by means of Carver and White's (1994) trait version of the Behavioral Inhibition Scale. We also assessed the trait version of Carver and White's Behavioral Activation Scale. Behavioral inhibition significantly predicted the decision to intervene in the footbridge dilemma, whereas no significant influence of behavioral inhibition was found in the trolley dilemma. In the footbridge dilemma, 20% of those with a low inhibition score pushed the person off the bridge, whereas none of the participants with a high behavioral inhibition score did so. In the trolley dilemma, 80% of both participants scoring above or below the median inhibition score threw the switch to save the five lives. The Behavioral Activation Scale (a global scale and three subscales) did not yield significant main or interaction effects on behavioral decisions to intervene in the moral dilemmas. Thus, dispositional behavioral inhibition and not behavioral activation was affecting participants' behavioral decisions in the footbridge dilemma.

The findings of our fifth experiment thus show that dispositional levels of behavioral inhibition (and not behavioral activation) significantly affect participants' behavioral decision to intervene in the footbridge dilemma (and not the trolley dilemma). The aim of our sixth experiment was to demonstrate the disinhibition hypothesis in a behavioral interaction context that was more involving for our participants and in which they directly experienced and responded to the following dilemma: In the experiment, participants worked together with another participant (in reality, a professional actor hired as a confederate) to complete as many correct answers on an intelligence test. The persons of the best-performing pair would each win an iPod. When the experimenter would leave the laboratory, the actor would suddenly pull out of his or her jacket an illegally obtained note with the correct answers on the intelligence test and fill out the intelligence test using these answers, thus confronting participants with a dilemma regarding how to respond toward now having a very good chance of obtaining an iPod but obtaining this desirable product by means of unethical or unfair means. When the experimenter came back, the participants were asked to fill in a questionnaire individually. The dependent variable was measured amid filler questions and assessed to what extent participants agreed with the statement "If I would win the iPod, I would be very pleased." After participants had been paid or had received course credit for their participation, they were thoroughly debriefed. During the debriefing procedure, all participants indicated that they did not object to the procedures used in the experiment.

Because most personal norms disapprove of unethical and unfair behaviors by others (Folger, 1984), and relatively few people have the resources needed to openly resist strong behavioral demands (e.g., Milgram, 1974), we predicted that when disinhibited behavior was not salient, participants would be inhibited to intervene in the other person's behavior and hence would indicate that they would be satisfied with the desirable product. When disinhibited behavior was salient, however, participants should find it easier to resist the other person's unethical behavior and hence to indicate that they were not that satisfied with the product obtained by unethical means. Thus, following our disinhibition hypothesis, we predicted that satisfaction with winning the iPod would be high when behavioral disinhibition was not salient and would be less high when disinhibited behavior was salient. This result is indeed what we found: Satisfaction with winning the iPod was high when disinhibited behavior was not salient and was less high when behavioral disinhibition was salient.

In sum, several experiments supported the hypothesis that people may feel inhibited to intervene in dilemmatic situations and that behavioral disinhibition therefore may help to overcome intervention inertia. Specifically, in the majority of our studies, we tested this disinhibition hypothesis by showing that relatively simple manipulations that remind people of having behaved in aggressive ways, holding powerful positions, or acting with less-than-normal constraints strengthen behavioral disinhibition, and this result leads to more positive interventions in moral dilemmas and to a greater dislike of unethical behavior. We also showed the influence of stable individual differences in the level of behavioral inhibition on behavioral decisions about whether to intervene in (at least some) moral dilemmas.

The robustness of these effects across different dependent variables is striking given that our salience manipulations were relatively brief tasks for participants to complete. One implication of these findings is that disinhibited conditions (such as aggression or power) that often have negative effects on human behavior may help to free people from behavioral constraints that prevent intervening in moral dilemmas and disliking unethical behavior. Taken together, our findings suggest that, contrary to conventional wisdom, aggression (e.g., Berkowitz, 1983), power (e.g., Kipnis, 1972), and disinhibited behavior (e.g., Le Bon, 1896/2001) can have positive consequences on human behavior.

The findings were obtained in various experiments in three different types of dilemmas, making it less likely that the results were caused by specifics of the studies. Of course, aggression and power can have effects other than reducing behavioral inhibition (see, e.g., Galinsky, Magee, Inesi, & Gruenfeld, 2006), and behavioral disinhibition certainly can have negative effects on human behavior (see, e.g., Keltner, Gruenfeld, & Anderson, 2003). Furthermore, saving the greater number of lives or having other reactions in moral dilemmas should not be equated with acting in a morally good way. The fact, however, that disinhibited behavioral states can overcome intervention inhibition and that this can have positive consequences such as a greater dislike of unethical behavior is a novel observation. This finding is also significant because recent theories have suggested that moral reasons may often follow, not drive, human reactions toward moral dilemmas (e.g., Haidt, 2001), making it relevant to know what psychological conditions influence reactions to moral dilemmas. Related to this are some interesting findings collected by Tomohiro Kumagai, showing that Japanese participants who first had their self-esteem boosted (using the implicit manipulation by Dijksterhuis, 2004) subsequently showed less inhibition to intervene in the footbridge dilemma, presumably because higher self-esteem increases people's tendencies to not care so much about what others think of their actions (Kumagai & Van den Bos, 2008).

Overcoming intervention inhibition is an important task, not only in moral dilemmas but also in other significant social situations such as when a society wants to reduce the bystander effect (e.g., Darley & Latané, 1968). The present chapter reviewed some first evidence for the novel proposition that behavioral disinhibition helps to free people from behavioral constraints that prevent them from intervening in dilemmas and disliking unethical behavior. One of the reasons we thought it was interesting to provide evidence for our disinhibition hypothesis is it can be contrasted with earlier observations noting the detrimental effects of behavioral disinhibition on people's reactions (e.g., Le Bon, 1896/2001). Our findings suggest that when disinhibited behavior is salient or when people have stronger individual predispositions toward behavioral disinhibition, people are more likely to overcome intervention inertia and behaviorally intervene in moral dilemmas and show a greater dislike of a desirable product they will receive by unethical means. An implication of our findings may be that disinhibited states may facilitate disliking or resisting injustice in the world, suggesting that behavioral disinhibition can be conducive for the greater good.

I started this chapter with the suggestion that the regulated and homeostatic self leads to calmer, better composed human beings. Related to this, conventional wisdom holds that behavioral disinhibition has negative effects on what humans do (e.g., Kant, 1785/1959; Le Bon, 1896/2001). In the second part of this chapter, however, I focused on the issue that the calm, regulated self may also lead to less desirable reactions. In particular, I proposed that calm, well-regulated people may perhaps not intervene in bystander and dilemmatic situations and that at least some public disinhibition (see, e.g., Latané & Nida, 1981) is needed to overcome people's inhibition to intervene in these kinds of situations. In other words, behavioral disinhibition may be bad, except when it is not.

Coda

Disruption of human homeostasis may lead to functional (sometimes dysfunctional) self-regulatory mechanisms that may help (respective to those that do not help) to fully or at least partially restore the homeostatic imbalance. Personal uncertainty and mortality salience are two important conditions that may lead to homeostatic imbalance and hence are important antecedents of self-regulation. A well-regulated individual is more likely to react in composed or calm ways to events or persons that threaten the individual's cultural worldviews. This is not to say, however, that well-regulated, rationalistic people always will do what is best in the social situation at hand. Moral dilemmas and bystander situations, for instance, are cases in which people may well feel flabbergasted or behaviorally inhibited to intervene in the situation. Research findings discussed in this chapter indeed suggest that behaviorally disinhibited behavioral states may help to overcome intervention inertia and that this, at least sometimes, may contribute to the greater good.

A substantial and growing body of research shows that people react more vigilantly to violations or bolstering of their cultural worldviews when they are coping with personal uncertainty (e.g., Hogg, 2007; McGregor, 2003; Van den Bos et al., 2005). Furthermore, it is known that personal uncertainty causes behavioral inhibition (Peterson & Flanders, 2002). Taken together this perhaps suggests that the vigilance as a result of personal uncertainty may be the result of activation of the behavioral inhibition system (Van den Bos & Lind, in press). Interestingly, Marigold et al. (in press) recently argued that activation of the behavioral inhibition system following goal disruptions constitutes essentially a state of aroused uncertainty. According to these authors, when the behavioral inhibition system is activated, approach motivation remains prominent, but inhibition processes join in: With activation of the behavioral inhibition system, the organism remains in an inhibited approach orientation, with aroused attention vigilantly focused on awareness of threat cues or tenable alternative goals. This inhibition-mediated vigilant distress persists until, for example, flight becomes clearly necessary (Gray & McNaughton, 2000). Perhaps these speculations suggest the following general hypothesis that may be tested in future research: Self-regulation and behavioral (dis)inhibition (and related processes such as vigilance and aroused attention) may well work together in intriguing ways (in all likelihood more subtly or more complexly than previously assumed) to predict justice judgments and moral behavior. I hope that proposing this hypothesis, together with the other issues put forward in this chapter, may contribute to the fascinating study of self-regulation, homeostasis, and behavioral disinhibition in social psychology and in normative judgments and behavior.

Author's Note

The work on this chapter was supported by a VICI innovational research grant from the Netherlands Organization for Scientific Research (NWO,

453.03.603). I thank Aaron Kay and Patrick Müller for their comments on earlier versions of this chapter.

Address correspondence to Kees van den Bos, Department of Social and Organizational Psychology, Utrecht University, Heidelberglaan 1, 3584 CS Utrecht, the Netherlands; e-mail: k.vandenbos@uu.nl.

Note

1. It is difficult to give prototypical examples of what participants wrote for each of the disinhibition manipulations briefly reviewed here (aggression, power, and behavioral disinhibition). Inspection of what participants wrote down seems to indicate that they were describing a variety of situations and reporting different experiences in these situations. More research is needed to carefully sort out what different disinhibition manipulations may trigger in different groups of participants.

References

Baumeister, R. F., & Leary, M. R. (1995). The need to belong: Desire for interpersonal attachments as a fundamental human motivation. *Psychological Bulletin, 117*, 497–529.

Baumeister, R. F., & Vohs, K. D. (Eds.). (2004). *Handbook of self-regulation: Research, theory, and applications*. New York: Guilford.

Beauchamp, T. L. (2001). *Philosophical ethics: An introduction to moral philosophy* (3rd ed.). Boston: McGraw-Hill.

Berkowitz, L. (1983). Aversively simulated aggression. *American Psychologist, 38*, 1135–1144.

Carver, C. S., & White, T. L. (1994). Behavioral inhibition, behavioral activation, and affective responses to impending reward and punishment: The BIS/BAS scales. *Journal of Personality and Social Psychology, 67*, 319–333.

Darley, J. M., & Latané, B. (1968). Bystander intervention in emergencies: Diffusion of responsibility. *Journal of Personality and Social Psychology, 8*, 377–383.

Dechesne, M., Janssen, J., & Van Knippenberg, A. (2000). Derogation and distancing as terror management strategies: The moderating role of need for closure and permeability of group boundaries. *Journal of Personality and Social Psychology, 79*, 923–932.

Descartes, R. (1644). *Principia philosophiae* [Principles of philosophy]. Amsterdam: Ludovicum Elzevirum.

DeWall, C. N., & Baumeister, R. F. (2007). From terror to joy: Automatic tuning to positive affective information following mortality salience. *Psychological Science, 18*, 984–990.

DeWall, C. N., Baumeister, R. F., Stillman, T. F., & Gailliot, M. T. (2007). Violence restrained: Effects of self-regulation and its depletion on aggression. *Journal of Experimental Social Psychology, 43*, 62–76.

Dijksterhuis, A. (2004). I like myself but I don't know why: Enhancing implicit self-esteem by subliminal evaluative conditioning. *Journal of Personality and Social Psychology, 86*, 345–355.

Eisenberger, N. I., & Lieberman, M. D. (2004). Why rejection hurts: A common neural alarm system for physical and social pain. *Trends in Cognitive Sciences, 8*, 294–300.

Eisenberger, N. I., Lieberman, M. D., & Williams, K. D. (2003). Does rejection hurt? An fMRI study of social exclusion. *Science, 302*, 290–292.

Folger, R. (Ed.). (1984). *The sense of injustice: Social psychological perspectives*. New York: Plenum.

Folger, R., & Cropanzano, R. (1998). *Organizational justice and human resource management*. Thousand Oaks, CA: Sage.

Gailliot, M. T., & Baumeister, R. F. (2007). The physiology of willpower: Linking blood glucose to self-control. *Personality and Social Psychology Review, 11*, 303–327.

Gailliot, M. T., Schmeichel, B. J., & Baumeister, R. F. (2006). Self-regulatory processes defend against the threat of death: Effects of self-control depletion and trait self-control on thoughts and fears of dying. *Journal of Personality and Social Psychology, 91*, 49–62.

Gailliot, M. T., Schmeichel, B. J., & Maner, J. K. (2007). Differentiating the effects of self-control and self-esteem on reactions to mortality salience. *Journal of Experimental Social Psychology, 43*, 894–901.

Galinsky, A. D., Magee, J. C., Inesi, M. E., & Gruenfeld, D. H. (2006). Power and perspectives not taken. *Psychological Science, 17*, 1068–1074.

Gray, J. A., & McNaughton, N. (2000). *The neuropsychology of anxiety: An enquiry into the functions of the septo-hippocampal system*. Oxford, UK: Oxford University Press.

Greenberg, J., Solomon, S., & Pyszczynski, T. (1997). Terror management theory of self-esteem and cultural worldviews: Empirical assessments and conceptual refinements. In M. P. Zanna (Ed.), *Advances in experimental social psychology* (Vol. 29, pp. 61–139). New York: Academic Press.

Greene, J. D., Sommerville, B., Nystrom, L. E., Darley, J. M., & Cohen, J. D. (2001). An fMRI investigation of emotional engagement in moral judgment. *Science, 293*, 2105–2108.

Haidt, J. (2001). The emotional dog and its rational tail: A social intuitionist approach to moral judgment. *Psychological Review, 108*, 814–834.

Hauser, M. (2006). *Moral minds: How nature designed our universal sense of right and wrong*. New York: Ecco.

Higgins, E. T., Herman, C. P., & Zanna, M. P. (Eds.). (1981). *Social cognition: The Ontario Symposium* (Vol. 1). Hillsdale, NJ: Lawrence Erlbaum.

Hofstede, G. (2001). *Culture's consequences: Comparing values, behaviors, institutions, and organizations across nations* (2nd ed.). Thousand Oaks, CA: Sage.

Hogg, M. A. (2000). Subjective uncertainty reduction through self-categorization: A motivational theory of social identity processes. In W. Stroebe & M. Hewstone (Eds.), *European review of social psychology* (Vol. 11, pp. 223–255). Chicester, UK: Wiley.

Hogg, M. A. (2005). Uncertainty, social identity and ideology. In S. R. Thye & E. J. Lawler (Eds.), *Advances in group processes* (Vol. 22, pp. 203–230). New York: Elsevier.

Hogg, M. A. (2007). Uncertainty-identity theory. In M. P. Zanna (Ed.), *Advances in experimental social psychology* (Vol. 39, pp. 70–126). San Diego, CA: Academic Press.

Kant, I. (1959). *Foundation of the metaphysics of morals*. Indianapolis, IN: Bobbs-Merrill. (Original work published 1785)

Keltner, D., Gruenfeld, D. H., & Anderson, C. (2003). Power, approach, and inhibition. *Psychological Review*, *110*, 265–284.

Kipnis, D. (1972). Does power corrupt? *Journal of Personality and Social Psychology*, *24*, 33–41.

Kumagai, T., & Van den Bos, K. (2008). [The effects of implicit self-esteem on intervention inertia in moral dilemmas]Unpublished raw data.

Landau, M. J., Johns, M., Greenberg, J., Pyszczynski, T., Martens, A., Goldenberg, J. L., & Solomon, S. (2004). A function of form: Terror management and structuring the social world. *Journal of Personality and Social Psychology*, *87*, 190–210.

Latané, B., & Nida, S. (1981). Ten years of research on group size and helping. *Psychological Bulletin*, *89*, 308–324.

Le Bon, G. (2001). *The crowd: A study of the popular mind*. Kitchener, Canada: Batoche Books. (Original work published 1896)

Leroy, D., & Brauer, M. (2007, September). *When power leads to behavioral disinhibition*. Paper presented at the Ninth Transfer of Knowledge Conference of the European Social Cognition Network, Brno, Czech Republic.

Leung, K., Tong, K.-K., & Ho, S. S.-H. (2004). Effects of interactional justice on egocentric bias in resource allocation decisions. *Journal of Applied Psychology*, *89*, 405–415.

Lieberman, M. D., & Eisenberger, N. I. (2004). The neural alarm system: Behavior and beyond. Reply to Ullsperger et al. *Trends in Cognitive Sciences*, *8*, 446–447.

Lind, E. A. (2001). Fairness heuristic theory: Justice judgments as pivotal cognitions in organizational relations. In J. Greenberg & R. Cropanzano (Eds.), *Advances in organizational behavior* (pp. 56–88). Stanford, CA: Stanford University Press.

Locke, J. (1997). *An essay concerning human understanding*. London: Penguin. (Original work published 1689)

Loseman, A., Miedema, J., Van den Bos, K., & Vermunt, R. (2009). Exploring how people respond to conflicts between self-interest and fairness: The influence of threats to the self on affective reactions to advantageous inequity. *Australian Journal of Psychology, 61,* 13–21.

Marigold, D. C., McGregor, I., & Zanna, M. P. (in press). Defensive conviction as emotion regulation: Goal mechanisms and interpersonal implications. In R. M. Arkin, K. C. Oleson, & P. J. Carroll (Eds.), *The uncertain self: A handbook of perspectives from social and personality psychology*. Mahwah, NJ: Lawrence Erlbaum.

Martin, L. L. (1999). I-D compensation theory: Some implications of trying to satisfy immediate-return needs in a delayed-return culture. *Psychological Inquiry*, *10*, 195–208.

McGregor, I. (2003). Defensive zeal: Compensatory conviction about attitudes, values, goals, groups, and self-definition in the face of personal uncertainty. In S. J. Spencer, S. Fein, M. P. Zanna, & J. M. Olsen (Eds.), *Motivated social perception: The Ontario Symposium* (Vol. 9, pp. 73–92). Hillsdale, NJ: Lawrence Erlbaum.

McGregor, I., Gailliot, M. T., Vasquez, N., & Nash, K. A. (2007). Ideological and personal zeal reactions to threat among people with high self-esteem: Motivated promotion focus. *Personality and Social Psychology Bulletin, 33*, 1587–1599.

McGregor, I., & Marigold, D. C. (2003). Defensive zeal and the uncertain self: What makes you so sure? *Journal of Personality and Social Psychology, 85*, 838–852.

McGregor, I., Zanna, M. P., Holmes, J. G., & Spencer, S. J. (2001). Compensatory conviction in the face of personal uncertainty: Going to extremes and being oneself. *Journal of Personality and Social Psychology, 80*, 472–488.

Miedema, J., Van den Bos, K., & Vermunt, R. (2006). The influence of self-threats on fairness judgments and affective measures. *Social Justice Research, 19*, 228–253.

Mikulincer, M., Florian, V., Birnbaum, G., & Malishkevich, S. (2002). The death-anxiety buffering function of close relationships: Exploring the effects of separation reminders on death-thought accessibility. *Personality and Social Psychology Bulletin, 28*, 287–299.

Milgram, S. (1974). *Obedience to authority: An experimental view*. New York: Harper & Row.

Murray, S. L., Holmes, J. G., & Collins, N. L. (2005). *The relational signature of felt security*. Paper presented at the conference of the Society of Experimental Social Psychology, San Diego, CA.

Olson, J. M., Herman, C. P., & Zanna, M. P. (Eds.). (1986). *Relative deprivation and social comparison: The Ontario Symposium* (Vol. 4). Hillsdale, NJ: Lawrence Erlbaum.

Peterson, J. B., & Flanders, J. L. (2002). Complexity management theory: Motivation for ideological rigidity and social conflict. *Cortex, 38*, 429–458.

Pyszczynski, T. A., Greenberg, J., & Solomon, S. (1999). A dual-process model of defense against conscious and unconscious death-related thoughts: An extension of terror management theory. *Psychological Review, 106*, 835–845.

Routledge, C., Arndt, J., & Goldenberg, J. L. (2004). A time to tan: Proximal and distal effects of mortality salience on sun exposure intentions. *Personality and Social Psychology Bulletin, 10*, 1347–1358.

Sedikides, C., De Cremer, D., Hart, C. M., & Brebels, L. (in press). Procedural fairness responses in the context of self-uncertainty. In R. M. Arkin, K. C. Oleson, & P. J. Carroll (Eds.), *The uncertain self: A handbook of perspectives from social and personality psychology*. Mahwah, NJ: Lawrence Erlbaum.

Simon, H. A. (1955). A behavioral model of rational choice. *Quarterly Journal of Economics, 69*, 99–118.

Solomon, S., Greenberg, J., & Pyszczynski, T. (1991). A terror management theory of social behavior: The psychological functions of self-esteem and cultural worldviews. In M. P. Zanna (Ed.), *Advances in experimental social psychology* (Vol. 24, pp. 93–159). New York: Academic Press.

Tajfel, H., & Turner, J. C. (1979). An integrative theory of intergroup conflict. In W. G. Austin & S. Worchel (Eds.), *The social psychology of intergroup relations* (pp. 33–47). Monterey, CA: Brooks/Cole.

Tajfel, H., & Turner, J. C. (1986). The social identity theory of intergroup behavior. In S. Worchel & W. G. Austin (Eds.), *Psychology of intergroup relations* (pp. 7–24). Chicago: Nelson-Hall.

Taubman-Ben-Ari, O., Florian, V., & Mikulincer, M. (1999). The impact of mortality salience on reckless driving: A test of terror management mechanisms. *Journal of Personality and Social Psychology, 76*, 35–45.

Thomson, J. J. (1976). Killing, letting die, and the trolley problem. *The Monist, 59*, 204–217.

Thomson, J. J. (1986). *Rights, restitution and risk.* Cambridge, MA: Harvard University Press.

Towler, R. (1984). *The need for certainty: A sociological study of conventional religion.* London: Routledge & Kegan Paul.

Tyler, T. R., & Lind, E. A. (1992). A relational model of authority in groups. In M. P. Zanna (Ed.), *Advances in experimental social psychology* (Vol. 25, pp. 115–191). San Diego, CA: Academic Press.

Tyler, T. R., & Smith, H. J. (1998). Social justice and social movements. In D. Gilbert, S. T. Fiske, & G. Lindzey (Eds.), *The handbook of social psychology* (4th ed., Vol. 2, pp. 595–629). Boston: McGraw-Hill.

Van den Bos, K. (2001). Uncertainty management: The influence of uncertainty salience on reactions to perceived procedural fairness. *Journal of Personality and Social Psychology, 80*, 931–941.

Van den Bos, K. (in press). The social psychology of uncertainty management and system justification. In J. T. Jost, A. C. Kay, & H. Thorisdottir (Eds.), *Social and psychological bases of ideology and system justification.* New York: Oxford University Press.

Van den Bos, K., Ham, J., Lind, E. A., Simonis, M., Van Essen, W. J., & Rijpkema, M. (2008). Justice and the human alarm system: The impact of exclamation points and flashing lights on the justice judgment process. *Journal of Experimental Social Psychology, 44*, 201–219.

Van den Bos, K., Heuven, E., Burger, E., & Fernández Van Veldhuizen, M. (2006). Uncertainty management after reorganizations: The ameliorative effect of outcome fairness on job uncertainty. *International Review of Social Psychology, 19*, 75–86.

Van den Bos, K., & Lind, E. A. (2002). Uncertainty management by means of fairness judgments. In M. P. Zanna (Ed.), *Advances in experimental social psychology* (Vol. 34, pp. 1–60). San Diego, CA: Academic Press.

Van den Bos, K., & Lind, E. A. (in press). The social psychology of fairness and the regulation of personal uncertainty. In R. M. Arkin, K. C. Oleson, & P. J. Carroll (Eds.), *The uncertain self: A handbook of perspectives from social and personality psychology.* Mahwah, NJ: Lawrence Erlbaum.

Van den Bos, K., Müller, P. A., Beudeker, D. A., Cramwinckel, F. M., Kumagai, T., Ruben, S., Smulders, L., & Van der Laan, J. (2008). On the Role of behavioral disinhibition in reactions to moral dilemmas. Manuscript submitted for publication.

Van den Bos, K., Poortvliet, P. M., Maas, M., Miedema, J., & Van den Ham, E.-J. (2005). An enquiry concerning the principles of cultural norms and values: The impact of uncertainty and mortality salience on reactions to violations and bolstering of cultural worldviews. *Journal of Experimental Social Psychology, 41*, 91–113.

Van Prooijen, J.-W., Karremans, J. C., & Van Beest, I. (2006). Procedural justice and the hedonic principle: How approach versus avoidance motivation influences the psychology of voice. *Journal of Personality and Social Psychology, 91*, 686–697.

Watson, D., Clark, L. A., & Tellegen, A. (1988). Development and validation of brief measures of positive and negative affect: The PANAS scales. *Journal of Personality and Social Psychology, 54*, 1063–1070.

Yavuz, H., & Van den Bos, K. (2008). *Effects of uncertainty and mortality salience on worldview defense reactions in Turkey.* Manuscript submitted for publication.

CHAPTER **10**

The Psychology of Punishment

Intuition and Reason, Retribution and Restoration

JOHN M. DARLEY and DENA M. GROMET

Princeton University

Abstract: Why should researchers care about how ordinary people think about punishment? First, people are citizens, and their opinions about what the criminal justice system should condemn have a claim to be considered when laws are passed. Second, when societies create legal codes that deviate from citizens' moral intuitions, citizens often move toward disrespect for the credibility of the legal codes. They no longer feel that the laws are a good guide to right and wrong. The first question to ask is whether criminal codes sometimes contradict people's sense of right and wrong. The authors review research about the crime of "attempt" that demonstrates the discrepancies between legal codes and the intuitions of citizens. Next, they summarize investigations on people's motives for punishment, which illustrates that people's punishment decisions are products of their intuitions rather than carefully reasoned decisions. Furthermore, the typical response to learning about a significant moral transgression is one of moral outrage, based on information about what offenders justly deserve for the wrongs committed. Put simply, people have a retributive response to crimes. The authors' final studies, however, demonstrate that people also are willing to allow restorative justice procedures that are designed to restore harmony and are willing to make reductions in punishments inflicted on the transgressor if these restorative goals are met.

Keywords: punishment, just deserts, morality, retribution, restoration

Introduction

For some time, we have been doing research on one particular aspect of the psychology of justice. Our concern has been with what sort of justice it is that people seek for a case in which an individual intentionally inflicts an unjustified harm on another person. The paradigmatic cases here are the ones people think of as the core of criminal codes, such as violent crimes such as rape and murder; robberies, thefts, and other takings of private property; and other crimes similar to those prototypic crimes. Much philosophical theorizing has been done on the various *justifications* for punishing offenders, such as inflicting a punishment that the offender justly deserves or inflicting a punishment sufficient to deter the offender (or others) from committing similar crimes in the future. Much less attention has been paid, however, to what it is that ordinary persons seek to achieve when assigning punishments. So we begin by investigating the "naïve psychology of punishment."

First, we consider why researchers should care about ordinary people's judgments about punishments for transgressions and the motives that underlie these decisions. We think it is a topic of great intrinsic interest, but it is also a question of social policy interest. One policy question focuses on whether citizens' judgments about the legality of various actions correspond to what legal codes stipulate. If criminal codes and citizens' intuitions deviate from one another, we suggest that this discrepancy may cause citizens to lose faith in the moral credibility of the law. Next, we demonstrate that in certain central doctrinal matters, code and community intuitions do in fact deviate from one another and that these discrepancies begin to generate contempt for the law.

We then discuss evidence from research on the naïve psychology of punishment, which indicates that many punishment judgments are made intuitively rather than based on chains of reasoning. Therefore, because the determinants of intuitions are not consciously accessible, it is necessary to use research methods that do not rely on self-reports of various punishment motives. Several studies we review lead us to conclude that punishment judgments are generally driven by information relevant to a just deserts punishment stance and therefore are not generally driven by utilitarian considerations such as creating deterrence or incapacitating dangerous offenders.

We then examine more deeply the implications of the claim that punishment decisions are often intuitive ones. We suggest that differentiating between intuitions and reasons should be taken as a dual process model.

Following a transgression, just deserts intuitions automatically come to mind and register as a feeling of moral outrage. Other considerations or cues can bring the reasoning system into play, and this can cause the perceiver to modify the retributive punishment assignments fueled by moral outrage. We demonstrate in several studies that people, even when assigning punishments to moderately or highly severe crimes, will wish to include elements that allow for restorative practices designed to deal with victim and community needs when their attention is drawn to these concerns.

Community Sentiments and Criminal Codes

There are a number of reasons to investigate the naïve psychology of punishment and the implications it has for both individual and societal outcomes. First, of course, to psychologists, exploring people's thinking about retributive justice is an intrinsically fascinating task. The "urge to punish" seems a deeply felt impulse that people hold, and exploring its rules of application and its determinants is therefore an important psychological project. But second, there are a number of social-policy-relevant reasons to explore this area. In a democracy, citizens' opinions have a claim to be represented in setting social policies, and this is certainly true in the realm of criminal laws and criminal justice system practices. Therefore, research determining the contours of the lines that citizens draw between allowable actions and those that should be criminalized are important inputs into informed law-making processes. Politicians take this into account in one way. They often are concerned with enhancing their chances of reelection by voting in accordance with voters' preferences in criminal justice matters, and voters' preferences have been assumed to be for more protection against crime, achieved by harsher punishments for criminals. This concern in the United States has lead to a remarkably rapid increase in the duration of prison sentences since the 1970s, along with a remarkable increase in the number of individuals incarcerated in prisons (Robinson & Darley, 2004).

A third reason to be concerned with the goodness of fit between legal codes and citizens' beliefs is closely related to the second one. Laws that contradict citizens' moral intuitions can challenge or even destroy citizens' core belief in the legitimacy of legal codes as a guide to right and wrong conduct. The efficacy of the criminal justice system depends to a considerable extent on citizens' beliefs that the laws are a legitimate guide to what actions are morally appropriate, as well as to what actions are morally wrong ones, which are not to be committed. To achieve this legitimacy, the lines that the criminal laws draw between innocent actions and culpable ones should be in broad concordance with the lines drawn by the moral thinking of the citizens.

One example of the consequences of discordance between laws and citizens' views is the American experiment in Prohibition. The manufacture and consumption of alcoholic beverages was criminalized, thus rendering punishable activities that large segments of the population considered relatively innocent. Many people initially evaded these restrictions, saw others evading them also, and came to believe that the laws in question were undue invasions of their rights. Scholars commented that this specific contempt for one aspect of the law spread to a more generalized distrust in the degree to which the legal codes were appropriate guides to allowable or wrong behaviors, which in turn lead to a watering down of the symbolic weight of the criminal sanction in the minds of citizens. A recent study (Nadler, 2005) generates evidence for this tendency to be willing to flout the law. Respondents reported being more willing to take actions that disobeyed laws on learning of past instances in which legal codes have criminalized apparently innocent actions or have failed to criminalize actions that most thought were shockingly immoral. (Here is an example of an actual instance of the legal codes failing to criminalize actions that seemed immoral to citizens. Two college-age men were in each other's company and arrived at a casino. One grabbed a young girl, dragged her to a secluded location, and molested her. The other, although generally aware that this was happening, did nothing to stop the incident. In the state in which this occurred, there was no law criminalizing this failure to intervene.)

Do Code and Community Discrepancies Exist?

Are there discrepancies between criminal codes and the relevant moral principles of the communities they govern? Putting this another way, are there reasons to worry about the psychological distancing reactions that communities would have if such discrepancies existed? Obviously the answer depends on the communities under discussion. In many small societies, the codes of the society are simply the shared moral intuitions of the societal members, perhaps existing in memory as specific instances of transgressions. In larger societies, more formal criminal codes are likely to have emerged. In common law countries, heirs to British judicial traditions, it is often thought that the laws that emerged and were enforced by courts were largely codifications of the thinking of the community about various offenses, perhaps systematized by judges, through a process of stare decisis involving comparisons between prior and present cases. In other countries, particularly European ones, the criminal laws are statuary laws put into effect by some legitimately constituted body of lawgivers.

The current state of sources of criminal codes in the United States is complicated. The criminal laws are largely set at a state level, and in the middle years of the 20th century, legal scholars noticed what seemed to them to be a good many different approaches to similar issues taken by different states, some of which were contradictory. Thus, in 1962 the American Law Institute, after some years of work, produced the *Model Penal Code* as a guide to legislatures who wished to update and rationalize the criminal codes of their states. The code has been adopted in whole or in part in a majority of states, and the various states generally have a statutory structure to their criminal codes. One influence on the code, however, was the customary practices of communities. The model code drafters were instructed to take the point of view of "contemporary reasoned judgment" in code drafting. Thus, one thread of influence in code drafting has always been the moral reasoning of the communities governed by the code.

Sometimes, however, the code drafters missed their mark. That is, research (Finkel, 1995; Robinson & Darley, 1995) has discovered that criminal codes and the moral intuitions of citizens governed by those codes frequently differ in important ways. One illustration will make this point, and it concerns when, exactly, a person has committed the crime of "attempt." The common law approach to this question focused on the actual occurrence of the harm of the event and generally asserted that a person had committed the crime of attempt when he or she was in "dangerous proximity" to committing the actual harm. The dangerous proximity condition was fulfilled when, for instance, the actor had entered the place of business in which the safe containing the valuables was located. This traditional view focused on how close the actor had come to the commission of the crime. It is "objectivist" in orientation in that it focuses on the objective harm of the crime and whether the actor's conduct was close to bringing this harm about. The code drafters thought that this was the incorrect approach. Instead, they focused on the actor's subjective culpability in attempting to bring the act about. Their question was whether the actor had moved far enough to demonstrate that his or her intention to commit the crime was "fully formed and resolute," which was demonstrated by the person's taking a "substantial step" toward the commission of the crime. This has been called the "subjectivist" stance toward criminalization, because it focuses on the intent of the actor. It is this intent that is demonstrated when the actor takes the substantial step toward the commission of the crime.

This stance also settled the appropriate penalty for the offense. From the subjectivist perspective, there is little significance attached to whether the harm of the offense actually occurs. Culpability lies in forming the settled intent to commit the crime. This made it an easy task for the code drafters to specify the penalty for the attempt: The penalty in the *Model Penal Code* for attempt is "of the same grade and degree as the most serious offence which is attempted" (American Legal Institute, 1985, s. 5.05(1)). In contrast to this, common law crimes of attempt were assigned penalties that were generally considerably reduced from the sentence given the completed offense.

As the reader will realize, the subjectivist stance criminalizes actions that are farther away from the crime than does the dangerous proximity test of the criminal law. The difference is great enough so that some jurisdictions actually identified a point intermediate between the two as the appropriate step to criminalize; this involved an unequivocal action, such as constructing a tool to open the safe.

These two definitions of what constituted criminal actions are importantly different, and it seemed to us important to determine which best approximated the thinking of citizens about this issue. The experiment is easily designed. The experimenter creates a scenario core about a crime that an actor moves toward committing and has respondents assign punishments to various versions of the scenario that represent a substantial step toward the crime, an unequivocal action, a case of dangerous proximity, and finally the completed crime. And rather than doing this for just one completed crime, the experimenter does it for two or more crimes to see the general shape of the punishment function.

We did several versions of this study (Darley, Sanderson, & LaMantia, 1996; Robinson & Darley, 1995, 1998), and the results for attempt crimes were similar. First, and not surprising, a more serious crime, murder as compared to robbery, draws longer duration prison sentences. Second, in terms of the *graduation* of punishment for the various degrees of movement toward the completed crime, the older, common law pattern is what the respondents imposed, including a higher sentence for the actual offense than for being in dangerous proximity to it. Notice, however, that there is a slight penalty imposed on the actor who takes a substantial step toward the offense and slightly higher penalties imposed on actors who have taken further steps toward the offense but not reached the dangerous proximity test.

Summarizing this, respondents took the common law, objectivist stance toward the grading of the punishment for the offense, in sharp contrast to the high and constant punishment that would be imposed by the subjectivist for all attempts from settled intent to completed crime. One subjectivist result was found, in that respondents were in accord with the subjectivist stance in that they thought that forming the settled intent to commit the crime and demonstrating that intent through taking a substantial step toward the crime was worth a minor penalty.

To review our arguments to date: There is reason to worry about what happens to respect for the law when laws are inconsistent with what citizens think they should be. Furthermore, some quite important legal doctrines promulgated by the *Model Penal Code*, when investigated empirically, were in fact inconsistent with community sentiment. We concluded from this finding that it is worth further investigating whether there are other central doctrines contained in the *Model Penal Code* that are in important disagreement with community sentiments.

Do Code Drafters Promulgate Their Enactments?

Recall that we said that a possible consequence of codes contradicting community sentiments is that the community will become alienated from the legal system and will obey laws only when they fear the punishments that would follow disobedience. We pointed out that, therefore, when the legal system enacts laws that are in disagreement with community sentiments, at a minimum the authorities should engage in a campaign of persuasion seeking to convince the community of the moral correctness of the laws or at least publicize the laws so that people can govern their conduct accordingly. We have some evidence that the code drafters are failing in this task. In one of the studies discussed earlier (Darley et al., 1996), after the respondents indicated what punishments they thought were the appropriate ones to assign to the various attempts toward a crime, we asked them to tell us what punishments they thought the legal system in effect in their state of residence actually assigned to those various attempts. Their state of residence was a state with a *Model Penal Code*. But the respondents reported the pattern we reported previously, which much more resembled the dangerous proximity standard. Probably the respondents were generating their views of what the laws were from what they thought that the laws should be. But they certainly showed no signs of having been successfully educated as to what the laws are.

We explored this issue in one other study (Darley, Carlsmith, & Robinson, 2001). Paul Robinson, our criminal law expert, identified four states that had mainly adopted the *Model Penal Code*, but legislators in those states had found it important to modify one provision of the law to better fit with what they thought the right laws for their state should be. The modifications, of course, differed by state. For example, one state criminalized a failure to rescue somebody in distress if the rescue could be achieved without much inconvenience, whereas another state allowed the use of deadly force in defense of property. Yet another required a person to report the location of a felon if he or she became aware of that location. We then asked the citizens of these four states to respond to scenarios in which, depending on the law of the specific state, the actor was guilty or innocent. We asked respondents to tell us whether they themselves thought the action described in the scenario should be against the law and also to tell us whether they thought the law of the state actually criminalized the action. We also measured various punishment attitudes designed to predict whether the respondents themselves thought the specific action in question should be criminalized. Citizens in all the states responded to all scenarios. Therefore, the states that followed the *Model Penal Code* could be compared to the state that held the deviant version on the particular law in question for each of the four states.

Overall, there was no evidence that citizens of the states in which the legislatures had passed the deviant laws accurately knew what the deviant law held. Respondents whose attitudes suggested they would be in favor of these deviant laws often were in favor of them, but that was also true in the control states. The background attitudes predicted what the respondents thought the law should be, which in turn generally predicted what they thought the law was.

A possible reason for this ignorance emerged. We scanned capital-city newspapers of those five states during the time periods that legislatures were considering adopting the *Model Penal Code*, and in each instance when they were modifying it, we could not detect any stories discussing the specific modifications that the relevant committees were considering. It seems unlikely that the states engaged in the publicity campaigns that would have been necessary to educate the public about the specific state of the laws. They thus failed to enable the citizens to guide their behavior with advance knowledge of the shape of the laws. Of course, it may well have been the case that the legislatures thought the changes they were making were the morally appropriate ones, and the citizens of the states would certainly intuit these laws, but they were wrong about this.

One last point, and we will leave the topic. We have not found among our acquaintances, whom we have casually questioned, any memories of attempts to educate them in either the specifics of the laws or the general doctrines that generate the laws. It is mysterious to us just how it is that our culture expects that the law can fulfill what is called its "ex ante" function without some education in the laws that govern our behaviors as ordinary undifferentiated citizens. We notice that in the training of advanced students who are about to go into the professions or of students studying to go into various trades, there is normally a good deal of attention paid to educating them in the laws that govern the practice of their profession or trade. It must be that we assume that citizens can intuit what the law allows and forbids about their ordinary affairs. This is yet another reason why it is important to investigate the fit between criminal codes and the intuitions of the community governed by those laws.

Those sorts of investigations continue, by both others and us. We now turn, however, to describing our findings in one more area in which psychology can illuminate an issue that is foundational to criminal practice. This concerns the motives that people have for inflicting punishments on rule transgressors.

The Psychological Motives for Punishment

Imagine that someone, a person who lives a few blocks away from you, is mugged at gunpoint on his way home from work, and you learn about it.

What would this event make you feel? People characteristically report feeling an immediate and strong desire to punish the mugger. But why would they, and you, feel this way?

Self-Reports of Motives

Generally, if people are asked to explain why they want to see the mugger punished or asked how they would justify punishing the offender, they name a number of reasons. People may say that punishment is important to deter the mugger and other would-be offenders from committing more crimes in the future and that it is important to have the offender in prison to at least ensure the safety of the community, as the offender cannot commit other acts of violence while in prison (this latter motive is referred to in the criminal justice literature as "incapacitation"). People often add that the mugger "deserves to be punished" for his actions. Indeed, a number of studies have demonstrated that when people are asked to report on why they wanted to punish an offender or why they punished an offender to the extent that they did, they indicate all of these reasons, as well as many others, suggesting that multiple motivations underlie their punishment decisions (Carlsmith, Darley, & Robinson, 2002; Ellsworth & Ross, 1983; Graham, Weiner, & Zucker, 1997; McFatter, 1982; Orth, 2003; Oswald, Hupfeld, Klug, & Gabriel, 2002; Reyna & Weiner, 2001; Warr & Stafford, 1984).

Experimental Investigations of Motives for Punishment

Generalizing from these sorts of self-reports, it appears that the desire to punish an offender may be multiply determined. Psychologists could suggest, however, that people's self-reports about what factors are driving their punishment decisions are somewhat suspect. As Nisbett and Wilson (1977) famously showed, people are likely to provide reasons for their behavior that are available to them and that conventional wisdom suggests as causes for their behavior, which may or may not reflect what did in fact influence their punishment decisions. On this account, self-report is useful more in understanding what motives are common within the culture's discourse about punishment than in determining the actual motives that people draw on to make their punishment decisions. Therefore, as some studies have demonstrated, when people report on motives for their decisions, they may state that they are fulfilling goals that, in practice, do not have an effect on their punishment decisions (Carlsmith et al., 2002; Darley, Carlsmith, & Robinson, 2000; Ellsworth & Ross, 1983).

Given these issues with self-report, we have used a "policy-capturing" methodology to investigate which motive underlies people's punishment decisions for individual crimes (Carlsmith, 2006; Carlsmith et al., 2002; Darley et al., 2000; for a more detailed review of these studies, see

Carlsmith & Darley, 2008). In this procedure, the features of the offense (that are derived from different motives) are varied, and then people are asked to indicate how severe they think the punishment should be. This procedure allows investigators to determine which motives influence people's sentencing decisions without relying on what participants *think* drive their decisions. The logic here is simple: A respondent reads a scenario about an offender who commits a crime. The specifics of the crime are held constant (e.g., the offender embezzles $11,460 from a company). Some respondents read a version of the scenario in which they discover that he had committed similar crimes in the past, and some read a version in which he had not done so. This manipulation of information should influence one's desire to assign higher prison sentences to individuals who are more likely to reoffend and thus is a manipulation of the incapacitation motive to punish. If this manipulation causes a difference in sentence duration assigned by the two test groups, then we infer that incapacitation is a goal that the respondents had that drove their sentence assignment. In ways that the reader can easily intuit, this two-cell design can be expanded to a multidimensional analysis of variance design in which manipulations of scenarios that should affect other motives for punishment can be included. In summary, this design enables a "functional measurement" approach in which the degree of influence of one sort of motive can be compared to the degree of influence of another motive.

We have done studies that compared the motive of retribution (just deserts) to the utilitarian motives of deterrence and incapacitation. For intentional instances of wrongdoing, we have found that people punish offenders for retributive purposes. In other words, people want to see offenders get the punishment that they deserve, not the punishment that would best serve deterrent or safety concerns generated by the offense. For instance, when participants were presented with information related to just deserts (e.g., offender intent) and information related to deterrence (e.g., ease of detection), it was only the just deserts factors that influenced the severity of the punishment that people assigned. In fact, people's perceptions of offense severity and the feelings of the moral outrage elicited by the offense mediated the relation between the just deserts factors and the severity of punishment (Carlsmith et al., 2002). In another study (Darley et al., 2000) with a similar design, we contrasted just deserts motives with incapacitation motives and found that the duration of the prison sentence assigned was driven by just deserts rather than incapacitation concerns. This result, too, was mediated by the respondents' ratings of the degree of moral outrage that they felt in response to the crime scenario. And when people must choose which information to receive about an offense to make a judgment about punishment, they seek information about the moral gravity of the offense (e.g., amount of harm, offender intent) prior to information about either deterrence (e.g., offense frequency) or incapacitation (e.g., offender's prior record; Carlsmith, 2006).

This set of studies was concerned with demonstrating the dominance of retributive over utilitarian concerns in people's punishment assignments, but a number of studies had previously demonstrated that people pay attention to the moral severity of the offense, which is one of the key retributive factors, when assigning a prison sentence. People assign more severe prison sentences as the seriousness of an offense increases (Hogarth, 1971; McFatter, 1978, 1982; Warr, Meier, & Erickson, 1983). This attention to offense severity does not disappear even when participants are required to use a sentencing strategy other than retribution (Darley et al., 2000; McFatter, 1978). Importantly, these studies have also demonstrated that features that should be central to other motives do not influence sentencing decisions. For instance, it has been shown that people do not adjust their sentencing recommendations based on deterrence factors such as ease of detection, the publicity that the crime received, and the frequency with which the crime occurred (Carlsmith et al., 2002; Warr et al., 1983).

The primacy of the retributive impulse is also supported by findings demonstrating that, by default, people act like "retributionists," rather than "deterrentists" or "incapacitators," when assigning punishment to offenders (Carlsmith et al., 2002; Darley et al., 2000; McFatter, 1978). That is, when people receive no instruction about what sentencing strategy to adopt, they spontaneously produce sentences that are responsive to retributive factors. Their default sentencing strategy differs, however, from both an incapacitation (Darley et al., 2000) and a deterrence (Carlsmith et al., 2002) orientation, as people change the severity of the punishment they assign from their default responses when induced to use one of the utilitarian strategies.

These results indicate that there is a disjunction between people's self-reports (a number of motives underlying punishment decisions) and what in actuality influences the extent to which people punish (just deserts concerns). Indeed, studies have directly shown that there are differences in what people say and what people do with regard to the motives that underlie their sentencing decisions. The motives that people support generally are not able to predict what sentences they assign for individual transgressions (Carlsmith et al., 2002; Graham et al., 1997). For instance, we found that participants endorsed just deserts and deterrence motives equally, but their sentences varied only with regard to just deserts concerns.

People frequently cite deterrence in particular as a motivation behind the punishments they assign. Correspondingly, a number of studies have shown that people believe that deterrence is more important to their sentencing decisions than is in fact the case. McFatter (1982) found that people rated deterrence to be the most important punishment motive, whereas retribution was the most important motive in determining the sentence they assigned. In addition, we found that although participants stated they would give more resources to preventing future crimes (deterrence) than catching the offender (just deserts), they responded *only* to the just deserts manipulations for allocating resources (Carlsmith et al., 2002).

Punitive Judgments as Intuitions

This disjunction between why people punish and how they justify it appears to be a divide between the intuitive response people feel when faced with wrongdoing (retribution) and their reasoning about what should or what is likely to influence their desire to punish. The general intuitive versus reasoned response has received much attention from dual process theorists (Chaiken & Trope, 1999). In particular we will focus on Kahneman and colleagues' (Kahneman, 2003; Kahneman & Frederick, 2002) two reasoning systems, System 1 (the intuitive system) and System 2 (the reasoning system). System 1 is responsible for quick, intuitive judgments. These judgment processes are difficult for people to articulate, as these processes are automatic and not available to conscious reflection. System 2 deals with judgments that are controlled and require conscious effort. These judgment processes are available to the judgment makers, and people can articulate their System 2 reasoning to others. People's default responses originate in System 1, and the influence of System 2 is seen only if an error in the System 1 judgment is detected.

The System 1–System 2 framework provides a way to understand why people punish to achieve retribution but claim to punish wrongdoers to achieve a variety of other motives. When people are asked to consider what is the appropriate punishment for an individual offense, System 1 governs people's responses. People assign punishments that are proportionate to the severity of the offense (retribution). When people are asked why they punished, however, their System 1 reasoning is not available to them. People then move to System 2 and recruit possible reasons for their decision. These reasons, however, may not correspond with what guided their intuitive judgments in System 1, which is why people believe that other considerations beyond retribution (such as deterrence and incapacitation) also influenced their judgment.

This dual process framework may account for previous research that demonstrates that people provide post hoc explanations for a number of moral issues, including support for the death penalty (Ellsworth & Ross, 1983) and disgust-eliciting actions (Haidt, Koller, & Dias, 1993). In particular, Haidt's social intuitionist model (Greene & Haidt, 2002; Haidt, 2001) outlines how moral judgment is dictated not by reasoning but rather by quick, intuitive, affective responses to the situation. Then, when people have to explain why an action was morally wrong (or right), they look for reasons that will support their position. Returning to the assignment of punishment, people can cite a number of motives that would all explain why they punished, although it is only just deserts that underlies the severity of their punishments for individual transgressions.

Neural Imaging Results Support the Dual Process Account

Recent neural imaging research done by a group led by Josh Greene (Greene, Nystrom, Engell, Darley, & Cohen, 2004; Greene, Sommerville, Nystrom, Darley, & Cohen, 2001) has discovered brain-imaging evidence that supports the claim that moral judgments of the sort we have been discussing are dual process judgments and that these dual processes take place in somewhat different brain locations. In these studies, a respondent, whose brain processes are being scanned, reads a short scenario that poses a moral question. Would it be morally appropriate for the actor in the scenario to take a certain action that would inflict harm on another? Some of these moral dilemmas, for instance whether a teenage single mother who found her baby a great burden should kill her infant, invoked rapid responses of "no." These cases were decided rapidly, and typically the response was that the action was morally inappropriate. This we suggest was an intuitive system response. Imaging results showed heightened activity in brain areas involving social cognition and emotional responding (specifically, the medial prefrontal cortex, posterior cingulate/precuneus, and superior temporal sulcus/temporoparietal junction).

A few cases, however, elicited quite a different pattern of results. One such case is the famous "smother baby" story:

> Enemy soldiers have taken over your village. They have orders to kill all remaining civilians. You and your townspeople have sought refuge in the cellar of a large house. Outside you hear the voices of those who have come to search the house for valuables. Your baby begins to cry loudly. You cover his mouth to block the sound. If you remove your hand from his mouth, his crying will summon the attention of the soldiers who will kill you, your child, and the others hiding out in the cellar. To save yourself and the others, you must smother your child to death. Is it appropriate for you to smother your child to save yourself and the other townspeople?

Here imaging work demonstrated first the previously described pattern of heightened activity in brain areas involving social cognition and emotional responding. We suggest that this was the intuitive reaction triggered by contemplating terrible harm to the baby. But there is a dilemma built into this story, and respondents' attention is called to it by the last sentence of the story. The choice is not between killing and not killing, it is between killing one so the entire group will not be killed.

This decision takes considerably longer to make, and respondents are split about the awful choices they face. Brain activation patterns now show both the fast response in the social reasoning and emotional areas (specifically, the medial prefrontal cortex, posterior cingulate/precuneus, and superior temporal sulcus/temporoparietal junction) and a slower response in both abstract reasoning areas (regions in the dorsolateral prefrontal cortex) and other areas activated when there is cognitive conflict and also the engagement of cognitive control (specifically the anterior cingulate cortex, a brain region associated with cognitive monitoring and integration). Some respondents, characterized by relatively strong brain activation patterns in social cognition and emotional brain areas, report that the baby must not be killed. Others respond in a more utilitarian matter, judging the ordinarily forbidden killing of the baby to be acceptable because the baby's life is forfeit, along with the lives of others, if this action is not taken. These respondents are characterized by relatively stronger activations of the higher order reasoning areas.

We interpret these results as suggesting that there are dual processes involved in making moral judgments. One is the intuitive system response, which is driven by relatively rapid reactions, produced by brain areas that register social cognitive and emotional processes. The second system involves abstract reasoning areas of the brain, ones that developed evolutionarily later than did the social cognitive and emotional brain areas, and is not always triggered into action. This, we suggest, is the reasoning system. It is sometimes the case that the reasoning system will be in conflict with the intuitive system, and conflict will be experienced. As we suggest in the next section, however, it is sometimes the case that the way in which moral issues are presented may preferentially cause reasoning processes to be engaged. Specifically, when questions are posed at the level of social policies, people may reason about their answers and report support for policies that they might oppose if the question accessed their intuitive judgments.

Dual Process Theories and Policy-Level Judgments

The two systems have implications for justice judgments at the policy level as well. People may engage in reasoning processes when thinking about how they would like wrongdoing handled on at the societal level. That is, people may endorse using a deterrent-based policy to respond to wrongdoing, although they would disagree with the application of this policy to an individual case, as it would not correspond to a just deserts philosophy. Indeed, a number of investigators (us included) have posited that whereas retribution is the dominant motive when people are assigning individual prison sentences, utilitarian motives may take over when people think about punishment on the policy level (Carlsmith et al., 2002; Darley et al.,

2000; Tyler, Boeckmann, Smith, & Huo, 1997; Warr et al., 1983). As utilitarian motives such as deterrence and incapacitation tend to lead to assignments of increasing punishment severity (Carlsmith et al., 2002; Darley et al., 2000; McFatter, 1978, 1982), this disjunction leads to a conflict between people's intuitive responses to individual cases of intentional wrongdoing (the punishment fitting the crime) and their reason-driven analysis that calls for harsher punishment when thinking about crime in general (for recent empirical research on this issue, see Carlsmith, 2008).

In sum, there are a number of ways in which people can justify why they punish wrongdoers, which capture both utilitarian (e.g., deterrence) and moral (e.g., retribution) concerns. When people are confronted with a specific crime, however, it is a desire for retribution that underlies their punishment decisions. This distinction between people's intuitive System 1 response and their System 2 approach to thinking about the reasons that might underlie their punishment responses illuminates a conflict between how people *feel* that individual wrongdoers should be punished and how people *think* about what should drive punishment decisions.

Is Justice Only Punishment?

Returning again to the example of your neighbor who was mugged on his way home from work, we have previously discussed the retributive impulse that you would feel upon learning this news. But would a desire for punitive action against the offender be your only response? You may learn of your neighbor's distress at being mugged and desire that your neighbor be compensated for his losses and be able to feel better about what happened to him (i.e., feeling closure and being able to move on with his life). You may also feel that there are issues created by the crime that affect the community in which you live, and you may desire to repair any harm caused to your community and to reinforce the moral values and boundaries of the community.

As much of the research on people's responses to intentional wrongdoing has focused on people's punitive responses (reviewed earlier), we know very little about justice concerns that focus on the victim and the community in which the crime occurred (Darley & Pittman, 2003). We propose that when people are confronted with wrongdoing, they are concerned with the victim and the community in addition to the punishment of the offender. Although these concerns can influence people's punitive responses toward the offender (Oswald et al., 2002), they may also cause people to be concerned with nonpunitive responses that impact the victim and the community. One such response is restorative justice, which focuses on repairing the harm that results from wrongdoing for all of the actors involved (victim, offender, and community). Ideally, restorative justice allows for the victim to be restored to where he or she was before the crime occurred, the offender to be rehabilitated and reintegrated into society, and the community in which the crime occurred to be restored as well (Marshall, 2003). In recent years, restorative justice has received considerable attention as an alternative to the current criminal justice system that prioritizes punishing the offender and neglects the other justice concerns associated with the victim and the community (Bazemore, 1998; Braithwaite, 1989, 2002; Sherman, 2003; Tyler, 2006).

The growing interest in, and testing of, restorative justice has motivated investigations into whether people in fact find restorative justice to be an effective way to achieve justice. One key to this discussion is the severity of the offense. We have conducted research that investigated how acceptable people find a restorative response to crime for offenses that range in severity (Gromet & Darley, 2006). We presented participants with different crimes and had them assign the case to one of three procedures that they believed would best achieve justice: a restorative conference, the standard court process leading to a prison sentence, or a restorative conference and a prison sentence. Consistent with previous research (e.g., Doble & Greene, 2000), we found that participants chose to use a restorative justice option over a more punitive, retributive one when handling offenses that are low in severity. For crimes that were high in severity, however, we found that participants selected the mixed procedure (a restorative conference followed by a court proceeding that assigned a prison sentence) over either the restorative conference or the prison sentence options on their own. In addition, these participants who used the mixed procedure assigned lower prison sentences than a control group of participants who could use only prison to handle the same offenses.

It appears, then, that people are open to accomplishing both retribution and restoration for serious offenses. If people must select either a restorative or a retributive option to handle a serious offense, however, then people select the retributive option over the restorative option (Doble & Greene, 2000; Roberts & Stalans, 2004). Also, people show high levels of support for punitive responses to wrongdoing, such as the death penalty, when they are provided with no alternatives that allow them to inflict both retributive and restorative measures (Doob & Roberts, 1988; McGarrell & Sandys, 1996). These findings indicate that retribution is a dominant concern when people respond to serious criminal violations and is able to trump restorative concerns. Therefore, although people are open to the fulfilling of additional justice goals beyond punishing the offender, we argue that these considerations are not part of people's intuitive reactions to wrongdoing. People's intuitive response is to punish the offender in proportion to the severity of the offense, and this intuitive judgment will dominate people's reactions. This intuitive response, however, can be adjusted if people are given the opportunity to think about how the victim and the community have been impacted by the offense.

We can understand this process through the System 1 versus System 2 framework discussed previously. System 2 oversees the judgments of System 1 and steps in to amend, or override, System 1 judgments that are determined to be erroneous. If System 2 does not detect an error or if System 2 is impaired because of time, cognitive load, or motivation constraints, then the intuitive judgment of System 1 is likely to remain. Applying this to people's justice judgments, if people have time and cognitive resources and their attention is drawn to how additional justice concerns (e.g., restoring the victim and the community) are not addressed by their initial retributive reaction, then they should incorporate these additional justice concerns into their responses to wrongdoing.

In recent research that we have conducted, we found evidence that people do tend by default to focus on the offender but that they are responsive to victim and community concerns when their attention is drawn to these targets (Gromet & Darley, 2009). When participants are left to their own devices about how to achieve justice for a particular crime, they tend to select sanctions that focus on the offender and his or her punishment (as opposed to rehabilitation). And participants show the same pattern if they are instructed to think about the offender when confronted with a crime, which indicates that participants were, by default, focusing on the offender when they received no instruction. If participants have their attention drawn to how the crime affected the victim or the community when confronted with wrongdoing, however, they choose more sanctions that will impact the victim and the community, respectively. Although these participants showed heightened concern with the victim and the community in these conditions, they continued to desire sanctions that would punish the offender. These results indicate that although punishment is the default and an essential response to intentional wrongdoing, people are open to other justice responses that target the victim and the community in which the crime occurred.

Justice, then, does not necessarily stop at the punishment of the offender. People do think other responses are just, and they will use these responses, either alone or in combination with retribution, to achieve justice. These justice considerations beyond punishing the offender, however, may not appear as strongly in people's intuitive reactions to wrongdoing as the desire for retribution. People will need attention and motivation to consider these additional responses to wrongdoing that focus on the victim and the community.

Conclusion

This chapter has explored the convergence and divergence between existing criminal codes and theories about how people respond to criminal violations and how people desire to punish offenders in actuality, as well as their broader conceptualizations of what constitutes just responses to wrongdoing. People's conceptions of what should be punished do differ in important and systematic ways from what legal codes (particularly the *Model Penal Code*) dictate is punishable behavior. This discrepancy has implications for overall compliance with the law, as laws that lack moral validity will decrease perceptions of the legitimacy of those laws. In addition, although people may cite a number of motivations driving their desire to see offenders punished, it is their just deserts concern with proportional retribution that determines the extent to which they punish offenders. This disjunction presents difficulties with regard to determining criminal law policies, as policies that people would endorse at the macro, societal level (punishments usually determined by utilitarian motives such as deterrence and incapacitation) would clash with the punishments that people consider just for individual cases (punishments usually determined by the retribution motive). Finally, although Western justice systems focus on the punishment of the offender, there is evidence that people, if given the opportunity, can think about justice in broader terms than punishment, including the restoration of crime victims, offenders, and communities. By highlighting these discrepancies between current criminal codes and current practices in legal systems and what people view as just responses to criminal violations, we have attempted to demonstrate how the psychological understanding of these differences can inform theory development in this area and create possibilities for sensible modifications in criminal justice practices that would be accepted by the community those practices govern.

References

American Legal Institute. (1985). *Model Penal Code and commentaries.* Philadelphia: Author.

Bazemore, G. (1998). Restorative justice and earned redemption: Communities, victims and offender reintegration. *American Behavioral Scientist, 41*, 768–813.

Braithwaite, J. (1989). *Crime, shame, and reintegration.* Cambridge, UK: Cambridge University Press.

Braithwaite, J. (2002). *Restorative justice and responsive regulation.* Oxford, UK: Oxford University Press.

Carlsmith, K. M. (2006). The roles of retribution and utility in determining punishment. *Journal of Experimental Social Psychology, 42*, 437–451.

Carlsmith, K. M. (2008). On justifying punishment: The discrepancy between words and actions. *Social Justice Research, 21*, 119–137.
Carlsmith, K. M., & Darley, J. M. (2008). Psychological aspects of retributive justice. In M. Zanna (Ed.), *Advances in experimental social psychology* (Vol. 40, pp. 193–236). London: Academic Press/Elsevier.
Carlsmith, K. M., Darley, J. M., & Robinson, P. H. (2002). Why do we punish? Deterrence and just deserts as motives for punishment. *Journal of Personality and Social Psychology, 83*, 284–299.
Chaiken, S., & Trope, Y. (1999). *Dual process theories in social psychology*. New York: Guilford.
Darley, J. M., Carlsmith, K. M., & Robinson, P. H. (2000). Incapacitation and just deserts as motives for punishment. *Law and Human Behavior, 24*, 659–683.
Darley, J. M., Carlsmith, K. M., & Robinson, P. H. (2001). The ex ante function of the criminal law. *Law and Society Review, 35*, 165–189.
Darley, J. M., & Pittman, T. S. (2003). The psychology of compensatory and retributive justice. *Personality and Social Psychology Review, 7*, 324–336.
Darley, J. M., Sanderson, C. A., & LaMantia, P. S. (1996). Community standards for defining attempt: Inconsistencies with the Model Penal Code. *American Behavioral Scientist, 39*, 405–420.
Doble, J., & Greene, J. (2000). *Attitudes towards crime and punishment in Vermont: Public opinion about an experiment with restorative justice*. Englewood Cliffs, NJ: John Doble Research Associates.
Doob, A. N., & Roberts, J. (1988). Public punitiveness and public knowledge of the facts: Some Canadian surveys. In M. Hough & N. Walker (Eds.), *Public attitudes to sentencing* (pp. 111–133). Aldershot, UK: Gower.
Ellsworth, P. C., & Ross, L. (1983). Public opinion and capital punishment: A close examination of the views of abolitionists and retentionists. *Crime and Delinquency, 29*, 116–169.
Finkel, N. (1995). *Commonsense justice: Jurors' notions of the law*. Cambridge, MA: Harvard University Press.
Graham, S., Weiner, B., & Zucker, G. S. (1997). An attributional analysis of punishment goals and public reactions to O. J. Simpson. *Personality and Social Psychology Bulletin, 23*, 331–346.
Greene, J., & Haidt, J. (2002). How (and where) does moral judgment work? *Trends in Cognitive Sciences, 6*, 517–523.
Greene, J. D., Nystrom, L. E., Engell, A. D., Darley, J. M., & Cohen, J. D. (2004). The neural bases of cognitive conflict and control in moral judgment. *Neuron, 44*, 389–400.
Greene, J. D., Sommerville, R. B., Nystrom, L. E., Darley, J. M., & Cohen, J. D. (2001). An fMRI investigation of emotional engagement in moral judgment. *Science, 293*, 2105–2108.
Gromet, D. M., & Darley, J. M. (2006). Restoration and retribution: How including retributive components affects the acceptability of restorative justice procedures. *Social Justice Research, 19*, 395–432.
Gromet, D. M., & Darley, J. M. (2009). Punishment and beyond: Achieving justice through the satisfaction of multiple goals. *Law and Society Review*.
Haidt, J. (2001). The emotional dog and its rational tail: A social intuitionist approach to moral judgment. *Psychological Review, 108*, 814–834.

Haidt, J., Koller, S. H., & Dias, M. G. (1993). Affect, culture, and morality: Or is it wrong to eat your dog? *Journal of Personality and Social Psychology, 65*, 613–628.

Hogarth, J. (1971). *Sentencing as a human process*. Toronto, Canada: University of Toronto Press.

Kahneman, D. (2003). A perspective on judgment and choice: Mapping bounded rationality. *American Behavioral Scientist, 58*, 697–720.

Kahneman, D., & Frederick, S. (2002). Representativeness revisited: Attribute substitution in intuitive judgment. In T. Gilovich, D. Griffin, & D. Kahneman (Eds.), *Heuristics and biases: The psychology of intuitive judgment*. New York: Cambridge University Press.

Marshall, T. F. (2003). Restorative justice: An overview. In G. Johnstone (Ed.), *A restorative justice reader: Texts, sources, and context* (pp. 28–45). Portland, OR: Willan.

McFatter, R. M. (1978). Sentencing philosophies and justice: Effects of punishment philosophy on sentencing decisions. *Journal of Personality and Social Psychology, 36*, 1490–1500.

McFatter, R. M. (1982). Purposes of punishment: Effects of utilities of criminal sanctions on perceived appropriateness. *Journal of Applied Psychology, 67*, 255–267.

McGarrell, E. F., & Sandys, M. (1996). The misperception of public opinion toward capital punishment: Examining the spuriousness explanation of death penalty support. *American Behavioral Scientist, 39*, 500–513.

Nadler, J. (2005). Flouting the law. *Texas Law Review, 83*, 1398–1441.

Nisbett, R. E., & Wilson, T. D. (1977). Telling more than we can know: Verbal reports on mental processes. *Psychological Review, 84*, 231–259.

Orth, U. (2003). Punishment goals of crime victims. *Law and Human Behavior, 27*, 173–186.

Oswald, M. E., Hupfeld, J., Klug, S. C., & Gabriel, U. (2002). Lay-perspectives on criminal deviance, goals of punishment, and punitivity. *Social Justice Research, 15*, 85–98.

Reyna, C., & Weiner, B. (2001). Justice and utility in the classroom: An attributional analysis of the goals of teachers' punishment and intervention strategies. *Journal of Educational Psychology, 93*, 309–319.

Roberts, J. V., & Stalans, L. J. (2004). Restorative sentencing: Exploring the views of the public. *Social Justice Research, 17*, 315–334.

Robinson, P. H., & Darley, J. M. (1995). *Justice, liability, and blame: Community views and the criminal law*. Boulder, CO: Westview.

Robinson, P. H., & Darley, J. M. (1998). Objectivist versus subjectivist views of criminality: A study in the role of social science in criminal law theory. *Oxford Journal of Legal Studies, 18*, 409–447.

Robinson, P. H., & Darley, J. M. (2004). Does criminal law deter? A behavioural science investigation. *Oxford Journal of Legal Studies, 24*, 173–205.

Sherman, L. W. (2003). Reason for emotion: Reinventing justice with theories, innovations, and research. *Criminology, 41*, 1–37.

Tyler, T. R. (2006). Restorative justice and procedural justice: Dealing with rule breaking. *Journal of Social Issues, 26*, 307–326.

Tyler, T. R., Boeckmann, R. J., Smith, H. J., & Huo, Y. J. (1997). *Social justice in a diverse society*. Boulder, CO: Westview Press.

Warr, M., Meier, R. F., & Erickson, M. L. (1983). Norms, theories of punishment, and publically preferred penalties for crimes. *The Sociological Quarterly*, *24*, 75–91.

Warr, M., & Stafford, M. (1984). Public goals of punishment and support for the death penalty. *Journal of Research in Crime and Delinquency*, *21*, 95–111.

CHAPTER **11**

Legitimacy and Rule Adherence

A Psychological Perspective on the Antecedents and Consequences of Legitimacy

TOM R. TYLER

New York University

Abstract: This chapter argues that legitimacy plays an important role in shaping people's willingness to adhere to group-based rules. It further suggests that legitimacy is the product of judgments about the fairness of group procedures. Taken together these two findings suggest the potential value of a new model of social regulation—the value-based approach.

Keywords: deterrence, legitimacy, compliance, rule adherence, procedural justice, moral values

Overview

I want to focus on one objective of the law: rule adherence. Rule adherence includes compliance in people's everyday lives and compliance in response to the decisions made by legal authorities such as judges and police officers. In either case, a key societal concern is with our ability to bring people's behavior into line with the law. The law has little ability to be effective in helping groups to manage their interactions and resolve conflicts if people do not obey it.

Having the ability to bring people's behavior into line with the law is not something that societies can take for granted. Irrespective of whether we are talking about general rule following or the acceptance of particular decisions, contentious issues are involved. In the case of everyday compliance, the government faces many situations where securing widespread compliance is difficult. Drug laws are a classic example. And more recently there have been ongoing struggles over efforts to control the illegal downloading of music, the illegal copying of movies, and so on. Wherever people do not generally support the law, it is difficult to enforce it. Similarly, in particular instances in which people are personally dealing with legal, political, managerial, or familial authorities, people often resist or ignore judicial orders and defy the police (Tyler, 2006c). The authorities can never take compliance for granted and often find securing it to be a formidable challenge.

The traditional method for securing compliance in the United States is deterrence. This approach is based on the belief that people will shape their behavior based on their fear of possible punishment if they are caught breaking rules. This is the dominant model of social control within the American legal system (Blumstein, Cohen, & Nagin, 1978). My primary critique of this model, which I will elaborate in this chapter, is that it is costly and minimally effective.

The problems associated with deterrence derive from the difficulties of creating and maintaining a credible threat of punishment. People try to hide their behavior because they do not want to be punished, so some type of surveillance system is needed to detect wrongdoing, and that system has to be credible. People have to believe that if they or others break the law, they are likely to be caught and punished. Except in situations where people are easily monitored, the problems of surveillance are formidable and can be resolved only by the large and continuing provision of resources to create and deploy agents of social control.

The problem of establishing credibility is made more difficult by findings suggesting that it is the probability of punishment, not its severity, that most strongly shapes rule following. As a consequence, authorities cannot motivate widespread rule following by administering a few draconian punishments. They need to spend the money required to create a general perception that those who break rules are likely to be caught and punished, for example, by hiring and deploying a police force or by developing other effective ways of monitoring behavior.

Despite these limitations, the literature on deterrence argues that it does work, that is, the perceived likelihood of punishment influences rule following. Even studies that show such a deterrence effect suggest, however, that the magnitude of the influences found is small. For example, MacCoun (1993) found that variations in the certainty and severity of punishment associated with drug use accounted for approximately 5% of the variance in the rate of drug use. Although drug use is in some ways not a typical crime, this finding is typical of studies in this area, which either find no significant influence or find an influence that occurs but is small in magnitude.

In addition, the deterrence approach has troubling side effects. One is the creation of a vast prison population. The United States today is one of the world leaders in the proportion of its population in jail or prison, with 1 in 100 U.S. adults currently incarcerated (*One in 100*, 2008). In addition to creating major fiscal costs, such high rates of incarceration have been damaging to poor and minority communities, because large numbers of people have been removed from those communities for long periods of time.

Although it is natural to assume the costs of punishment, for example, the many costs of being imprisoned, are borne by poor and minority group members, the extension of the deterrence model to broken windows approaches to order maintenance has meant that many middle-class people have been incarcerated for lesser periods of time for committing minor crimes. Broken windows approaches argue that people's initial minor crimes if unpunished escalate into more serious crimes. As a consequence, it is important for the police to sanction people for minor crimes so that they do not come to believe that crimes are not punished. For example, over one half of a million people per year were incarcerated for minor crimes as part of New York's effort to use broken windows approaches to maintain order. Those people spend some time in custody, and many paid fines for offenses such as drinking beer on their front steps or other similar crimes.

Deterrence approaches are particularly problematic because they create their own pernicious social dynamic. As already noted, research has suggested that it is probability of punishment that shapes behavior, not severity. As a result it is not possible to enact effective strategies of social control without allocating substantial resources. Society cannot simply capture a few offenders and give them heavy punishments as an approach to deter others from breaking the rules. Instead, it is necessary to spend the resources needed to create a widespread sense that there is a credible risk of punishment for wrongdoing. Furthermore, because deterrence approaches do not create supportive values, these costs are ongoing, and when there are no surveillance mechanisms in place, people cease rule following.

As a consequence of these financial pressures, despite the research pointing to the limited effectiveness of severity of punishment in shaping law-related behavior, there is consistent political pressure against such high levels of allocations and a search for a "cheap" form of effective deterrence. This political pressure inevitably leads to a focus on ever-harsher punishments, in the effort to have a low-cost, high-effectiveness model of deterrence. Because increases in severity have little actual impact on people's behaviors, however, the pressure for ever-greater sanctions increases over time as the system continually fails to deliver expected changes in public behavior. And although increasing the severity of punishment is low in effectiveness, it is also low in cost relative to the costs of raising the probability of punishment. Once a person is arrested and convicted of a crime, he or she can easily be sentenced to longer terms. Although those more severe punishments have their own additional costs, the costs are pushed into the future and are less obvious.

Beyond these problems, the use of a punishment-based strategy has social costs. One such cost is that it undermines the trust and confidence that people have in the legal system. Legal authorities become associated with punishment, and people come to dislike and avoid them. This is ironic, because studies have indicated that the primary reason people come into contact with legal authorities is because they seek their help (Tyler & Huo, 2002). It is a consequence of the salience of their role as enforcers of the law, a salience linked to deterrence. This negative framing of legal authorities is especially troubling because recent studies have emphasized the value of public cooperation with legal authorities in a joint effort to manage problems of social order within communities (Sampson, Raudenbush, & Earls, 1997). Low trust and confidence lead to low cooperation, thus undermining the ability of the authorities to effectively manage crime (Tyler & Fagan, 2008).

In addition to associating legal authorities with the delivery of sanctions, a deterrence approach generally frames people's relationship to the legal system in terms of rewards and costs, which has the consequence of crowding out other possible motivations for following the law, motivations such as personal values (Deci, 1975; Frey, 1994; Frey & Oberholzer-Gee, 1997). People become more focused on the potential risk of being caught and punished for wrongdoing, and they evaluate their actions in these instrumental terms. They become less focused on issues of obligation and responsibility, and these value-based considerations become less central to their decisions about how to behave.

Summary

This discussion outlines some of the problems associated with deterrence approaches to creating and maintaining social order. In my experience, legal authorities are aware of and acknowledge the problems associated with deterrence approaches to social control. They indicate, however, that there is no better system. And deterrence approaches have the virtue of delivering desirable behaviors, even if minimal influence is delivered at high cost. Hence, authorities often decide to live with the troubling side effects. The issue I want to address is the assumption underlying this approach—the argument that there is not a better system. I will argue that there is a better system, and I will base my argument on psychological research. That research develops out of a long-standing tradition of social psychology rooted in the work of Lewin.

Cooperation in the Lewinian Tradition

Cooperation, in the form of rule following, is conceptualized in the tradition of motivational research begun by Lewin (Gold, 1999) and central to the Research Center for Group Dynamics inspired by that research. In Lewin's classic studies, the focus of concern is the behavior of groups of boys, that is, whether their behavior is consistent with group rules. Various types of behavior are considered, including the performance of designated group tasks (making theatrical masks) as well as inappropriate behavior such as aggression toward others in the group. In the studies, leaders sought to encourage or discourage these behaviors by using a variety of styles of motivation, including authoritarian and democratic leadership. Lewin focused his own attention primarily on issues of aggression and scapegoating, whereas the focus here will be on the performance of group tasks that reflects rule-following behavior. The focus on group performance carried forward as an important aspect of the agenda of the Research Center for Group Dynamics inspired by the work of Lewin and his students. This discussion will focus on rule adherence rather than the motivation of productivity, although as noted the ability to motivate community members to work on behalf of legal authorities is recognized as equal in importance to the ability to secure rule adherence.

A key experimental distinction introduced in the Lewinian research approach was between behavior while the leader was present and behavior when the leader was absent. It was found that when the leader of an autocratic group left the room, the behavior of the boys changed. When the leader was democratic, this change did not occur. Lewin argued that democratic leadership, which was participatory, engaged the internal motivations of the boys, so their behavior was no longer linked to the presence of the external forces represented by the leader. Instead, behavior flowed from internal motivations. Furthermore, and central to Lewin's concerns, the boys were not suppressing negative feelings in the presence of the democratic leader, so there was less motivation for them to be aggressive toward other boys when the leader left.

In other words, field theory identified two sources of motivation. The first was external and reflected the contingencies in the environment. Lewin recognized that the environment shaped behavior by altering the costs and benefits associated with various types of behavior. The second type of motivation was internal and was shaped by the traits, values, and attitudes of the person. These were the motivational forces developing from sources within people and reflecting their own desires. When external contingencies are strong, individual differences in behavior do not emerge. Conversely, in the absence of strong external pressures, behavior reflected people's attitudes and values.

As was the case in the Lewinian approach, the argument here is that the engagement of internal, or value-based, motivations is especially key when the type of cooperation of concern is behavior that will occur outside of the surveillance of authorities. Such behavior is voluntary in the sense that it is not a reflection of the contingencies of the external environment. The leader is not present and cannot either reward or punish behavior. Hence, the behavior that occurs is a more direct reflection of the internal attitudes and values of the boys. Similarly, in legal settings, much of the behavior that occurs is outside the observation of legal authorities and, hence, is not a response to the fear of punishment.

The present analysis is broadly framed using the field theory model in two ways. First, as is true with the work of Lewin, this analysis of people's actions views behavior as a reflection of two factors: external (instrumental) and internal (value-based) motivations. Second, and again as articulated by Lewin, the key issue is the mix of these motivations. Finally, this analysis distinguishes between those behaviors that are and those that are not voluntary, that is, behaviors that do and do not occur in settings in which behavior is being observed and those who engage in it are aware that incentives and sanctions will be shaped by their actions.

Societal Implications

The Lewinian approach to studying social settings developed following the Second World War and was directly influenced by the contrast between authoritarian and democratic styles of group leadership. In the aftermath of that war, the dangers of autocratic leadership were apparent, and Lewin and his group of researchers sought to emphasize the virtues of democracy. The core such virtue is the ability to motivate voluntary action because people are psychologically involved in groups and act based on their own internal attitudes and values to assist those groups.

An argument in favor of the virtues of voluntary engagement does not fit well with the emphasis on command-and-control strategies that is widely found within writing about law and management or with the dominance of the deterrence approach to social control in the area of law and the corporate world. Hence, one benefit that is derived from adopting the broader framework of motivation articulated here is that it draws attention to the benefits of engaging people in groups in ways that support voluntary cooperation.

Such an argument fits well with the increasing recognition that groups, organizations, and communities benefit from more active engagement than can be motivated by incentives and sanctions. Within law and public policy, the virtues of citizen participation are recognized, whereas in political science the importance of social capital is noted. In management many studies focus on the need for extra-role behavior that moves beyond doing one's job. And in education the centrality of intrinsic motivation to active learning is a key issue. All of these literatures point to the value of having a broader conception of motivation.

Value-Based Approaches to Regulation

The value-based model of regulation draws on psychological studies of legitimacy and morality (Tyler, 2006a, 2006c). These studies show that values can shape people's rule-following behavior, which suggests that it is possible to motivate rule following in a different way. The core issue is whether these findings do, indeed, make a compelling case that a value-based model will be effective.

A model of regulation based on values focuses on legitimacy and morality. Legitimacy refers to the view that "the actions of an entity are desirable, proper, or appropriate within some socially constructed system of norms, values, beliefs and definitions" (Suchman, 1995, p. 574), and thus feelings of legitimacy are expected to be related to adherence to rules and policies. A typical statement used to operationalize legitimacy is "If a judge makes a decision, you should follow it." Moral values are important through moral value congruence. Moral value congruence is the belief that the behavior the law proscribes is consistent with or inconsistent with moral values concerning right and wrong. We might, for example, ask people to agree or disagree with the statement "Your moral values and the values in the law are usually quite similar."

The focus of this chapter is on the first of these two values: legitimacy. Legitimacy is especially important to authorities because it provides them with some degree of freedom of action (Tyler, 2006a). If they are legitimate, they are deferred to whatever their decisions are, within a particular scope of authority. With moral values the authorities are most likely to gain value-based deference when their actions accord with people's prior moral values. As I will argue next, however, both values play an important role in a value-based, self-regulatory strategy.

The value-based model represents an alternative approach to encouraging rule following because it focuses on peoples' intrinsic motivations. It identifies rule following as originating with an individual's intrinsic desire to follow rules and not with external contingencies in the environment that are linked to rule following. That desire is the motivation to behave in accord with one's values.

Summary

As noted earlier, the value-based approach examines the influence of two judgments regarding one's group, community, or organization: (a) the perceived legitimacy of rules and authorities and (b) the congruence of those rules with one's moral values. Congruence between rules and an individual's moral values should also motivate adherence, as people strive to follow their inclinations to do what they feel is morally right. The value-based model argues that the concerns embodied in these two judgments intrinsically motivate people to feel a personal responsibility and desire to bring their behavior into line with the rules and policies of a group or organization. Furthermore, these values are suggested to be more important than the instrumental issues of risk that have been outlined.

The second argument is that the procedural justice of groups, organizations, and societies shapes values. These groups typically have hierarchical structures, with authorities and institutions to make and enforce rules. Procedural justice is the fairness through which these authorities and institutions create rules and make decisions. Research suggests that the justice of the procedures used to exercise authority is the key factor shaping the legitimacy of leaders and institutions. These procedural issues are argued to be more powerful than the evaluations of personal outcomes.

The procedural justice argument suggests that issues of rule of law are central to legitimacy (Tyler, 2000, 2006a, 2006c). Two core elements of procedural justice are also central to discussions of the rule of law. First, decisions should be made in neutral, transparent, and rule-based ways. Second, respect should be accorded to people and to their rights. When authorities exercise their authority following these principles, they are widely viewed as legitimate and entitled to be obeyed. In addition, research has suggested that when authorities follow fair procedures, people are more likely to believe that their decisions and policies are consistent with their moral values (Tyler & Blader, 2000, 2005; Tyler, Callahan, & Frost, 2007).

Empirical Studies of Values

To explain the role of values in shaping compliance, I will first discuss studies of everyday compliance with the law among people who have personal experiences with police officers and judges. I examined this question in a study using interviews with a random sample of the residents of Chicago (Tyler, 2006c). In the interviews people were asked about their law-related values, their perception of the risk of being caught and punished for rule breaking, and their everyday compliance with the law. The results suggest that both risk and values shape compliance. Furthermore, both moral value congruence and legitimacy are more influential than risk assessments. These findings are typical of those from studies of this type, and they suggest both that values shape compliance and that their influence is equal to and even greater than the influence of risk judgments.

The second argument is that procedures influence legitimacy and facilitate decision acceptance by legitimating the solutions reached. To examine this argument, we can look at studies of personal experiences with legal authorities. Such experiences are explored in the already noted Chicago study as well as in Tyler and Huo (2002). Tyler and Huo (2002) examined people's experiences in a sample of 1,656 personal interactions with legal authorities in Oakland and Los Angeles. They found that people were most likely to deal with the authorities because they had contacted them to solve a problem. Irrespective of whether they contacted the authorities or vice versa, however, undesired outcomes occurred approximately 30% of the time. Our concern is with people's willingness to defer to the decisions of those authorities, particularly when the outcomes of the experience were negative.

In Tyler and Huo (2002), people were interviewed about their experience—the justice of the procedures, the desirability of the outcome, and their deference to the decisions made. The results suggest that the legitimacy of the authorities and people's willingness to accept their decisions were primarily a consequence of the fairness of the procedures the authorities used to make their decisions. The valence of the decision had very little influence on people's willingness to accept it (also see Tyler, 2006a). Subsequent panel studies extended this finding and showed that legitimacy actually increased preexperience levels when people received an undesirable outcome through a fair procedure (Tyler & Fagan, 2008).

This procedural justice argument has been supported in other studies of people's general rule-following behavior, as well as in research on decision acceptance. A study of New Yorkers, for example, found that the primary factor shaping both judgments of police legitimacy and the willingness to work with the police to fight crime was an assessment of the fairness of the procedures used by the police within the respondent's respective communities (Sunshine & Tyler, 2003).

These findings suggest that we can use fair procedures as a way to increase both rule following and decision acceptance. Fair procedures lead people to view authorities as more legitimate, leading to rule adherence. The effects of procedural justice are particularly important because they are found to encourage voluntary deference rules, as well as deference that is long term in nature. Because the rule adherence that flows from procedural justice and legitimacy is motivated by values, it is not shaped by the risk of detection but maintained by people's own internal values.

Interestingly, these studies also suggest that when authorities use fair procedures, people evaluate those decisions and policies as being more consistent with their own moral values. Hence, people's second relevant value—their own sense of what is right and wrong—is also shaped by the fairness of the procedures used by authorities.

Values as a Basis for Social Regulation

The arguments underlying value-based regulation can be used as a model for addressing several important regulatory issues that have emerged in recent years. One example is the regulation of employees. Another is the regulation of agents of social control (the police, the military). In each case the issue is the same: How can we best encourage rule adherence?

Employees

During the past several years, issues of corporate wrongdoing have led to increasing concern about how it is possible to get employees to follow organizational policies and rules of appropriate conduct. I have studied this issue among several groups of employees, including a sample of New York–based employees (Tyler & Blader, 2000), a study of corporate bankers (Tyler & Blader, 2005; Tyler, Dienhart, & Thomas, 2008), and a national sample of employees (Tyler & Blader, 2005; Tyler et al., 2008). In each case the basic approach was the same: We interviewed employees about their values, the characteristics of their workplace, and their rule-related behavior in work settings.[1]

What do these studies tell us about the possibility of motivating employees to follow workplace rules by means of their views about the legitimacy of management and the morality of company policies? The studies find that rule following is consistently shaped by judgments about both legitimacy and moral value congruence. Furthermore, these value-based influences are both stronger than the influence of the risk of being caught and punished for rule breaking and stronger for voluntary deference than for compliance. Relative influences on rule adherence are shown in Table 11.1.

Table 11.1 The Influence of Values on Rule Adherence in Work Settings

	Corporate Bankers	Sample of U.S. Employees
Beta weights		
Legitimacy	.29***	.46***
Moral value congruence	.19***	.17***
Incentives/sanctions	.01	.15***
Adjusted R-squared	15%***	36%***
N	540	4,430

Note: DV = rule adherence.
***$p < .001$.

Table 11.2 Procedural Justice and Values in Work Settings

	Corporate Bankers		Sample of U.S. Employees	
	Values	Deference	Values	Deference
Beta weights				
Procedural justice	.24***	.28***	.55***	.31***
Distributive justice	.02	.03	.05**	.08***
Outcome favorability	.00	.03	.06**	.01
Adjusted *R*-squared	19%	7%	40%	10%
N	540	540	4,430	4,430

Note: DV = values; deference.
**$p < .01$, ***$p < .001$.

These results suggest that employee judgments about legitimacy and moral value congruence have an important influence on employee behavior in the workplace. Legitimacy matters! If people feel that they have a personal responsibility to accept and follow rules and think those rules are morally right, they will follow those rules voluntarily.

To test the argument that procedures matter, we can contrast procedural justice to nonfairness-related judgments of outcome valence (outcome favorability, outcome fairness). Does procedural justice matter? The results shown in Table 11.2 suggest that it does. It shapes both values and deference to rules. So the justice of organizational procedures does shape employee values and influence employee willingness to defer to rules. Those employees who believe that their organization is managed with fair procedures think its managers are more legitimate and entitled to be obeyed and its policies are more moral. And, of particular importance, its employees follow the rules more frequently. In particular, they are more likely to voluntarily defer to the rules.

Finally, we can examine which procedural elements shape overall evaluations of procedural justice. On the basis of Blader and Tyler (2003a, 2003b), we identify two elements of procedures—quality of decision making and quality of interpersonal treatment—and two levels on which each might be important—in one's immediate group and for the overall organization. The results are shown in Table 11.3. They indicate that people distinguish between the two issues and two levels. Table 11.3 indicates that all four elements of procedures make an independent contribution to assessments of the overall fairness of the workplace.

Agents of Social Control

We can extend the consideration of how to motivate rule adherence to a second arena that has also proved problematic in recent years—agents of social control (the police, the military). The core issue is similar in the

Table 11.3 The Components of Procedural Justice in Work Settings

	Single Corporation Study	Multicompany Study
Beta weights		
Quality of decision making (org.)	.19*	.26**
Quality of decision making (super.)	.22*	.31**
Quality of treatment (org.)	.43***	.24*
Quality of treatment (super.)	.12*	.25*
Adjusted R-squared	66%***	79%***

Note: DV = procedural justice of the workplace.
$^{*}p < .05$, $^{**}p < .01$, $^{***}p < .001$.

case of both the police and the armed forces. Society creates authorities to enforce rules. But who then controls those authorities and brings their behavior into line with rules and policies? When employees deviate from rules, financial scandals ensue. When agents of control violate the rules, people are injured or killed. Hence, in this case society has the same interest in understanding how to motivate rule adherence as it does with employees in general, but with this particular class of employee, the consequences of noncompliance are sometimes more immediate and dire.

Tyler et al. (2007) used a sample of law enforcement officers and soldiers as the basis for a study of these issues. In the study, 419 social control agents completed questionnaires about their values, the characteristics of their workplace, and their rule adherence behavior.

As with employees in general, a comparison of the influence of legitimacy, morality, and risk on policy adherence suggests that legitimacy and morality are the central factors shaping policy and rule following (see Table 11.4). If people feel that they have a personal responsibility to accept and follow rules, then they will follow those rules voluntarily. In this setting it is legitimacy that is the central value of importance, and the influence of moral value congruence is weaker.

Table 11.4 Values and Rule Adherence Among Agents of Social Control

	Comply	Defer
Beta weights		
Legitimacy	.40***	.44***
Moral value congruence	.06	.14**
Probability of detection	.18***	.18***
Adjusted R-squared	26%***	37%***

Note: DV = rule adherence.
$^{**}p < .01$, $^{***}p < .001$.

Table 11.5 Procedural Justice, Values, and Voluntary Behavior Among Agents of Social Control

	Values	Voluntary Behavior
Beta weights		
Procedural justice	.42***	.38***
Distributive justice	.14**	.06
Outcome favorability	.05	.07*
Adjusted *R*-squared	31%***	22%***

Note: DV = values, rule-following behavior.
*$p < .05$, **$p < .01$, ***$p < .001$.

Does procedural justice shape values in this setting? The results shown in Table 11.5 suggest that it does. So the justice of organizational procedures shapes values and influences willingness to defer to rules among agents of social control. Those police officers, federal agents, or members of the military who believe that their organization is managed with fair procedures think its managers are more legitimate and entitled to be obeyed and its policies are more moral. And, of particular importance, its members follow the rules more frequently. Of special interest is the finding that they are more likely to voluntarily defer to the rules.

Finally, we can examine which procedural elements shape overall evaluations of procedural justice. The results are shown in Table 11.6. They indicate that people distinguish between two issues and two levels. The issues are quality of decision making and quality of interpersonal treatment. The levels are their immediate workgroup and the larger organization. Table 11.6 indicates that all four elements of procedures make an independent contribution to assessments of the overall fairness of the workplace.

Table 11.6 Components of Procedural Justice Among Agents of Social Control

	Procedural Justice
Beta Weights	
Quality of decision making (org.)	.16*
Quality of decision making (super.)	.23**
Quality of treatment (org.)	.24**
Quality of treatment (super.)	.34***
Adjusted *R*-squared	69%

Note: DV = procedural justice.
*$p < .05$, **$p < .01$, ***$p < .001$.

Discussion

Earlier studies in the area of everyday law-related behavior highlight the important role of social values in encouraging citizen compliance with the law (Tyler, 2006c). It has been shown that people are more likely to comply with laws when they feel that legal authorities are legitimate and ought to be obeyed. The findings outlined here support this argument and extend it to two different arenas—employees and the agents responsible for maintaining social control. This extension is especially striking because the work arena is one in which the influence of values has traditionally been downplayed in favor of alternative instrumental or "rational" approaches.

The current findings also extend previous work by considering the social value not only of legitimacy but also of moral value congruence (i.e., the match between the person's moral values and those of the organization). When those within an organization feel that the values of their organization are congruent with their own, their own motivation to behave morally leads them to follow organizational rules out of their intrinsic motivation to behave appropriately. These findings are parallel in studies of general work settings and with agents of social control. Both types of studies show that employees' views about the congruence of organizational policies with their personal moral values shape their rule following (Tyler & Blader, 2000, 2005).

In addition to the empirical support for the utility of the value-based strategy reported here, such an approach has additional benefits over a command-and-control strategy. For instance, it prevents organizations from expending resources on creating and maintaining credible systems of surveillance to enforce rules. These problems are typical of any efforts to regulate conduct using incentive- or sanction-based strategies. Exacerbating this problem, such strategies actually encourage people to hide their behavior and thus make it necessary to have especially comprehensive and costly surveillance systems.

A core issue within organizational psychology is the distinction between selection and training. The selection approach suggests that the key to organizational viability is to hire people with values that fit the organization. The training approach argues that the experiences people have within the organization can shape their attitudes, values, and behaviors. Those experiences can potentially develop from people's general experiences with the structures and processes of the organization or can result from particular training programs. The results outlined here support the argument that the general structures and processes of organizations shape the people within those organizations. Even in the case of moral values, I suggest the structures and processes of organizations have an impact.

The findings suggest the importance of one particular aspect of organizations: the justice of their rules and procedures. Consistent with prior findings in the literature on procedural justice, rule following and, in particular, deference to rules are linked to the fairness through which authorities act. In contrast to the many studies showing procedural justice effects among the general public (Tyler, 2006c; Tyler & Huo, 2002), however, this study shows that the manner in which both employees in general and agents of social control in particular experience their own work organizations shapes their behavior in relationship to social rules. The findings, in fact, suggest general support for a value-based model of cooperation (Tyler & Blader, 2000).

Consider two approaches that authorities might potentially use to try to secure rule adherence. One approach would be to stress the punishments associated with being caught engaging in rule breaking or the benefits of adhering to the rules in terms of promotion. A second approach would be to focus on creating fair procedures within the organization so that people feel that decision making follows just procedures and that people are treated with dignity and respect. These results clearly point to the value of the second approach.

Of course, we also have to ask if we might try both approaches at the same time. One persistent difficulty has been the finding by psychologists and economists that the use of incentives and sanctions undermines values. Frey (1994) used the term "crowding out" to describe the finding that people who become focused on issues of reward and cost are less strongly influenced by their values. To the extent this is true, then the two approaches are not easily compatible.

A second way to combine the two approaches is to consider them as a funnel (Ayres & Braithwaite, 1992). From this perspective we approach everyone on a values basis, seeking to engage them through this model of rule following. Most people can and will be engaged. A small group, however, will lack the values on which to base rule following, and those people will persist in rule breaking when opportunities arise. This group must then be dealt with by means of sanctions and rewards. Because a small group has been identified, however, instrumental models can be more easily deployed to deal with their behavior. From this perspective the problem with beginning with sanctioning is that it creates a relationship with everyone that is based on risk perceptions, leading authorities to have to manage their entire regulatory approach on this basis.

The argument advanced here is for a broader view of the antecedents of rule-following behavior for everyone, that is, people in neighborhoods, employees in work settings, agents of social control, and so on (Tyler, 2007). This approach looks at the influence of both instrumental and value-based motivations in shaping rule-following behavior. The results presented suggest that the consideration of both models together better explains such behavior than is possible with either model taken alone.

The view presented here includes not only the motivations traditionally studied, motivations that are linked to sanctions and incentives, but also value-based motivations for following group rules. These value-based motivations are linked to concerns about acting in fair ways in work settings. The case for this broader model rests on the finding that in two different studies, employees were found to be motivated in their rule following by their social values concerning legitimacy and morality. These findings suggest that we would be better able to understand rule-following behavior in work organizations, as well as other settings, if we adopted a broader model of human motivation that added an account of value-based motivations to our models of behavior.

Within the context of this message, it is especially important to note the breadth of the findings. The first setting studied involved both high-level corporate banking employees and a broad sample of workers. The second setting considered agents of social control—police officers, federal agents, and members of the armed forces. In all of these contexts, values mattered, and procedural justice shaped values.

Legal institutions are designed based on the assumption that behavior is shaped by the risk of sanctioning. As a result, there is a fundamental misalignment of the organization, in this case the legal system, and models of motivation, leading the system to be less efficient and effective than might potentially be the case.

One approach that might potentially be taken to the limits outlined is to shore up the effectiveness of instrumental approaches. A strategy that is often used is to increase the magnitude of gains and losses. For example, law enforcement authorities often increase the severity of punishment for wrongdoing. Such strategies are shown by research to be ineffective, because the probability of punishment has a more important influence on wrongdoing than does the severity of punishment. Authorities may view this implementation model as desirable, however, because, like the instrumental approach itself, increasing severity is something that is under the control of authorities. To increase the likelihood of punishment, authorities would need to be able to deploy more resources, which they may not be able to do.

Another approach is to more effectively deploy available resources. Again, using the example of deterring wrongdoing, police resources can be deployed in response to crime risks in a more direct manner than is commonly the case (Tyler, 2007). Rather than deploying the police based on political or economic considerations, police can be deployed based on actual crime rates.

Similarly, management models such as pay for performance may increase productivity by more directly linking productivity and the delivery of incentives. Although it might seem obvious that that linkage would exist, one common critique of CEO compensation is that it is not linked to company performance, with social factors such as connections between the CEO and a member of the board better accounting for CEO compensation. And discussions of public sector employees such as the police commonly note the lack of connection between performance and compensation in a sector where pay is heavily based on seniority and union-negotiated pay scales.

The approach to increasing the motivation to cooperate taken in this chapter is not to strengthen instrumental approaches but instead to broaden the conception of what motivates employees. By including value-based motivations in the overall motivational framework, researchers can better explain why people cooperate. The implication for organizational design is that there needs to be a focus on creating the organizational conditions conducive to promoting social motivations.

This argument is based on the distinction in utility functions between expectancy and value. The judgment and decision-making literature has made clear in the past several decades that there is a great deal to be gained by exploring the individual's thought processes, that is, by developing the expectancy aspect of utility. Hence, we are not much more aware of the heuristics and biases that can distort the decision making of people seeking to act in their own immediate material interests.

This chapter suggests that there is a similar benefit to developing the second aspect of the utility model—our understanding of what people value. In other words we need to create an expanded version of what it is that motivates people in social settings. Although people are motivated by material incentives, such as opportunities for pay and promotion, and seek to avoid losses, such as sanctions for rule breaking, they are also motivated by a broader set of issues, issues loosely collected here and labeled "value-based motivations."

Value-based motivations are distinct from instrumental motivations, conceptually, and as a consequence they have distinct strengths and weaknesses. A distinct strength is, as has been noted, that they do not require organizational authorities to possess the ability to provide incentives for desired behavior or to be able to create and maintain a credible system of sanctions. At all times groups benefit from having more resources available that can be directed toward long-term group goals. If everyday group actions are shaped by self-regulating motivations, groups have more discretionary resources.

And as the findings outlined make clear, value-based motivations are important because they are more powerful and more likely to produce changes in cooperative behavior than are instrumental motivations. Hence, value-based motivations are both more powerful and less costly than are incentives and sanctions. Of course, this does not mean that value-based motivations can be immediately and automatically deployed in all situations.

A weakness of value-based motivations is that they cannot be quickly activated within any social context. A CEO with a million-dollar war chest can create an incentive system to motivate behavior in desired directions overnight. Conversely, a city can shift its police patrols around to vary the nature of the threat faced by community residents. Such flexibility is a major advantage of the instrumental system. Value-based motivations must be developed over time as the culture of an organization is created. Hence, a long-term strategy is needed to build an organization based on value-based motivations.

A strategy based on social motivation also has the disadvantage of taking control away from those at the top of the social hierarchy. If a group relies on voluntary cooperation, its leaders need to focus on the attitudes and values of the people in the group. For example, they have to create work that people experience as exciting. Furthermore, they have to pursue policies that accord with employees' moral values. These aspects of value-based motivation create constraints on the actions of leaders.

It is natural that some leaders would prefer a strategy in which they are the focus of attention, irrespective of its effectiveness, to one in which they focus their attention on employees. Yet within business organizations, a focus on the customer is a widely institutionalized value. Similarly, the concept underlying democratic processes is that within communities, policies ought to be a reflection of the values of the members of those communities. Hence, it is hardly a radical suggestion that organizations benefit when they develop their policies and practices in consultative ways that involve all of the relevant stakeholders, including leaders, group members, and external clients such as customers. And, of course, most leaders may simply prefer the most effective strategy, which is a value-based approach.

Aspects of procedural justice feed directly into the need to make group policies and practices consistent with the attitudes and values of group members. Participatory decision making and consultation at all levels are mechanisms through which people's views are represented.

Ironically, those constraints may often have additional value for groups. The era of corporate excess makes clear that the power of those in high management, when unchecked, does not always end up serving the interests of the company. Hence, the need to be accountable to others within the organization may have valuable benefits for the group and may check the tendency of leaders to engage in unwise actions. Just as "checks and balances" is frequently held out as one of the primary desirable design features of U.S. government, the balancing of policies and practices among stakeholders has the benefit of restricting any tendency toward excesses.

Images of Human Nature

The key argument in this chapter is that people are motivated by value-based motivations. This message emerges clearly from the analyses reported. Across variations in the form of the analysis, value-based motivations are consistently found to have an influence on cooperation, intention to remain in the group, and well-being. Furthermore, this influence is typically larger in magnitude than is the influence of instrumental influences. As a consequence, the effective design of organizations will be enhanced if that design is based on an awareness of the organizational factors encouraging social motivations.

The value of value-based motivations emerges across the forms of cooperation considered. The gains of value-based motivation, however, are particularly relevant given the need for voluntary cooperation. Within the study of law and regulation, for example, it has been increasingly recognized that we want more from people than compliance. We also want active cooperation, with people working both with the police and with others in their community to fight crime. In management, the virtues of voluntary extra-role behavior are touted in firms ranging from Silicon Valley start-ups to large corporations such as Apple and Microsoft. As work becomes less physical and more intellectual and social in nature, it is increasingly important that people move beyond simply doing their jobs to being creative and innovative in work settings. And it is important that we recognize the need for people to work actively with others to develop consensual community-based policies and to create the social capital in communities to implement them. Given the problems that authorities are seeking to solve in modern organizations, social motivations are especially central, because those problems require voluntary cooperation on the part of the members of groups, organizations, and communities.

Note

1. Of course studying the rank and file does not address all issues of misconduct, because in some settings the issue is top management. It does address those issues in which there is widespread misconduct that involves breaking rules of appropriate conduct.

References

Ayres, I., & Braithwaite, J. (1992). *Responsive regulation: Transcending the deregulation debate.* Oxford, UK: Oxford University Press.

Blader, S., & Tyler, T. R. (2003a). A four-component model of procedural justice: Defining the meaning of a "fair" process. *Personality and Social Psychology Bulletin, 29,* 747–758.

Blader, S., & Tyler, T. R. (2003b). What constitutes fairness in work settings? A four-component model of procedural justice. *Human Resource Management Review, 12,* 107–126.

Blumstein, A., Cohen, J., & Nagin, D. (1978). *Deterrence and incapacitation.* Washington, DC: National Academy of Sciences.

Coglianese, C. (1997). Assessing consensus: The promise and performance of negotiated rulemaking. *Duke Law Journal, 46,* 1255–1349.

Deci, E. L. (1975). *Intrinsic motivation.* New York: Plenum.

Frey, B. S. (1994). How intrinsic motivation is crowded in and out. *Rationality and Society,* 6, 334–352.

Frey, B. S., & Oberholzer-Gee, F. (1997). The cost of price incentives. *American Economic Review, 87,* 746–755.

Gold, M. (1999). *The complete social scientist: A Kurt Lewin reader.* Washington, DC: American Psychological Association.

MacCoun, R. J. (1993). Drugs and the law: A psychological analysis of drug prohibition. *Psychological Bulletin, 113,* 497–512.

One in 100: Behind Bars in America 2008. (February, 2008). The Pew Center on the States.

Sampson, R. J., Raudenbush, S. W., & Earls, F. E. (1997). Neighborhoods and violent crime. *Science, 277,* 918–924.

Suchman, M. C. (1995). Managing legitimacy: Strategic and institutional approaches. *Academy of Management Review, 20,* 571–610.

Sunshine, J., & Tyler, T. R. (2003). The role of procedural justice and legitimacy in shaping public support for policing. *Law and Society Review, 37,* 513–548.

Tyler, T. R. (2000). Social justice: Outcome and procedure. *International Journal of Psychology, 35,* 117–125.

Tyler, T. R. (2006a). Legitimacy and legitimation. *Annual Review of Psychology, 57,* 375–400.

Tyler, T. R. (2006b, Winter). What do they expect? New findings confirm the precepts of procedural fairness. *California Court Review,* 22–24.

Tyler, T. R. (2006c). *Why people obey the law.* New Haven, CT: Yale University Press.

Tyler, T. R. (2007). *Psychology and the design of legal institutions.* Nijmegen, the Netherlands: Wolf Legal.

Tyler, T. R., & Blader, S. L. (2000). *Cooperation in groups: Procedural justice, social identity, and behavioral engagement.* Philadelphia: Psychology Press.

Tyler, T. R., & Blader, S. L. (2005). Can businesses effectively regulate employee conduct? The antecedents of rule following in work settings. *Academy of Management Journal, 48,* 1143–1158.

Tyler, T. R., Callahan, P., & Frost, J. (2007). Armed, and dangerous(?): Can self-regulatory approaches shape rule adherence among agents of social control. *Law and Society Review, 41,* 457–492.

Tyler, T. R., Dienhart, J., & Thomas, T. (2008). The ethical commitment to compliance: Building value-based cultures that encourage ethical conduct and a commitment to compliance. *California Management Review, 50*, 31–51.

Tyler, T. R., & Fagan, J. (2008). Legitimacy and cooperation: Why do people help the police fight crime in their communities? *Ohio State Journal of Criminal Law, 6*, 231–275.

Tyler, T. R., & Huo, Y. J. (2002). *Trust in the law*. New York: Russell Sage Foundation.

CHAPTER 12

Justice in Aboriginal Language Policy and Practices

Fighting Institutional Discrimination and Linguicide

STEPHEN C. WRIGHT
Simon Fraser University

DONALD M. TAYLOR
McGill University

Abstract: Nearly two decades of research in Arctic Quebec (Nunavik) has yielded strong evidence of the costs of dominant-language-only education in the development of Inuit children. The authors' research shows that dominant-language-only education, compared to a heritage-language program, negatively affects Inuit children's self-esteem, undermines heritage-language development, and has little or no long-term benefit for second-language acquisition. In this chapter, the authors use this research as the foundation for considering the broader justice implications of the continued failure to recognize the critical role of language practice in the education of Aboriginal children. The authors consider social psychology's general lack of attention to discrimination on the basis of language and conclude, on the basis of their research evidence, that continued use of dominant-language-only programs represents as a case of institutional language discrimination that, in the case of Aboriginal children, can also be justly described as part of an ongoing process of linguicide.

Precolonial North America boasted hundreds of unique and vibrant First Nations, Inuit, and Métis groups, each with their own unique language. From the many Algonquian groups in the East, to the Cree-speaking communities spanning the middle of the continent, to the Wakashan groups along the Pacific coast, to the Inuit scattered over the vast Arctic, what is now Canada was a land of stunning linguistic diversity. Sadly, the Aboriginal linguistic landscape of today is much less vibrant. Some languages are entirely gone, with the last known speakers passing on years, even decades, ago. Many others hang on the brink, with the number of fluent speakers dwindling and the use of the language in daily life reduced to a minimum. Despite heroic efforts by many First Nations, Inuit, and Métis groups to preserve and strengthen what remains, the decline in language vibrancy continues (see Statistics Canada, 2007), and the future remains bleak. Without clear changes to current trends, most experts foresee only one or two Aboriginal languages having a reasonable chance of long-term survival (see Norris, 2007).

This picture of language loss is cause for grave concern. Language is not only a vehicle for communication but the medium by which cultural knowledge is transmitted. But more than this, a language holds in its vocabulary, structure, and delivery the traditions, views, and ways of thinking that *are* a group's cultural uniqueness. Aboriginal groups have recognized this deep connection, claiming, for example, that their "languages embody the past and the future … carrying unique, irreplaceable values and spiritual beliefs that allow speakers to relate with their ancestors and to take part in sacred ceremonies" (Task Force on Aboriginal Languages and Cultures, 2005, p. 22). Although a person can certainly be Inuit and not speak Inuktitut, many Aboriginal communities recognize heritage language can be a critical component of identity. A lack of language competence can be experienced as a personal loss, a community loss, and ultimately a loss for the entire group.

The collective nature of language makes its loss in Aboriginal communities even more worrisome. When Italian, Greek, or Portuguese families immigrate to Canada, and their children lose the heritage language, it is a loss for the child and the family. When Aboriginal children lose their heritage language, it is not only a loss for the child and the family but a cultural tragedy. Many linguistic groups can withstand some language loss in one geographic location because they are part of a larger, linguistically vibrant group. Italy, Greece, and Portugal, as nation-states, ensure the vitality of their languages. For Aboriginal groups, Canada is the cultural and linguistic homeland. If the language is lost here, it is entirely lost. The authors of the *Atlas of the World's Languages in Danger of Disappearing* (Wurm, 1996) asserted that a language can be considered in serious danger of disappearing if it is not the mother tongue for at least 30% of the children in the community. If this assessment is correct, the claims of Aboriginal leaders (e.g., Assembly of First Nations, 1988, 1990; Task Force on Aboriginal Languages and Cultures, 2005) that they are involved in an 11th-hour struggle to pull their languages back from the brink are not exaggerated. According to 2006 Canada Census data, only approximately 24% of First Nations claim an Aboriginal language as their mother tongue, and these Aboriginal language speakers are highly concentrated in a small number of languages (Statistics Canada, 2007).

The story of disappearing languages should be a source of disquiet not only for those who speak them but also for all who lament the loss of unique cultural contributions. A language is certainly one of a society's most complex, unique, and important cultural contributions. Its destruction represents the elimination of what is arguably the apogee of cultural achievement.

But what does this have to do with justice? Specifically, what does this have to do with the social psychology of justice? First, the loss of Aboriginal languages is clearly part of the broader colonization project that accompanied European settlement of North America. Like Aboriginal cultures generally, language was directly targeted as part of the assimilationist agendas of governments, churches, and schools. This agenda was implemented with draconian and violent tactics, including, but not restricted to, forced removal of children from their communities and placement in residential schools, severe corporal punishment for using their heritage languages, and a host of coercive and abusive means of ensuring they spoke only English. The decline of Aboriginal languages has been no accident, no benign or inevitable cultural evolution, but rather it is to a great degree the result of injustices perpetrated by a colonial oppressor, made obvious by the recent formal apology made to Aboriginal people by Canada's prime minister. This moves the issue of Aboriginal language loss into the realm of social justice.

We argue, however, that the connection between Aboriginal language loss and social justice does not rest only in the past. The relevance of social justice did not end with the closing of residential schools. We will use data from a longitudinal research project in an Inuit community in Arctic Quebec (Nunavik) to argue that current educational policies and practices have a significant impact on the development of Aboriginal children and that failure to recognize the impact of harmful policies and to alter them should be understood as an issue of social justice.

Kurt Lewin is often credited with introducing the idea that social psychology can often bring a valuable focus to real-world social issues while at the same time considering real-world social issues can expose shortcomings in current theoretical and empirical work. We argue that the current discussion is an excellent case of this Lewinian perspective. Bringing concepts and theoretical perspectives from social psychology to bear on the issue can help to focus attention on processes that continue to undermine the vibrancy of Aboriginal, and perhaps other minority, languages. Specifically, the concept of institutional discrimination can be effectively applied to school policies affecting Aboriginal children. In addition, the concept of "institutional language discrimination" can provide a valuable frame for understanding the impact of mainstream society and its agents on Aboriginal children's language development. At the same time, we believe that investigating language loss in Aboriginal communities, framed in social justice terms, also points to some curious gaps in the social psychological literature on discrimination and justice.

Language Discrimination

Social Psychology's Blind Spot

Prejudice and discrimination have been central themes in social psychology for most of its history. Although discussions of language rights and linguistic discrimination can be found in political science (e.g., Kymlicka & Patten, 2003), anthropology (e.g., Mascia-Lees & Lees, 2003), and applied linguistics and education (e.g., Brutt-Griffler, 2002; Skutnabb-Kangas, 2002), social psychology, has virtually ignored prejudice and discrimination against groups defined by linguistic differences. Often distinctiveness in terms of language is subsumed under distinctions in terms of ethnicity, race, nationality, or immigrant status (e.g., Latinos in the United States, Turks in Germany), and these other bases for categorization are judged more relevant. We study racism or ethnocentrism, and limitations on linguistic rights or action and policies that undermine the strength or even the existence of a language are not described as similar to actions that might undermine religious rights or cultural practices. The term *language discrimination* is decidedly absent in the social psychological literature.

Social psychology's inattention to discrimination on the basis of language mirrors the apparent popular perception of language. For example, imagine the public response had the state of California mandated that only Christianity or European history and culture could be taught in public schools. This form of blatant disregard for religious or cultural rights would be readily recognized as discrimination and immediately judged as unacceptable in the current climate of tolerance and diversity. In 1998, however, California outlawed classroom instruction in any language other than English. Proposition 227 passed with a clear majority, and the resulting legislation virtually ended bilingual education in the state. Similar initiatives have followed in other states (e.g., Arizona). Supporters of this form of English-only legislation describe it as beneficial for minority children.

They contend that mandated English-only instruction speeds minority children's acquisition of, and proficiency in, English and that fluency in English has many economic benefits and allows for increased participation in mainstream culture. The veracity of some of these claims will be a critical issue for this chapter. Let's imagine for a moment, however, that these claims are true and that there are economic and acculturative benefits to English fluency and that English-only instruction increases English acquisition. By this logic, if conversion to Christianity or adoption of European culture would reduce impediments to economic advancement and hasten acculturation to the mainstream, then it follows that it is in minority children's interests to pass the legislation mandating Christian- and European-only education in our schools. Interestingly, arguments such as these were common in the 19th and the first half of the 20th centuries and were used to support, for example, mandatory residential schooling of Aboriginal children in the United States, Canada, and Australia. Given recent formal apologies for these schools from the Canadian and Australian governments, however, it appears the popularity of these kinds of assimilationist arguments has diminished. At the same time, rejection of these "benefits of assimilation" arguments appears not to have extended to language. Mandated language assimilation seems acceptable even in these supposedly more tolerant times.

In Canada, we might expect that the national policy of bilingualism and multiculturalism would lead to a different perspective on language. Indeed, the province of Quebec has enacted laws to protect French, and lawmakers highlighted the minority status of the language in Canada (and North America) as the foundation for these special rules. Despite opposition to the specifics of these laws, there appears to be general recognition of using language as a defining characteristic of group identity and protecting the rights of its speaker as a legitimate concern. In fact, opponents of Quebec's laws often describe how these laws violate the language rights of English speakers.

Nonetheless, language assimilation and language loss remain the reality for many ethnolinguistic groups in Canada. As described earlier, most Aboriginal languages in Canada have been badly eroded, and members of many immigrant groups see their heritage language replaced by English or French within one or two generations. Few schools in Canada provide significant instruction in languages other than English or French, even in areas where non-English or non-French speakers are the majority. Thus, although formal assimilationist groups such as the English-only movement in the United States are rare in Canada, normative practices seem to demonstrate little concern for the language loss experienced by many linguistic minorities.

In summary, language, compared to other markers of ethnic or cultural distinctiveness, is generally not recognized by social psychologists or the general public to be an aspect of identity that can be a target of discrimination. Explicitly limiting the use of minority languages and attempting to replace it with the dominant language and labeling speakers "linguistically disadvantaged" or "language deficit" are common practices. Similar devaluation of other aspects of social or cultural identities would almost certainly be interpreted as discrimination.

Why might this be? Perhaps languages are considered only in terms of their functional utility as a tool for communicating and gaining access to material and social rewards. If language is simply a skill or resource, which language one knows and uses should be determined by its economic utility compared to other alternatives. This kind of economic model, however, entirely ignores a language's psychological meaning and its role as an aspect of cultural identity. Nonetheless, if language were a relatively unimportant aspect of a group's identity, efforts to assist the group in replacing it with a "more useful" one might not be seen as discrimination.

Language can be a critical and highly visible aspect of collective identity, however. Research on youth suicide supports this claim in the most graphic of terms. In some Aboriginal communities, including some Arctic Inuit communities, youth suicide rates have reached crisis levels, with some communities reporting rates 10 times the Canadian national average. Chandler and Lalonde (1998), however, showed that a critical but often overlooked point is the enormous variation across communities, with some having no incidence of suicide at all. In a review of communities in the province of British Columbia, Hallet, Chandler, and Lalonde (2007) found that a simple measure of the percentage of the community population who spoke the heritage language was a key predictor of suicide rates. In fact, youth suicide was virtually nonexistent in the few communities in which over half the band members reported conversational proficiency in the heritage language. Importantly, this measure of language proficiency had predictive power over and above six other cultural continuity factors identified in previous research as excellent predictors of youth suicide rates in these same communities. Community competence in the heritage language may be a critical aspect of cultural identity because it contributes directly to what Chandler and Lalonde (2008) described as a sense of ownership of the collective past and commitment to future prospects. Clearly, when this powerful psychological significance of language is ignored in favor of purely economic concerns, it should qualify as discrimination.

Institutional Language Discrimination

Like other forms of discrimination, language discrimination can take several forms. Discrimination can be perpetrated and experienced as primarily *interpersonal.* Here, actions directed at harming or excluding a person because of the language he or she speaks are perpetrated by individuals (alone or in small groups). Ridiculing an individual or refusing to hire someone because of his or her accent or heritage language would be examples of *interpersonal language discrimination.*

We focus on *institutional language discrimination,* however, wherein policies and/or practices enacted by institutions impact negatively on a particular language group. The concept of institutional discrimination is usually described as involving a dominant group introducing structures or procedures that sustain intergroup inequality by encouraging (mandating) behavior that conforms to the current status relations (Sidanius & Pratto, 1999; Wright & Taylor, 2003). Thus, institutional discrimination is not directed at a specific individual, but rather it undermines the opportunities, status, or well-being of all members of the target group who come in contact with the institution. Some discussions have used the term *structural discrimination* to describe cases where the offending policy was "neutral in intent" but nonetheless harmed the minority group (Pincus, 1996, p. 186). Although this distinction is interesting, it is often the case that once a policy or normative practice is in place, those who continue to enact it may be unaware of the reasons for its implementation. The practice continues simply because "this is how we do things around here" or the original prejudice is replaced by much more benign and acceptable legitimizing beliefs (see Sidanius & Pratto, 1999). Thus, whether a policy was inspired by negative intentions may not matter, as institutional discrimination often renders the original negative attitudes unnecessary for continued inequality of treatment.

Thus, one criterion for labeling a practice "institutional language discrimination" is that it must cause demonstrable harm that is disproportionately experienced by members of a particular language group. Even evidence of unequal harm may not be adequate, however, if the harm serves to reduce or prevent an even greater harm that members of the group would almost certainly face. In addition, if there exists no alternative practice that caused less harm, then it may not be appropriate to label the practice "discrimination." Thus, to claim institutional language discrimination, one would need to show, first, that a particular language group is disproportionately harmed by an institutional practice and, second, that there is no long-term benefit conferred by the current practice that could not be conferred by a less harmful alternative.

Is English-Only Instruction Institutional Language Discrimination?

Given this definition, is there a basis for labeling English-only instruction (or instruction exclusively in any societally dominant language) institutional language discrimination? Why might this harm minority-language children in ways not experienced by dominant-language speakers? In many cases, minority-language children in classrooms designed for majority-language speakers receive relatively little support. They are expected to "sink or swim" in what has been labeled "submersion" education. Their limited understanding of the classroom activities, materials, and teacher's requests are likely to lead to greater frustration and anxiety. They may feel greater detachment or exclusion because their group's experiences, values, and culture are less prevalent in (or absent from) the curriculum (Cummins, 1994; Wright & Taylor, 1995). These feelings may hinder their academic performance, their acquisition of the dominant language (e.g., Bougie, Wright, & Taylor, 2003), and even their engagement in school. They can be stigmatized as being less intelligent than dominant-language children, which at times even results in their being overrepresented in special education classes (Cummins, 1994).

One solution that has been offered is ESL (English as a second language) programs. These programs involve a variety of strategies to develop minority-language children's English skills before submerging them in a dominant-language classroom (Baker, 2001). Instruction is usually exclusively in English, but the curriculum is modified to better serve children who do not speak English and to focus on English acquisition. Another solution is training culturally sensitive teachers (Villegas & Lucas, 2002) to reduce the degree to which classrooms reflect only the dominant ethnolinguistic culture. Some have also championed the use of cooperative education, in which children from both cultural groups are encouraged to learn with, and from, each other (Baker, 2001), and parents are included as coeducators (Cummins, 1994). These tactics can all help improve minority-language children's learning experiences. Where there is a large group of minority children, however, others have advocated for heritage-language education (e.g., Cummins, 1989; McLaughlin, 1989; Wright, Taylor, & Macarthur, 2000).

Heritage-language education can take many forms, and a distinction is often made between *transitional* and *maintenance* programs (Baker, 2001). Transitional programs use a combination of the heritage language and the dominant language for a short time, with the intent that children move into dominant-language-only classrooms as soon as possible. The aim is linguistic assimilation with little concern for the fate of the heritage language. Maintenance programs, however, involve more extensive and continued use of the heritage language. The aim is full bilingualism where the child becomes fluent and literate in both languages.

The dominant popular political ideology often dictates which model will be implemented (see Baker & Prys Jones, 1998). In the case of California's

Proposition 227, an exclusive focus on the economic imperative to learn English, accompanied by a belief that submersion in English would facilitate children's learning it, fueled opposition even to transitional heritage-language programs. On the other hand, a strong endorsement of an ideology of multilingualism might lead to a preference for programs that encourage maintenance of the heritage language.

All these programs are intended to reduce the harm done to minority-language children who would otherwise be forced to sink or swim in a mainstream classroom. Any program that uses only the dominant language and even transitional heritage-language programs, however, openly ignores the value of minority languages. These programs imply (or even express) that competence in the minority language is at best irrelevant and at worst a direct impediment to the pursuit of "what really matters." We contend that this direct devaluation of the minority language and the failure to give it full recognition as a medium for educating children who speak it cause harm. In addition, we propose that in the case of Aboriginal languages, programs designed to maintain and enhance heritage-language proficiency provide a reasonable alternative, one that not only reduces this harm but leads to much more positive social and linguistic outcomes for the child and for the community. If this is true, dominant-language-only instruction of Aboriginal language speakers would meet the criteria to be labeled "institutional language discrimination."

The Nunavik Language Project

Evidence in support of this claim of institutional language discrimination comes from a program of research in a region of Canada's Eastern Arctic known as Nunavik (Bougie et al., 2003; Taylor, Caouette, Usborne, & Wright, in press; Taylor & Wright, 2002; Wright & Bougie, 2007; Wright & Taylor, 1995; Wright et al., 2000). A group of Inuit children was followed through the first 8 years of schooling to investigate the differential impact of dominant-language-only and heritage-language instruction on collective and personal self-esteem, maintenance and development of the heritage language (Inuktitut), and the acquisition of a second language (English or French).

The Context

Nunavik is a vast Arctic region on the northern tip of the Canadian province of Quebec. The geographic isolation of 14 small communities scattered over this region and the recency of direct intrusion of mainstream Canadian and U.S. culture accounts, in part, for the vibrancy of traditional culture and of the Inuktitut language in this region. Inuktitut has been described as one of a few North American indigenous languages with the potential for long-term survival (Foster, 1982; Priest, 1985). Despite this optimistic claim, research in Nunavik shows clear evidence of a growing intrusion of English and French into the daily lives of the Inuit (Crago, Annahatak, & Ningiuruvik, 1993; Dorais, 1996; Taylor & Wright, 1989).

This research was conducted in the largest community in Nunavik. At the time, the population was roughly 1,400; approximately 80% were Inuit, 12% were Francophone, and 8% were Anglophone, and more than 90% of the Inuit spoke Inuktitut as their first language (Taylor & Wright, 1989). In 1975, the Inuit in Nunavik signed the landmark James Bay Agreement with the Quebec government, securing them considerable economic, cultural, and educational autonomy. Over the next decade, the independent Kativik School Board was created, with a strong Inuit presence in its administration. One of its key aims was to have Inuit culture reflected in the educational process. Committed to the maintenance of the heritage language, the board designed programs to foster expertise in Inuktitut while also attempting to prepare students to participate in mainstream French and English society. The result was a program whereby students in kindergarten, Grade 1, and Grade 2 received instruction exclusively in Inuktitut. From Grade 3 through to the end of secondary school, students chose either a French or an English stream of education.

At the outset of our research, this program was just being introduced, and parents were initially given the option to have their children participate in the new Inuktitut program or to attend existing French or English programs. This situation and community provided a unique research opportunity that addresses many of the criticisms of comparative field research. First, we could take advantage of a naturally occurring quasi experiment that allowed for direct comparison of Inuit children being schooled in Inuktitut or one of two dominant languages. Second, the size and geographic isolation of the community creates a homogeneity of social experiences (outside the classroom) among Inuit children that is impossible in most other field settings. Inuit children in the three language programs were indistinguishable on the basis of neighborhood, socioeconomic status, or cultural history. Although exposure to mainstream television (primarily English) was considerable, there was relatively little variation in the amount of exposure to TV across children in the three programs (Taylor, Wright, Ruggiero, & Aitchison, 1992). Third, all three programs were offered in the same school. Fourth, heritage-language instruction is compared to instruction in two distinct dominant languages (English and French). This allows for increased confidence that differences between heritage-language education and dominant-language education do not result from the unique historical or linguistic characteristics of one particular dominant language. Fifth, comparisons between the three groups of Inuit children are supplemented with comparisons with both mixed-heritage children and a sample of White children living in the same community. Sixth, a longitudinal design (totaling 12 test occasions over 8 years) and multiple cohorts (children entering kindergarten in 4 successive years) greatly reduces potential confounds associated with specific teachers, classroom groups, or pedagogical and historical peculiarities that might arise when a single measurement occasion or cohort is used. Twice a year from kindergarten through Grade 3 and then at the end of each school year from Grade 4 through Grade 7, children completed measures of personal and collective self-esteem and language ability in each of the three languages (Inuktitut, French, and English).

Language of Instruction and Self-Esteem

Supporters of heritage-language education contend that heritage-language education strengthens the child's personal and social (collective) identity, whereas education in only a dominant language undermines self-esteem (Wright & Taylor, 1995). People's personal identity includes aspects of the self that make each person unique (e.g., personal attributes, skills, experiences). Personal self-esteem is each person's evaluation of these self-aspects. Social identity refers to those aspects of the self that connect people to others—group memberships (Taylor, 2002). An evaluation of these self-aspects determines collective self-esteem (Crocker & Luthanen, 1990; Cross, 1987).

The challenges faced by minority-language children in dominant-language classrooms are considerable. The negative feelings and poorer performance that can result from these challenges can undermine confidence and reduce personal self-esteem (see Cummins, 1989). Moreover, social comparisons with dominant-language children, who appear more successful, are likely to generate negative self-evaluations. Dominant-language education can also harm collective self-esteem. Cultural discontinuity between the school and home environments may lead children to believe that their language, and by extension those who speak it, are not valued (Williamson, 1987). In contrast, heritage-language instruction, by conferring status to the minority language, can secure and reinforce children's cultural identity (Baker, 2001; Cummins, 1989; Hornberger, 1987; Huang, 1995; Wright & Taylor, 1995). For Aboriginal languages where the survival of the language itself is at stake, maintaining positive collective self-esteem of the speakers is crucial (Taylor, 2002), as it largely determines the extent to which children will speak their language outside the classroom. We tested these claims at two critical points in our sample of Inuit children's early education. The first was during their earliest school experience—kindergarten. The second was at the crucial transition point from heritage-language to second-language instruction in midelementary school (Bougie et al., 2003).

The Kindergarten Study We (Wright & Taylor, 1995) measured the personal and collective self-esteem of Inuit children in the three language programs (Inuktitut, French, and English) at both the beginning and the end of their kindergarten year. The measure was a picture-sorting task allowing for easy administration in any language. In response to a series of prompts (e.g., "Pick all the children who are smart"), the children selected from a set of nine photos, including one of themselves, four of Inuit children, and four of White children. The number of times they selected their own picture provided a measure of personal self-evaluation, and the pattern of selection of Inuit and White children provided a measure of collective self-evaluation.

Children in the three language programs had identical levels of personal self-evaluation when they entered kindergarten. At the end of the year, however, children in the Inuktitut program selected their own photo significantly more often in response to positive prompts than did Inuit children in the French and English programs. Although children in the Inuktitut program showed a marked increase in positive self-appraisal from the beginning to the end of the year, children in the dominant-language programs showed no increase, remaining as concerned about their self-worth at the end of the year as they were when they entered school (see Figure 12.1).

In addition, children in the Inuktitut program showed a pattern of selection of photos of Inuit and White children consistent with a healthy

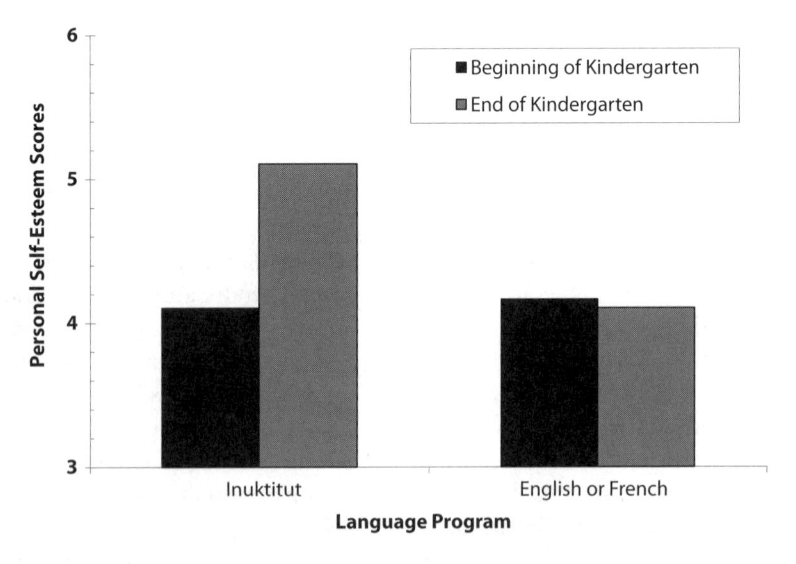

Figure 12.1 Inuit children's personal self-esteem scores, comparing children in Inuktitut and dominant-language (English or French) kindergarten classes.

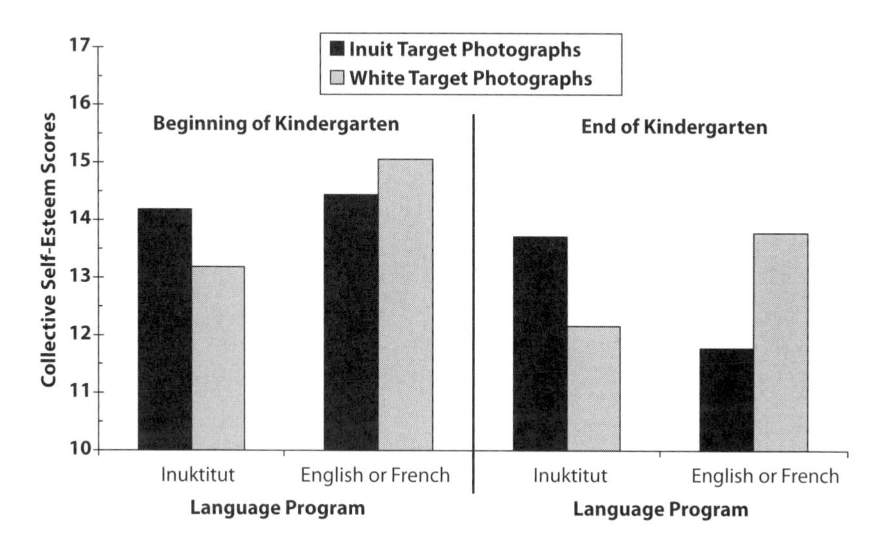

Figure 12.2 Inuit children's collective self-esteem scores, comparing children in Inuktitut and dominant-language (English or French) classes at the beginning and the end of their kindergarten year.

collective self-esteem. They showed a mild preference for Inuit targets over White targets, a preference that increased somewhat by the end of kindergarten. Children in English and French programs, however, showed the opposite pattern. When tested in the first month of kindergarten, this group showed no reliable difference in their evaluations of White and Inuit targets. By the end of the year, however, they showed a preferential positive evaluation of Whites over Inuit targets (see Figure 12.2).[1]

The Grade 3 Transition Study The Kativik School Board's original program resulted in children facing an abrupt transition from heritage-language to second-language instruction. At Grade 3, children who had received almost all of their instruction in Inuktitut were moved into an almost entirely English or French classroom environment, usually taught by White teachers who spoke little or no Inuktitut. Using the same procedures as in the kindergarten study, we explored the impact of this abrupt shift on children's self-esteem (Bougie et al., 2003). Because submersion in a dominant-language program appeared to undermine Inuit children's self-esteem during kindergarten, the question was whether 3 years of Inuktitut instruction might serve as an inoculation, minimizing the threat to self-esteem posed by this shift to a dominant-language-only curriculum.

The results, however, did not support this inoculation hypothesis. The shift was associated with a significant drop in personal self-esteem.

Inuit children previously enrolled in the Inuktitut program enjoyed high levels of personal self-esteem at the beginning of Grade 3. The shift to a dominant-language program, however, resulted in significantly lower levels of self-esteem at the end of the year. Children who had been in the English and French programs since kindergarten, and thus faced no change in classroom experience, showed no drop in self-esteem in Grade 3. Furthermore, the drop in self-esteem during the transition year was the only significant loss in self-esteem recorded during any period of our study. Thus, submersion in a dominant-language program may challenge children's sense of personal worth, even when a heritage-language program has effectively nurtured their self-esteem for the three initial years of schooling.

In terms of collective self-esteem, Inuit children's evaluations of White and Inuit targets showed a consistent pro-White bias at both the beginning and the end of Grade 3. Inuit children previously enrolled in the Inuktitut program, and those previously enrolled in dominant-language instruction, demonstrated a pattern consistent with negative collective self-esteem. It appears, therefore, that although Inuktitut instruction from kindergarten through Grade 2 was effective in building a positive image of the ingroup in the early years (kindergarten and Grade 1), this experience did not enable Inuit children to overcome the more powerful implicit message that the mainstream White culture, and those who represent it, hold more status than their own ethnic ingroup.

Conclusions Inuit children's submersion in a dominant-language program can have damaging effects on the children's self-esteem, and heritage-language education appears to provide a partial solution. Given that personal and collective self-esteem are critical psychological resources that can influence subsequent success and opportunities, continuing a program that is associated with less positive self-evaluations, when an alternative exists that has primarily positive effects, appear to meet the criteria for institutional discrimination. Some of the benefits of a heritage-language program, however, can be undermined by a later abrupt submersion in the dominant-language program. Thus, a true maintenance program with continued use of the heritage language as part of children's instruction well after Grade 2 may be needed. In addition, it appears that even a very robust heritage-language program cannot shield Aboriginal children from the strong cultural messages of out-group superiority that can lead to preferential evaluations of White children.

Language of Instruction and Subtractive Bilingualism

In addition to having adverse effects on self-esteem, dominant-language-only education can contribute to a pattern of *subtractive bilingualism*

(Wright et al., 2000). Subtractive bilingualism occurs when increasing acquisition of a dominant language is associated with a corresponding slowing or reversing of heritage-language development. In contrast, *additive bilingualism* occurs when a second language is added to one's linguistic repertoire, without reducing proficiency in one's heritage language (Genesee, 1987; Lambert & Taylor, 1983; Wong Fillmore, 1991).

Educational practices that lead to subtractive bilingualism should be labeled "institutional discrimination" because they undermine a critical component of cultural identity. In addition, undermining children's ability in their heritage language diminishes their ability to connect with, and learn from, their community and elders; it deprives them of the knowledge and social support that comes from language-based interactions with family and community. Moreover, undermining children's ability to converse in their heritage language robs them of the specific perspectives and cognitive advantages enjoyed by speakers of that language. At the group level (especially when a language is spoken by a relatively small group, which is the case for all Aboriginal languages), undermining children's language ability will hasten the extinction of the language itself, permanently robbing the group of a key element of cultural identity (Nettle & Romaine, 2000). Thus, if dominant-language-only instruction can be shown to contribute to subtractive bilingualism and heritage-language education does not, it would provide another basis for labeling dominant-language-only instruction a clear case of institutional language discrimination.

Heritage-Language Development Studies We tested the hypothesis that education exclusively in a dominant language would negatively affect heritage-language development by comparing the Inuktitut skills of Inuit children in the English and French programs to the skills of those children in the Inuktitut program. In consultation with specialists at the Kativik School Board, we developed an extensive test of oral language and basic literacy appropriate for this population. All three groups began school with exactly the same ability in Inuktitut. By the end of kindergarten, however, children in the English and French programs scored significantly lower on the Inuktitut tests than children in the Inuktitut program, and the gap grew dramatically wider over the next 2 years. This gap was found for both oral language (e.g., oral story comprehension) and literacy skills, although the gap was larger for literacy. In addition, those starting school in the English and French programs remained deficit in their Inuktitut skills compared to those in the Inuktitut program well after all children were in the English- and French-only programs in Grade 3. In fact, this deficit remained 5 years later at our final test occasion at the end of Grade 7 (see Figure 12.3).

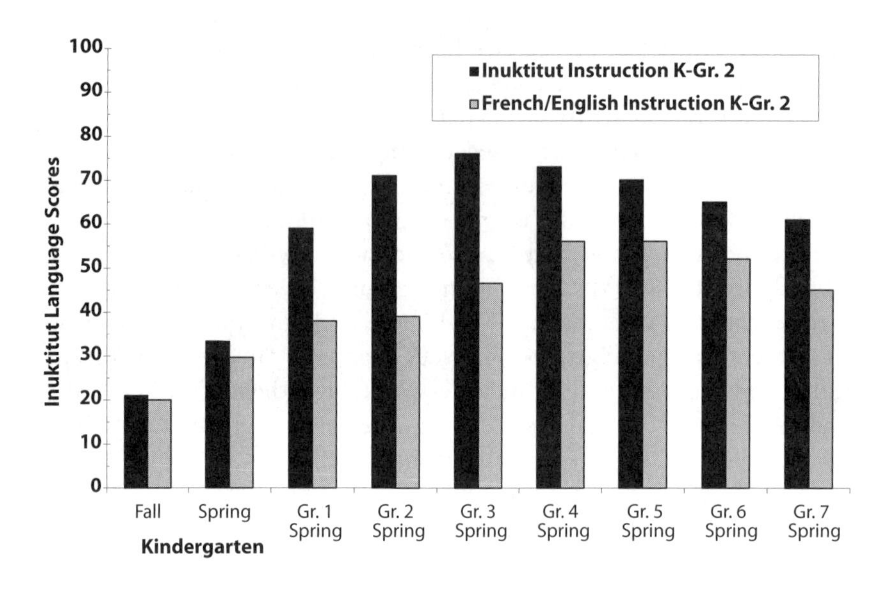

Figure 12.3 Inuit children's Inuktitut proficiency scores from kindergarten through Grade 7, comparing children who were in Inuktitut and dominant-language (English or French) programs from kindergarten through Grade 2.

Further evidence for the benefit of heritage-language instruction is found in comparisons between Inuit children and groups of mixed-heritage (Inuit and White) and White Francophone children living in the community. Francophone children were educated in French (their heritage language). Most mixed-heritage children received instruction in their first or home language (English). The language tests were carefully constructed so content and difficulty were equated across the three languages. Thus, it was possible to compare the performance of all of these groups in their own heritage language from kindergarten through Grade 2. All groups showed steady improvement in their respective heritage-language abilities, but, more important, Inuit children in the Inuktitut program showed a pattern almost identical to that of both the mixed-heritage and Francophone children, whereas Inuit children in the English and French programs stood out for their clear lack of ability in their heritage language. That is, in kindergarten through Grade 2, Inuit children in Inuktitut instruction demonstrated levels of heritage-language proficiency equal to or exceeding the English proficiency of mixed-heritage children and the French proficiency of Francophone children educated in their heritage languages. Inuit children educated in English or French were the clear outliers, showing poorer heritage-language skills.

Language of Instruction, Language Transfer, and Second-Language Acquisition

One of the key arguments used to defend dominant-language-only education has been that it is necessary to ensure that minority-language children develop strong English (or French) skills so that they can compete successfully in a society dominated by English (or French). The basic idea is that early English and more English means better English. This view is often accompanied by claims that too much time spent on the heritage language results in confusion that impedes the acquisition of English. Thus, the logic is that although dominant-language-only education may not be good for heritage-language development, it is the price that must be paid to ensure strong English acquisition. Therefore, submersing children in English is the best way to ensure that English will quickly replace the heritage language as the primary academic language.

Although this logic has some intuitive appeal, there is also good reason to believe that the exact opposite might be true, that dominant-language-only instruction will lead to *slower* and *less* complete acquisition of the dominant language (Cummins, 1989; Lambert & Taylor, 1983; Louis & Taylor, 2001; Skutnabb-Kangas, 1981). The explanation for this involves the concept of "language transfer," whereby initial mastery (including literacy) of the heritage language provides the basic cognitive skills and linguistic strategies that can then be transferred to the second language (Baker, 2001; Durgunoglu, Nagy, & Hancin-Bhatt, 1993; Odlin, 1989; Royer & Carlo, 1991). The argument is that when children develop strong academic language skills in the heritage language, they construct a platform for building second-language skills. Skills in one language support, rather than disrupt, the development of skills in a second language. If this is the case, dominant-language-only instruction will indeed slow mastery in the dominant language, which could have repercussions for Aboriginal children's academic achievement, likelihood of high school graduation and university attendance, and future employment opportunities in mainstream society.

Evidence from our research provides some support for the language transfer model and clearly contradicts the claim that early dominant-language-only education will ensure rapid and better skills in the dominant language. It is not surprising that when comparing the English and French skills of children in the Inuktitut program to those of children in the English and French programs at the end of Grade 2, we find that instruction in French or English does improve children's skills in French or English. Children in the English program have stronger English skills at the end of Grade 2 than do children in the French or Inuktitut programs. Similarly, those in the French program acquire more French than those in the Inuktitut or English

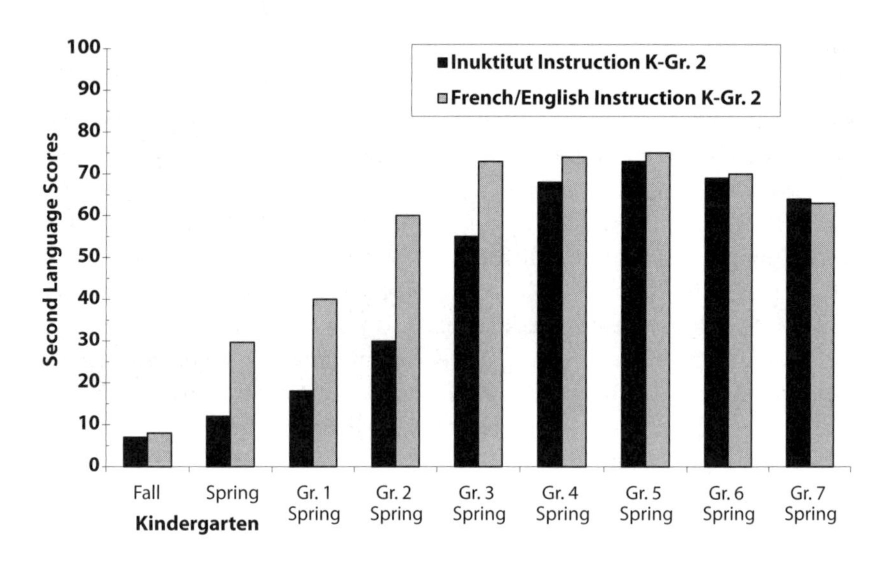

Figure 12.4 Inuit children's second-language proficiency scores from kindergarten through Grade 7, comparing children who were in Inuktitut and dominant-language (English or French) programs from kindergarten through Grade 2.

programs. These initial positive results, however, are completely erased when we consider the pattern of second-language acquisition from Grade 3 through Grade 7 when all the children are now in a French or English program. As shown in Figure 12.4, by the end of Grade 5 there is no difference in the second-language proficiency of children who spent their first 3 years receiving instruction in the second language and children who spent their first 3 years receiving Inuktitut instruction. Simply put, children educated from kindergarten to Grade 2 in Inuktitut catch up in English or French in just 3 years. Thus, for this group of Inuit children, early submersion in a dominant-language-only classroom had no benefit at all in terms of long-term proficiency in the dominant language. Also, this pattern of results was identical for both the English and the French programs. Although French proficiency is always lower than English proficiency (likely the result of greater "out-of-school support" for English, as English plays a much larger role in community life; see Taylor & Wright, 1989), the pace at which those educated in Inuktitut catch up to their classmates in the French program is identical to the pace at which they catch up to their classmates in English. Thus, these findings are not an artifact of one particular dominant language, lending credibility to their generality.

In addition, regression analyses were used to assess how well Inuktitut proficiency and English (or French) proficiency at the beginning of Grade 3 predicted English (or French) proficiency at the end of Grades 5, 6, and 7. The results (see Table 12.1) were consistent with the language transfer

Table 12.1 Standardized Betas for the Prediction of Second-Language Proficiency in Grade 5, Grade 6, and Grade 7 by Inuktitut and Second-Language Proficiency at the Beginning of Grade 3, Comparing Children Who Were in Inuktitut and Dominant-Language (English or French) Programs From Kindergarten Through Grade 2

Predictor	Children in Kindergarten through Grade 2 Inuktitut Programs	Children in Kindergarten through Grade 2 English and French Programs
	Predicting Second Language in Grade 5	
Grade 3 Inuktitut proficiency	.24*	–.10
Grade 3 second-language proficiency	.39**	.62**
	Predicting Second Language in Grade 6	
Grade 3 Inuktitut proficiency	.36**	–.12
Grade 3 second-language proficiency	.48**	.82**
	Predicting Second Language in Grade 7	
Grade 3 Inuktitut proficiency	.50**	–.09
Grade 3 second-language proficiency	.32*	.84**

Note: Grade 3 is the point where the Inuktitut program ends and all children are in English or French programs.
*$p < .01$, **$p < .001$.

hypothesis. When Grade 3 Inuktitut and English (or French) proficiency were both included as predictors, Inuktitut proficiency was a consistent and significant independent predictor of later English (or French) proficiency. This result was apparent, however, only for children in the Inuktitut program. For children in the English (or French) program, Grade 3 proficiency in a second language was a strong predictor of subsequent second-language skills, but Inuktitut skills played little or no role in predicting French or English proficiency later in the children's school career.

The most alarming evidence of dominant-language-only instruction's negative impact emerges, however, when we consider proficiency in heritage language and second language simultaneously. First, Inuit children educated exclusively in English or French do not show rapid development in their second language. Rather, their progress in English and French is slow. At the end of Grade 2, Inuit children in the dominant-language programs show English or French skills that are far below those of their English- or French-speaking classmates and equally far below the Inuktitut language skills of Inuit children in the Inuktitut program. Combining this with their poorer proficiency in their own heritage language, these children face the real possibility that they do not have adequate language skills *in any language* that would allow them to fully participate in the complex, abstract language activities required by the usual classroom curriculum.

To more effectively explore this concern, we divided the many tasks in our language tests into two groups based on the distinction between *conversational* and *academic* language proficiency. Conversational proficiency is required to carry on simple day-to-day conversations about concrete topics in contexts where situational cues assist comprehension. This measure included tasks such as learning simple vocabulary items; demonstrating sentence comprehension; naming colors, shapes, and numbers; and so on. Academic proficiency is required for rapid processing of complex information, when discussing abstract concepts, and/or when environment cues are minimal. This measure included tasks such as responding to questions about an oral story and demonstrating complex sentence comprehension and early literacy skills. At the end of Grade 2, scores on the measure of Inuktitut conversational proficiency were relatively high for Inuit children in all three language programs. Children in the French and English programs, however, scored much lower on the measures of Inuktitut academic proficiency than the children in the Inuktitut program.

This distinction between conversational and academic proficiency also unveils a disturbing finding for second-language proficiency. The slow development of English or French among children in the dominant-language programs was almost entirely due to poor academic proficiency. By the end of Grade 2, these children had fairly strong conversational proficiency in English or French but had extremely low academic proficiency. So although they could carry on simple conversations about concrete topics in English or French, they were largely unable to use these languages to participate in complex, intellectually challenging conversations.

These data depict a rather disheartening picture. At the end of Grade 2, Inuit children who have been educated exclusively in a dominant language appear to be able to carry on basic daily conversations in Inuktitut and their second language. They have not, however, developed the sophistication in either their heritage language or the second language that would allow them to fully participate in the complex linguistic interchanges expected of children their age (such as understanding the implications of a story told by an elder) or to master the difficult and complex tasks required to complete the normal Grade 2 curriculum. This pattern of language competence frames a story of current and future academic success that is bleak indeed.

Summary of the Findings

Our research considered three crucial outcomes of Inuit children's early educational experiences: their positive self-regard, their developing proficiency in their heritage language, and their acquisition of a dominant language. The results clearly contradict claims that early dominant-language-only education is helpful for Inuit children. Early submersion in a French or English program had negative effects on the children's personal and collective self-esteem and resulted in lower proficiency in Inuktitut that persisted until at least Grade 7, and any initial benefits in terms of conversational skills in English or French completely evaporated by Grade 5. In contrast, a heritage-language program provides a clear alternative that leads to a much more positive pattern. Early education in Inuktitut had positive effects on personal self-esteem and provided early support for collective identity. It results in strong development of the heritage language, equivalent to the first-language development of mixed-heritage English speakers and White French speakers. Finally, children in the Inuktitut program catch up to those in English- and French-only programs in terms of second language within 3 years of transferring to a dominant-language program.

Toward Justice in Language Policy

Returning to our central issue of social justice, it appears that our data provide strong support for the claim that educating Aboriginal children exclusively in a dominant language can be described as institutional language discrimination. Dominant-language-only instruction harmed Inuit children, and this harm cannot be justified in terms of any evidence of some larger benefit accrued through participation in these programs. Moreover, a heritage-language program provided a viable alternative that dramatically reduced the harm.

We think, however, that this analysis of institutional language discrimination also points out a number of issues related to social psychology's general understanding of discrimination. First, it makes clear the need for a more complete analysis of institutional discrimination. Although institutional discrimination is often alluded to briefly in reviews, it is seldom considered in detail. For example, it is often described as *differential treatment* of minority group members (e.g., Pincus, 1996). The current analysis questions the appropriateness, or at least the completeness, of this description. Our position is that institutional discrimination occurs precisely when minority-language speakers are given treatment that is identical to that given to majority-language speakers. Providing minority-language children with the exact same environment as their majority-language classmates usually means that everyone gets a dominant-language-only classroom. Our description of institutional discrimination refers not to differential treatment at all but rather to *differential harm*. Thus, *identical treatment* can be discriminatory when treating two groups the same leads to greater harm for one of them.

This discussion is also relevant to debates surrounding the *color-blind perspective*, which is the idea that the solution to discrimination is to ignore group-based difference and treat everyone as an individual. Of course, our call for heritage-language education is entirely inconsistent with a color-blind perspective in that it calls for clear recognition that members of language-minority groups must be recognized in terms of their group membership and for educational programs to be built with this unique group-shared attribute (language) clearly in mind.

In addition, our analysis points out the importance of recognizing that institutional discrimination can often be entirely unapparent, that the intentions of those enacting the discrimination may be entirely admirable, that the practice may be supported by apparently logical and persuasive arguments that serve as an ideological base for its continuation and that even describe the practice as being in the best interests of those it discriminates against. Dominant-language-only education is supported by a series of apparently logical arguments about how the practice will actually benefit minority-language children. Those people with clearly malicious intents may employ these arguments, but these arguments are also believed and promoted by people who believe they are helping the children they work with. Exposing the oversights and inaccuracies of these legitimizing arguments is a critical step in attacking this example of institutional language discrimination, and this is likely true of many other cases of institutional discrimination.

Finally, we believe that understanding the concept of institutional discrimination not only can be a valuable analytical tool for understanding the dominant-language–heritage-language debate but may also be a powerful political tool. Calling dominant-language-only programs *discrimination* has substantial rhetorical power. Even if opponents of dominant-language-only programs can show that these programs do not have the positive effects that supporters of the English-only movements claim, it provides only a small impetus to invest the resources and effort necessary to provide heritage-language education. If the practice of dominant-language-only education represents a clear case of discrimination against language-minority children, however, it provides a strong impetus to reevaluate its use and to consider alternatives, even if those alternatives require additional resources. Thus, a social psychological analysis in this case not only provides a clearer understanding of an important societal issue but also may provide tools for those who are motivated to promote social justice.

Authors' Note

The authors would like to acknowledge the contributions of Evelyne Bougie and Judy Macarthur to the research reported here. We would like to thank all the children who took part in our research; the Education Committee, principals, teachers, and staff at the school; Mary Elijassiapik, Qiallak Qumaluk, Annie Kudluk, Michelle Auroy, Claudette Baron, Linda Thessen, Gaston Cote, Sue McNicol, Nicole Allain, Betsy Matt, Siasi Clumas, Jason Annanhatak, Mary Kane, Julie Caouette, Esther Usborne, Mike King, Nina Jauenig, and Roxane de la Sablonniere who served for many hours as testers; and especially Doris Winkler and Mary Aitchison, who provided continuous invaluable assistance. We are grateful to the Kativik School Board for its ongoing support. Finally, our ongoing research in the Nunavik is also supported by funding from the Social Science and Humanities Research Council of Canada.

Note

1. When we first reported these findings (Wright & Taylor, 1995), we did not report the pattern of change from the beginning to the end of kindergarten. The lack of a statistically significant three-way interaction, Language Programs × Ethnicity of Target Photo × Time, led us to not elaborate on the combined effect of these three variables. We now believe that this was an overly conservative approach. A less conservative strategy yields a very different interpretation. At the end of the kindergarten year, the two-way crossover interaction between language program and selection of White and Inuit targets is obvious and statistically significant ($p = .03$). At the beginning of the kindergarten year, the interaction effect is very small ($F < 1.0$), and none of the t tests (arguably an overly liberal statistic for these comparisons) comparing differences between the four means are statistically significant.

References

Assembly of First Nations. (1988). *The Aboriginal language policy study.* Ottawa, Ontario, Canada: Assembly of First Nations Education Secretariat.

Assembly of First Nations. (1990). *Towards linguistic justice.* Ottawa, Ontario, Canada: Assembly of First Nations Education Secretariat.

Baker, C. (2001). *Foundations of bilingual education and bilingualism* (3rd ed.). Clevedon, UK: Multilingual Matters.

Baker, C., & Prys Jones, S. P. (1998). *Encyclopedia of bilingualism and bilingual education.* Clevedon, UK: Multilingual Matters.

Bougie, E., Wright, S. C., & Taylor, D. M. (2003). Early heritage-language education and the abrupt shift to a dominant-language classroom: Impact on the personal and collective esteem of Inuit children in Arctic Québec. *International Journal of Bilingualism and Bilingual Education, 6*(5), 349–373.

Brutt-Griffler, J. (2002). Class, ethnicity, and language rights: An analysis of British colonial policy in Lesotho and Sri Lanka and some implications for language policy. *Journal of Language, Identity, and Education, 1*(3), 207–234.

Chandler, M. J., & Lalonde, C. E. (1998). Cultural continuity as a hedge against suicide in Canada's First Nations. *Transcultural Psychiatry, 35*, 191–219.

Chandler, M. J., & Lalonde, C. E. (2008). Cultural continuity as a protective factor against suicide in First Nations youth. *Horizons, 9*(4), 13–24.

Crago, M. B., Annahatak, B., & Ningiuruvik, L. (1993). Changing patterns of language socialization in Inuit homes. *Anthropology and Education Quarterly, 24*, 205–223.

Crocker, J., & Luthanen, R. (1990). Collective self-esteem and intergroup bias. *Journal of Personality and Social Psychology, 58*, 60–67.

Cross, W. E. (1987). The two-factor theory of Black identity: Implications for the study of identity development in minority children. In J. S. Phinney & M. J. Rotheram (Eds.), *Children's ethnic socialization: Pluralism and development* (pp. 117–133). Newbury Park, CA: Sage.

Cummins, J. (1989). *Empowering minority students*. Sacramento, CA: California Association for Bilingualism Education.

Cummins, J. (1994). Knowledge, power, and identity in teaching English as a second language. In F. Genesee (Ed.), *Educating second language children: The whole child, the whole curriculum, the whole community* (pp. 33–58). Cambridge, UK: Cambridge University Press.

Dorais, L. J. (1996). The Aboriginal languages of Quebec, past and present. In J. Maurais (Ed.), *Quebec's aboriginal languages: History, planning, development* (pp. 43–100). Clevedon, UK: Multilingual Matters.

Durgunoglu, A., Nagy, W., & Hancin-Bhatt, B. J. (1993). Cross-language transfer of phonological awareness. *Journal of Educational Psychology, 85*, 453–465.

Foster, M. (1982). Indigenous languages: Present and future. *Language and Society, 7*, 7–14.

Genesee, F. (1987). *Learning through two languages: Studies of immersion and bilingual education*. Cambridge, MA: Newbury House.

Hallet, D., Chandler, M. J., & Lalonde, C. E. (2007). Aboriginal language knowledge and youth suicide. *Cognitive Development, 22*, 392–399.

Hornberger, N. H. (1987). Bilingual education success, but policy failure. *Language in Society, 16*, 205–226.

Huang, G. G. (1995). Self-reported biliteracy and self-esteem: A study of Mexican-American 8th graders. *Applied Psycholinguistics, 16*, 271–291.

Kymlicka, W., & Patten, A. (Eds.). (2003). *Language rights and political theory*. Oxford, UK: Oxford University Press.

Lambert, W. E., & Taylor, D. M. (1983). Language in the education of ethnic minority immigrants. In R. J. Samuda & S. L. Woods (Eds.), *Perspectives in immigration and minority education* (pp. 267–280). Washington, DC: University Press of America.

Louis, W., & Taylor, D. M. (2001). When the survival of a language is at stake: The future of Inuktitut in Arctic Québec. *Journal of Language and Social Psychology, 20*(1/2), 111–143.

Mascia-Lees, F. E., & Lees, S. H. (2003). Language ideologies, rights, and choices: Dilemmas and paradoxes of loss, retention, and revitalization. *American Anthropologist, 105*, 710–711.

McLaughlin, D. (1989). The sociolinguistics of Navajo literacy. *Anthropology and Education Quarterly, 20*, 275–290.

Nettle, B., & Romaine, S. (2000). *Vanishing voices: The extinction of the world's languages*. Toronto: Oxford University Press.

Norris, M. J. (2007). Aboriginal languages in Canada: Emerging trends and perspectives on second language acquisition. *Canadian Social Trends, 83*, 19–27.

Odlin, T. (1989). *Language transfer: Cross-linguistic influence in language learning*. New York: Cambridge University Press.

Pincus, F. (1996). Discrimination comes in many forms: Individual, institutional and structural. *American Behavioral Scientist, 40*(2), 186–194.

Priest, G. E. (1985). Aboriginal languages in Canada. *Language and Society, 15*, 13–19.

Royer, J. M., & Carlo, M. S. (1991). Transfer of comprehension skills from native to second language. *Journal of Reading, 34*, 450–455.

Sidanius, J., & Pratto, F. (1999). *Social dominance: An intergroup theory of social hierarchy and oppression*. New York: Cambridge University Press.

Skutnabb-Kangas, T. (1981). *Bilingualism or not: The education of minorities*. Clevedon, UK: Multilingual Matters.

Skutnabb-Kangas, T. (2002). Marvelous human rights rhetoric and grim realities: Language rights in education. *Journal of Language, Identity, and Education, 1*(3), 179–205.

Statistics Canada. (2007). *Aboriginal peoples in Canada in 2006: Inuit, Métis and First Nations*. Ottawa, Ontario, Canada: Minister of Industry.

Task Force on Aboriginal Languages and Cultures. (2005). *Towards a new beginning: A foundational report for a strategy to revitalize First Nation, Inuit, and Métis languages and cultures*. Ottawa, Ontario, Canada: Canadian Heritage, Aboriginal Affairs.

Taylor, D. M. (2002). *The quest for identity: From minority groups to generation Xers*. New York: Praeger.

Taylor, D. M., Caouette, J., Usborne, E., & Wright, S. C. (in press). Aboriginal languages in Québec: Fighting linguicide with bilingual education. *Divesite Urbanine*.

Taylor, D. M., & Wright, S. C. (1989). Language attitudes in a multilingual northern community. *Canadian Journal of Native Studies, 9*, 85–119.

Taylor, D. M., & Wright, S. C. (2002). Do aboriginal students benefit from education in their heritage language? Results from a ten-year program of research in Nunavik. *Canadian Journal of Native Studies, 22*, 141–164.

Taylor, D. M., Wright, S. C., Ruggiero, K. M., & Aitchison, M. C. (1992). Language perceptions among the Inuit of Arctic Quebec: The future role of heritage language. *Journal of Language and Social Psychology, 12*, 195–206.

Villegas, A. M., & Lucas, T. (2002). *Educating culturally responsive teachers: A coherent approach*. New York: State University of New York Press.

Williamson, K. J. (1987). Consequences of schooling: Cultural discontinuity amongst the Inuit. *Canadian Journal of Native Education*, *14*, 60–69.
Wong Fillmore, L. (1991). When learning a second language means losing the first. *Early Childhood Research Quarterly*, *6*, 323–346.
Wright, S. C., & Bougie, E. (2007). Reducing prejudice towards and the impact of discrimination against language minority groups: Heritage-language education, intergroup contact, and ingroup identification. *Journal of Language and Social Psychology*, *26*, 157–181.
Wright, S. C., & Taylor, D. M. (1995). Identity and the language in the classroom: Investigating the impact of heritage versus second language instruction on personal and collective self-esteem. *Journal of Educational Psychology*, *87*, 241–252.
Wright, S. C., & Taylor, D. M. (2003). The social psychology of cultural diversity: Social stereotyping, prejudice and discrimination. In M. A. Hogg & J. Cooper (Eds.), *Sage handbook of social psychology*. London: Sage.
Wright, S. C., Taylor, D. M., & Macarthur, J. (2000). Subtractive bilingualism and the survival of the Inuit language: Heritage- versus second-language instruction. *Journal of Educational Psychology*, *92*, 63–84.
Wurm, S. A. (Ed.). (1996). *Atlas of the world's languages in danger of disappearing*. Paris: UNESCO Publishing.

CHAPTER 13

The Antecedents, Nature, and Effectiveness of Political Apologies for Historical Injustices

KARINA SCHUMANN and MICHAEL ROSS

University of Waterloo

Abstract: Throughout history, governments have acted with prejudice and cruelty toward groups of people who differ on some identifiable dimension from the majority of their citizens. At the time, these discriminatory acts often result from deliberate political choice and are approved by legislatures, courts, and the majority of citizens. When the abuses finally end and those who perpetrated and endorsed the discriminatory acts pass from the scene, successive governments must decide how to respond to demands for redress from the victimized groups. In this chapter, the authors discuss when and how successive governments respond to historical injustices, as well as the psychological implications of their responses for members of the previously victimized group and the nonvictimized majority. In particular, the authors focus on political apologies for historical injustices, examining their antecedents, nature, and effectiveness.

Throughout history, governments have acted with prejudice and cruelty toward groups of people who differ on some identifiable dimension (e.g., skin color or religion) from the majority of their citizens. In Western democracies, these discriminatory actions are typically endorsed by parliaments, a majority of citizens, and the legal system (Backhouse, 1999;

Brooks, 1999). When the abuses finally end, governments often refuse to apologize or offer compensation. In many cases, members of the aggrieved group persist in demanding redress, decades or even centuries later (e.g., for African American slavery, see Brooks, 1999). Meanwhile, members of the government, legal system, and majority who perpetrated and condoned the discriminatory actions pass from the scene. In the current chapter, we discuss their successors' responses to the original injustice, as well as the psychological implications of their responses for members of the previously victimized group and the nonvictimized majority. In particular, we focus on political apologies for historical injustices, examining their antecedents, nature, and effectiveness.

How Do Governments Respond to Historical Injustices?

We use the term *historical injustice* to refer to discriminatory actions conducted by a previous government against a group of individuals sharing common characteristics (e.g., racial, ethnic, or national origins). The events are historical in the sense that they no longer occur—although their effects may persist—and members of the present government were not participants in the episodes. The former government's actions are unjust in that most people, as well as the legal system, would declare them to be legally and morally wrong if enacted today. Most of our examples of historical injustices concern harms committed against minorities living within the country in question, but some examples include harms committed against citizens of other nations (e.g., war crimes); still others include wrongs committed both within and outside of the country's borders (e.g., African American slavery).

When confronted with historical injustices, governments react much as individuals do to accusations of personal misdeeds, with a range of responses varying from denial to apology. At one end of the spectrum, governments reject claims that their predecessors committed an injustice. For example, Turkish officials typically deny the occurrence of the Armenian genocide of 1915 ("Armenian Genocide Dispute," 2006). Governments often bolster their disclaimers with explanations and justifications. The Turkish government acknowledges that Armenians were killed, but it disputes the magnitude and source of the carnage. According to their government reports, the killings occurred in the context of interethnic violence during World War I (WWI) and did not represent a systematic effort to destroy the Armenian population (i.e., genocide). Almost 100 years later, Turkish governments continue to restrict the availability of information about the episode to the general public. Orhan Pamuk, a celebrated Turkish novelist, was recently prosecuted for "insulting Turkishness" by writing about the episode ("Controversial Turkish Novelist Wins Nobel," 2006), although the charges were eventually dropped following international protests.

The concept of "insulting Turkishness" is intriguing from a social psychological perspective. Apparently the Turkish government supposes that the Turkish social identity is threatened by accounts of these rather ancient grievances. Relevant social psychological research provides some support for the concerns of the government. Reminders of historical injustices committed by one's own country can have a negative impact on social identity, just as reminders of past glories can have a positive impact, especially among individuals who identify highly with their country (Doosje, Branscombe, Spears, & Manstead, 1998; Sahdra & Ross, 2007). It is difficult for governments to manage information about such episodes, however. The events last too long, involve too many people, and are often confirmed by official documents. For many years, the Japanese government denied that it played any role in forcing women into sexual slavery for the Japanese Army during World War II (WWII). Eventually a Japanese professor proved otherwise by publicizing documents obtained from the government's own archives (Brooks, 1999).

Rather than denying, minimizing, and suppressing information about past injustices, governments sometimes acknowledge, condemn, and even apologize for the actions of their predecessors. Of the various responses to historical injustices, apologies are perhaps the most psychologically intriguing. Why apologize for events in the distant past that you did not commit? Some politicians and scholars argue that an apology is powerful medicine that yields a host of benefits ("Harper's Speech," 2006; Lazare, 2004; Minow, 2002). Most important, perhaps, apologies for historical injustices are hypothesized to promote reconciliation and forgiveness. Politicians offering public apologies suggest that their statements of remorse will heal past wounds ("Apology to Residential School Students," 2008), "turn the page" (e.g., "Harper's Speech," 2006), and allow groups to put the injustice behind them (e.g., "Apology for Study Done in Tuskegee," 1997). As a result, members of the majority and minority can look forward to a just and harmonious future (e.g., Motion of Reconciliation, 1999).

Why do some politicians and scholars suppose that apologies have such transformative powers? In part, they likely generalize their beliefs about the effects of everyday interpersonal apologies to the presumed effects of political apologies. There are at least four problems with such a generalization. First, there is a lot of speculation and anecdotal evidence but scant empirical support for the hypothesis that everyday interpersonal apologies are wonderfully effective. We are not arguing that interpersonal apologies are typically ineffective but simply saying that there is little compelling research evidence establishing the conditions in which interpersonal apologies are or are not helpful. Second, in everyday life, people typically apologize for fairly trivial transgressions (e.g., being late, bumping into someone; Schumann & Ross, 2008), whereas governments apologize for severe wrongs. Apologies for more serious harms may be less effective in promoting forgiveness than apologies for trivial wrongs. Third, interpersonal apologies often occur between people who generally trust and care for each other. Victims may forgive transgressions, in part, to preserve an existing strong affiliation. The relationship between a previously victimized group and a majority group or government may not be characterized by similar trust and caring.

Finally, a government apology for a historical injustice occurs in a very different social context than the typical interpersonal apology, although they are semantically similar. In everyday life, people apologize after they have personally committed a transgression. In Western cultures at least, individuals rarely apologize for another person's misdeeds unless they are directly responsible for the transgressors, as when parents apologize on behalf of their children. Also, individuals typically apologize to people whom they have directly harmed and do so shortly after the transgression. The situation is quite different when government officials apologize for historical injustices. They apologize for episodes that sometimes occurred before they were born. Also, they speak not only for themselves but also for the citizens they represent, some of whom may not agree that an apology is warranted. Furthermore, government apologies for historical injustices are often targeted at people who did not experience the injustice directly but have some connection to the original victims (e.g., descendents or members of the same religious or ethnic group). For example, the Canadian government recently apologized for a tax imposed on Chinese immigrants between 1885 and 1923. The apology was directed mainly at current members of the Chinese Canadian community, many of whom immigrated to Canada in the past few decades. It is not surprising, then, that opponents of government apologies for historical injustices often argue that these apologies are offered by the wrong people to the wrong people ("We Won't Pay," 2007).

When and How Do Governments Apologize?

Whether or not government apologies possess transformative powers, their frequency has increased dramatically in recent decades (Lazare, 2004). This groundswell of apologies has been variously labeled an apology "epidemic" (Thompson, 2002, p. viii), "the age of apology" (Brooks, 1999, p. 3), the "new international morality" (Barkan, 2000, p. ix), the "global trend of restitution" (Barkan, 2000, p. x), and "the apology phenomenon" (Lazare, 2004, p. 7). In the past couple of years alone, seven American state legislatures offered official apologies for slavery. These apologies came more than 140 years after the ending of African slavery in the United States and constitute the first official U.S. government apologies for slavery. Also in the past few years, the Australian government apologized for the Stolen Generations, and the Canadian government apologized for Canada's earlier mistreatment of various minority groups. The increase in apologies raises the following question: Why so many now? Have governments worldwide suddenly acquired a moral rhetoric? Are displays of government sentimentality simply a passing fad? To understand the roots of this new political morality, we examine factors that influence when governments act to redress earlier harms. As with interpersonal apologies, little experimental research exists on the motivations behind political apologies for historical injustices. We therefore base our analysis on actual government apologies that have occurred in the past two decades.

One important influence on how governments respond to historical injustices is political pressure from victimized groups. For example, by exhibiting passion, persistence, and cohesiveness in pursuing redress for their internment during WWII, Japanese Americans received one of the first of the modern political apologies (Brooks, 1999). Political pressure is magnified when victimized groups are represented by key individuals within the government (Lazare, 2004). Four Japanese American members of Congress fought vigorously for the Civil Liberties Act (redress for the Japanese American internment during WWII) when it was presented to the House and Senate, even providing narratives of their own war experiences. According to Leslie T. Hatamiya (1999), a lawyer and daughter of former Japanese American internees, this inside leadership was essential to the passage of the act. More recently, six of the seven U.S. states that apologized for slavery had substantial African American representation in the state legislature. This internal support provided the redress movement with the political power to reverse decades of inaction.

Demands by other governments may also influence whether redress is provided. Germany offered extensive reparations to Jewish victims of WWII partly in response to pressure from the Allies (Brooks, 1999). By comparison, successive German governments have granted little to other groups targeted by the Nazis for elimination, such as homosexuals and Romany people, who did not receive the same degree of external support (Brooks, 1999).

A second factor that can influence the likelihood of government redress for historical injustices is perceived resistance from the nonvictimized majority of citizens. In representative polls on slavery reparations, only 37% of White Americans supported a federal government apology for slavery ("Polling Report: Race and Ethnicity," 2008), and a full 90% opposed government cash payments to descendents of slaves (Viles, 2002). Former U.S. president Bill Clinton presented the objections of the White majority as a major reason for not supporting an official federal government apology for slavery (Brooks, 1999).

A third influencing factor is whether other governments have apologized for similar harms and not suffered ill consequences. An earlier government apology that is accepted by the targeted victimized group, does not lead to further costly demands, and is generally approved by the majority seems especially likely to encourage subsequent apologies for similar injustices. When the U.S. government apologized and offered compensation to survivors of the Japanese American internment in August 1988, the Canadian government provided an almost identical redress package to survivors of the Japanese Canadian internment less than 6 weeks later. Canada would almost certainly have been more hesitant to act if the American majority had responded negatively to the U.S. redress effort. Similarly, in 2007, the state of Virginia became the first American state to offer an official expression of regret for its role in slavery. Six other states apologized within the next year, and more states are expected to apologize shortly. In support of the inspirational role played by Virginia's apology, Senator Tony Rand described the precedent set by Virginia as a major contributing factor to his decision to place an apology bill before the North Carolina legislature (personal communication, April 22, 2008). It seems likely that this series of state apologies, combined with the absence of escalating demands from African Americans or a significant majority backlash, will motivate an official apology from the U.S. federal government.

Finally, governments seem less likely to apologize when either too little or too much time has passed since the injustice. Governments may be disinclined to apologize soon after an injustice, in part because their current members supported the discriminatory actions. If they participated in the injustice, they are perhaps more likely to justify or deny the wrongdoing than to express remorse. Also, it is likely that shortly after the injustice, the attitudes of the majority of citizens have not yet shifted from acceptance to revulsion. With the passage of time, members of the current government are no longer directly associated with the injustice, and the beliefs and values of the majority have changed. The government can apologize for the past injustice while dissociating itself from the individuals who perpetrated the act.

When much time passes, governments may be reluctant to apologize for what they deem to be historical curiosities. The effects of the injustice will have seemingly dissipated, and there are no victims who need or demand an apology (Starzyk & Ross, 2008). As the recent spate of apologies for slavery indicates, however, apologies occasionally occur long after the episode ends. One explanation for such delayed apologies may be that negative effects of the injustice are perceived to persist. If members of the previously victimized group regard the effects of the injustice as continuing, they may demand redress long after the injustice has supposedly ended.

These four factors—persistent victim demands, minimal majority opposition, precedents, and time lapse—at least partially explain the modern "age of apology." And with every new apology, the barriers to subsequent apologies will almost certainly continue to dissolve. Previously victimized groups will mobilize and persist in the hope of receiving similar redress, majority groups will acclimate to the process and offer less resistance, and governments will imitate the apologies of other governments.

The Content of Political Apologies

Not every apology is created equal; the extent to which members of a previously victimized group benefit from receiving a government apology may depend on its content (Lazare, 2004; Negash, 2006). Linguists and psychologists have suggested that a comprehensive apology contains as many as six elements (Bavelas, 2004; Lazare, 2004; Scher & Darley, 1997; Schlenker & Darby, 1981): (a) remorse (e.g., "I'm sorry"), (b) acceptance of responsibility (e.g., "It's my fault"), (c) admission of injustice or wrongdoing (e.g., "What I did was wrong"), (d) acknowledgment of harm and/or victim suffering (e.g., "I know you're hurt"), (e) forbearance or promises to behave better in the future (e.g., "I will never do it again"), and (f) offers of repair (e.g., "I will pay for the damages").

Theorists have hypothesized that a statement that contains more of these elements is superior to one that contains fewer elements (e.g., Bavelas, 2004; Lazare, 2004). There is little research, however, linking the comprehensiveness of an apology to forgiveness and reconciliation. Moreover, there probably needs to be a match between the severity of a misdeed and the comprehensiveness of an apology. A comprehensive apology for a very minor transgression may seem facetious. This matching of the comprehensiveness of the apology to the severity of the harm may help explain why over 90% of everyday interpersonal apologies consist of a simple "I'm sorry" or its equivalent (Meier, 1998).

Governments don't apologize for minor historical wrongs. As a result, their apologies are likely to be more comprehensive than the typical interpersonal apology. Each of the six elements in a comprehensive apology could address important psychological needs in members of a previously victimized group. A government that expresses remorse for the injustice demonstrates good conscience and genuine concern for the victims and their group (Scher & Darley, 1997). By assigning blame for the injustice outside the victim group (usually to the responsible government or leaders), a government asserts the innocence of the victims. An admission of injustice assures the victimized group that the current government upholds the moral principles that were violated (Lazare, 2004) and is committed to maintaining a legitimate and just social system. By acknowledging harm and victim suffering, a government validates the victims' pain and corroborates the victims' suffering for outsiders (Lazare, 2004). A promise of forbearance can help restore trust between groups; it indicates that the government values current members of the previously victimized group and is willing to work to keep them safe (Lazare, 2004). Finally, an offer of repair (e.g., financial compensation, land transfers, or memorials) substantiates the apology (Lazare, 2004). Repair demonstrates a sincere commitment by the current government to address the wrongs of the past and uphold justice (Minow, 2002). A government apology that includes these various elements should theoretically make the victimized group feel better about themselves, the majority group, their government, and their country. As we noted earlier, however, such effects need to be demonstrated, not assumed.

Research on social identity theory (e.g., Branscombe & Doosje, 2004; Tajfel & Turner, 1986) and justice motivations (e.g., Jost & Banaji, 1994; Lerner, 1980) has suggested that for political apologies to be effective, they will also need to include elements not typically associated with everyday interpersonal apologies. According to social identity theory, people are motivated to think highly of the groups to which they belong. Members of a group that has been subject to a long-standing historical injustice may feel devalued by society. This perception of low regard can damage the social identities of current group members (Branscombe & Doosje, 2004; Tajfel & Turner, 1986). Governments could use an apology as an occasion to offset the harmful psychological implications of prior injustices by emphasizing the important contributions of the victimized group to society. Such praise should satisfy the identity concerns of group members by affirming their positive qualities. An apology that calls for reconciliation also addresses the identity concerns of the victimized group. By communicating a desire for reconciliation, a government demonstrates high regard for the victimized group and an interest in repairing its relationship with that group. Perhaps most important, calls for reconciliation highlight the central goals of an apology: to promote forgiveness and healing, as well as improve relationship well-being.

Although most discussions of government apologies focus on their impact on members of the previously victimized group, apologies also have implications for the majority. An apology could threaten the social identities of members of the majority. Opinion polls indicate that members of a nonvictimized majority sometimes strongly oppose government apologies for historical injustices (e.g., Viles, 2002). Members of majority groups may feel that a government apology offered on their behalf implicates them in the injustice (Blatz, Ross, & Starzyk, 2008). To minimize this threat to social identity, governments could use the apology as an occasion for affirming the positive qualities of the majority group and explicitly absolving them of responsibility for the injustice. By doing so, governments may reduce opposition to the apology.

According to theories of justice motivation, people are motivated to believe that they live in a just country where people deserve what they get and get what they deserve (Jost & Banaji, 1994; Lerner, 1980). By acknowledging a major injustice, an apology could threaten this psychologically important belief. An apology that emphasizes the fairness of the current system and dissociates it from the system that perpetrated the injustice reduces this threat to beliefs in a just world. In their apology, government officials could also criticize the perpetrators for making unjust decisions and neglecting the needs of their people. By condemning the actions of past governments, the current government disconnects itself from the wrongdoing and demonstrates its commitment to justice.

There is also a downside to dissociating the past from the present. Members of the majority offer stronger support for redress when they view the victimized group as still experiencing negative effects of the injustice (Starzyk & Ross, 2008). It is difficult to establish objectively whether the effects of the injustice linger over decades or even centuries. Is the lower socioeconomic status of African Americans attributable to slavery and subsequent discrimination during the Jim Crow era, or are the origins more recent, and are African Americans themselves partly to blame? The answers to such questions depend, in part, on group membership. Relative to members of the majority, members of the previously victimized group perceive the effects of a historical injustice as lasting longer (Banfield, Blatz, & Ross, 2008). To encourage the majority's support for the apology and affirm the perspective of the victim group, governments may attempt a delicate psychological balancing act: They could note that the effects of the historical injustice persist, but they at the same time dissociate the current system from the original injustice.

How Governments Apologize

We have speculated about the elements that governments could include in their apologies to address the psychological concerns of the majority of their citizens and of the targeted victimized group. With the growing sample of political apologies for historical injustices, we can examine the content of political apologies for evidence of these elements. Next, we assess the degree to which a set of government apologies included the elements we proposed. To obtain our sample for analysis, we composed a list of federal government apologies offered by various countries for domestic and international injustices. We considered apologies for analysis if they (a) were verbal and available in English, (b) were offered for events that were intentional rather than accidental, (c) were offered by a federal government institution (e.g., parliament) or leader (e.g., president, prime minister, or sovereign) for events that occurred before the current government took office, (d) were offered to identifiable groups rather than individuals, and (e) included the core element of an apology, an expression of remorse (Meier, 1998; Scher & Darley, 1997). In the end, we compiled a list of 14 government apologies offered in the past two decades. These 14 apologies composed the first subset of our sample.

In the past year, seven U.S. states have apologized for their roles in slavery. These apologies are of special interest in light of the U.S. federal government's repeated refusal to apologize officially for slavery. These seven apologies composed the second subset of our sample. We analyzed the subsets separately, because there are intriguing differences in content of the apologies in the two sets.[1] Brief descriptions of the injustices are provided in Table 13.1. Two raters independently examined each of the apologies for the presence of the 12 elements (see Tables 13.2 and 13.3 for Subsets 1 and 2, respectively). Interrater reliability was high ($K = .86$). We present the results of this analysis in order of frequency of appearance of apology elements in Subset 1, with the most common elements presented first.

All apologies in Subset 1 and Subset 2 included expressions of remorse, such as "we apologize" or "we regret"—the presence of such statements was a criterion for inclusion in this sample. Similarly, all apologies in both subsets acknowledged that the acts committed against the victims were unjust. For example, President Bill Clinton declared in his apology to African American victims of the Tuskegee syphilis study, "You did nothing wrong, but you were grievously wronged" ("Apology for Study Done in Tuskegee," 1997). All apologies in both subsets also described the harms caused by the governments' actions and recognized the victims' suffering. Prime Minister Stephen Harper of Canada acknowledged the "tragic accounts of the emotional, physical and sexual abuse and neglect of helpless children and their separation from powerless families and communities" in his apology to former Aboriginal residential school students ("Apology to Residential School Students," 2008).

A promise of forbearance (e.g., "This will never happen again") was apparently more common in apologies from Subset 1 (93%) than in the states' apologies for slavery (43%). In apologizing for the Japanese army's abuse of comfort women during WWII, Prime Minister Tomiichi Murayama declared, "To ensure that this situation is never again repeated, the Government of Japan will collate historical documents concerning the former wartime comfort women, to serve as a lesson of history" (Ministry of Foreign Affairs of Japan, 1995a). The tendency to omit forbearance in the states' apologies may reflect a shared, though potentially unfounded, belief that there is little threat of slavery in the future.

Table 13.1 Included Apologies and Descriptions of the Injustices

Injustice	Apologizer	Description of Injustice
Internment of Japanese Americans	Congress (1988) George Bush (1991) Bill Clinton (1993)	In 1942, 110,000 ethnic Japanese (62% American-born citizens) were interned in relocation centers with inadequate housing, clothing, and food. Most experienced significant property losses.
Internment of Japanese Canadians	Brian Mulroney (1988)	In 1942, 22,000 Japanese Canadians (59% Canadian-born citizens) were expelled from homes in British Columbia and interned under poor conditions. Their property was sold off by the government to pay for internment. After the war, internees were forced to leave British Columbia.
Overthrow of Kingdom of Hawaii	Congress (1993)	In 1893, U.S. naval forces invaded the sovereign Hawaiian nation, took over government buildings, disarmed the Royal Guard, and declared a provisional government. In 1898, the U.S. Congress approved a joint resolution of annexation, creating the U.S. Territory of Hawaii.
WWII comfort women	Tomiichi Murayama (1995)	During WWII, an estimated 200,000 girls and women were taken from their homes in Korea, China, and other Japanese-occupied regions and placed in brothels to be used as sex slaves for the Japanese army.
Japanese WWII crimes	Tomichii Murayama (1995)	In the 1930s and 1940s, the Japanese military murdered between 6 and 10 million East Asian civilians.
Seizure of Maori land	Queen Elizabeth II (1995)	Under the New Zealand Settlement Act of 1863, over a million acres of Waikato land was confiscated. The Maori resisted the confiscation, and many died in the fighting that followed.
Tuskegee syphilis study	Bill Clinton (1997)	In 1932, the U.S. Public Health Service began a 40-year study of the progression of syphilis with 600 Black men. They were never told they had syphilis or treated for it, even when penicillin became available.

Table 13.1 Included Apologies and Descriptions of the Injustices (Continued)

Injustice	Apologizer	Description of Injustice
Australian Aboriginal Stolen Generations	John Howard (1999) Kevin Rudd (2008)	Between 1915 and 1969, approximately 100,000 Australian Aboriginal children were removed from their families by the government and church and placed in internment camps, orphanages, and other institutions. They were forbidden to speak their language, received little education, and lived under poor conditions. Physical and sexual abuse was common.
Chinese Canadian Head Tax and Exclusion Act	Stephen Harper (2006)	In 1885, the Canadian Government levied a head tax on all Chinese immigrants to restrict the number of Chinese entering Canada. The Exclusion Act barred all Chinese from entering Canada from 1923 to 1947.
British role in slave trade	Tony Blair (2006)	Between 1660 and 1807, over three million Africans were sent to the Americas in British ships. Many died during capture and transportation.
U.S. states' roles in slave trade	Virginia (2007) Maryland (2007) North Carolina (2007) Alabama (2007) New York (2007) New Jersey (2008) Florida (2008)	Between 1654 and 1865, slavery was legal in at least 23 U.S. states. By the 1860 census, the slavery population in the United States had grown to over 4 million. Slave owners often treated their slaves inhumanely.
Canadian residential schools	Stephen Harper (2008)	In the 1870s, the Canadian federal government funded church-run schools with the aim of assimilating Aboriginals into the dominant culture. Children were forcibly removed from their homes and isolated from their families and cultures. Children were prohibited from speaking their native languages. Many children were physically and sexually abused, and many died because of poor sanitation, lack of medical care, and tuberculosis.

Table 13.2 Elements Present in Subset 1

	Element Number											
Injustice	**1**	**2**	**3**	**4**	**5**	**6**	**7**	**8**	**9**	**10**	**11**	**12**
Internment of Japanese Americans												
Congress, Civil Liberties Act (1988)	✓		✓	✓	✓	✓						
George Bush (1991)	✓		✓	✓	✓	✓				✓	✓	✓
Bill Clinton (1993)	✓	✓	✓	✓	✓	✓						✓
Internment of Japanese Canadians (1988)	✓	✓	✓	✓	✓	✓		✓			✓	✓
Overthrow of Kingdom of Hawaii (1993)	✓	✓	✓	✓			✓	✓				
WWII comfort women (1995)	✓	✓	✓	✓	✓	✓				✓	✓	
Japanese WWII crimes (1995)	✓	✓	✓	✓	✓					✓	✓	✓
Seizure of Maori land (1995)	✓	✓	✓	✓	✓	✓		✓				
Tuskegee syphilis study (1997)	✓	✓	✓	✓	✓	✓	✓	✓				✓
Australian Aboriginal Stolen Generations												
John Howard (1999)	✓	✓	✓	✓	✓	✓	✓	✓	✓	✓	✓	✓
Kevin Rudd (2008)	✓	✓	✓	✓	✓	✓	✓	✓	✓	✓	✓	v
Chinese Head Tax and Exclusion Act (2006)	✓	✓	✓	✓	✓	✓	✓	✓	✓	✓		✓
British role in slave trade (2006)	✓	✓	✓	✓	✓	✓	✓				✓	✓
Canadian residential schools (2008)	✓	✓	✓	✓	✓	✓	✓	✓	✓			
Percentage of time element present (/14)	100	86	100	100	93	86	50	57	29	43	50	64

Note: Element 1 = remorse, 2 = acceptance of responsibility, 3 = admission of injustice or wrongdoing, 4 = acknowledgment of harm and/or victim suffering, 5 = forbearance, 6 = offer of repair, 7 = praise for victimized group, 8 = call for reconciliation, 9 = continued suffering, 10 = praise for majority group, 11 = praise for present system, 12 = dissociation of injustice from present system.

Table 13.3 Elements Present in Subset 2

Injustice	Element Number											
	1	2	3	4	5	6	7	8	9	10	11	12
U.S. states' roles in slave trade												
Virginia (2007)	✓	✓	✓	✓	✓		✓	✓			✓	✓
Maryland (2007)	✓	✓	✓	✓					✓			
North Carolina (2007)	✓	✓	✓	✓	✓		✓	✓				
Alabama (2007)	✓	✓	✓	✓			✓	✓	✓			
New York (2007)	✓	✓	✓	✓		✓						
New Jersey (2008)	✓	✓	✓	✓	✓		✓	✓	✓			
Florida (2008)	✓	✓	✓	✓				✓				
Percentage of time element present (/7)	100	100	100	100	43	14	57	71	43	0	14	14

Note: Element 1 = remorse, 2 = acceptance of responsibility, 3 = admission of injustice or wrongdoing, 4 = acknowledgment of harm and/or victim suffering, 5 = forbearance, 6 = offer of repair, 7 = praise for victimized group, 8 = call for reconciliation, 9 = continued suffering, 10 = praise for majority group, 11 = praise for present system, 12 = dissociation of injustice from present system.

Most apologies (86% in Subset 1 and 100% of the states' apologies for slavery) explicitly assigned responsibility for the injustice to governments and institutions. Prime Minister Kevin Rudd of Australia stated in his apology for the Stolen Generations, "The uncomfortable truth for all of us is that the parliaments of the nation, individually and collectively ... made the forced removal of children on racial grounds fully lawful" ("PM Rudd's 'Sorry' Address," 2008).

The final element of an interpersonal apology, an offer of repair, was present in most (86%) of the apologies in Subset 1 but in only one of the states' apologies for slavery. Repair came in the form of either individual or community-based compensation. In apologizing for the Chinese head tax, Prime Minister Harper stated, "Canada will offer symbolic payments to living head tax payers and living spouses of deceased payers" ("Harper's Speech," 2006). Rather than offering payments to specific individuals, Prime Minister Tony Blair announced that Britain would increase aid to Africa, launching an immunization facility that is projected to save the lives of 5 million African children a year ("PM's Article for the New Nation Newspaper," 2006). The tendency to omit repair in the states' apologies for slavery probably reflects the widespread belief that financial compensation to descendents of slaves is impractical and unnecessary (Brooks, 1999; Viles, 2002). We are surprised, however, that more states did not offer symbolic, economically feasible reparations, such as New York's establishment of a day to commemorate those who were enslaved.

Two of the apology elements seem designed to protect the present system from being tainted by the historical injustice: dissociation of the present system from the one in which the injustice occurred and praise for the current system. A majority (64%) of apologies from Subset 1 explicitly dissociated the present system from the one in which the injustice occurred, and 50% offered praise for the present system. In apologizing for the Chinese head tax, Prime Minister Harper emphasized that the tax "was a product of a profoundly different time" and "lies far in our past" ("Harper's Speech," 2006). In his apology to Japanese Canadian internment victims, Prime Minister Brian Mulroney praised Canada's current commitment to equality and fairness for all:

> We are tolerant people who live in freedom in a land of abundance ... a Canada that at all times and in all circumstances works hard to eliminate racial discrimination at home and abroad. (Japanese Internment National Redress, 1988, p. 19499)

Similar statements were less common in the states' apologies for slavery, with only one apology containing dissociation and direct praise for the current system. It is intriguing that it was the first apology, that by Virginia, that contained both of these elements.

A call for reconciliation was present in 57% of the apologies from Subset 1 and 71% of the states' apologies for slavery. Former Australian prime minister John Howard stated in his apology for the Stolen Generations that the House "reaffirms its wholehearted commitment to the cause of reconciliation between indigenous and non-indigenous Australians as an important national priority for Australians" (Motion of Reconciliation, 1999). The apology from Florida includes the following statement: "It is important that the Legislature express profound regret for the shameful chapter in this state's history, and, in so doing, promote healing and reconciliation among all Floridians" ("Florida Senate and House Express Profound Regret for Slavery," 2008).

We suggested earlier that governments could use apologies to affirm the social identities of members of both the previously victimized minority and the nonvictimized majority. Of the apologies from Subset 1, 50% contained praise for the minority and 43% included praise for the majority. In apologizing for Japanese war crimes, Prime Minister Murayama praised the majority by referring to the "wisdom and untiring effort of each and every one of our citizens" in rebuilding a peaceful and prosperous Japan (Ministry of Foreign Affairs of Japan, 1995b). Of the states' apologies for slavery, 57% offered praise for the victimized group. The apology from New Jersey acknowledged that colonial laws relegated "the status of Africans and their descendents to slavery, in spite of their loyalty, dedication, and service to the country" (Assembly Concurrent Resolution No. 270, 2008). It then recognized "the faith, perseverance, hope, and endless triumphs of African-Americans and their significant contributions to the development of this State and the Nation" (Assembly Concurrent Resolution No. 270, 2008). Not a single apology from any state included praise for the majority group.

In a majority of apologies in both subsets, governments avoided explicitly linking current social problems or suffering to the original injustice. Only 29% of the apologies from Subset 1 and 43% of the states' apologies for slavery made this linkage. As an example, Alabama described in detail the suffering that African Americans continue to endure because of slavery:

> African-Americans have found the struggle to overcome the bitter legacy of slavery long and arduous, and for many African-Americans the scars left behind are unbearable. ("Expressing Profound Regret for Alabama's Role in Slavery," 2007)

As we noted earlier, there are disadvantages and advantages to emphasizing the scars left behind. In most cases, governments seem to have decided that the former outweighs the latter.

Relative to the states' apologies for slavery, the apologies in Subset 1 were more likely to dissociate the injustice from the present system, praise the present system, and praise the majority group. We suggest two opposing interpretations for these differences. One possibility is that the individuals who drafted the states' apologies felt confident that slavery no longer threatened the current system or social identities of the majority group, and therefore they felt little need to protect against these threats. Alternatively, the U.S. states may have omitted these elements because racism and inequality remain in the system. Statements of praise for the system could therefore appear disingenuous to the victimized group and might even anger them. The more frequent inclusion of statements of continued suffering in the state apologies (43% vs. 29%) provides some support for this second interpretation.

In summary, the political apologies in our sample, especially those in Subset 1, included most of the six proposed elements of an interpersonal apology. As Tavuchis (1991) anticipated, these government apologies are far more comprehensive than the typical interpersonal apology recorded by linguists (Meier, 1998). In fact, 10 (71%) apologies from Subset 1 contained all six of the elements associated with an interpersonal apology. All but 2 apologies from Subset 1 and all apologies from Subset 2 included remorse, acknowledgment of wrongdoing, acknowledgment of harm, and acceptance of responsibility. Our analysis suggests that these four elements constitute the core of political apologies.

In addition to the six interpersonal elements, we proposed that government apologies could address psychological concerns for social identity and justice that are more specific to historical wrongs. We obtained some evidence for the proposed additional elements, but they tended to be included less frequently than the interpersonal elements. Continued research in this area will reveal whether inclusion of these additional elements psychologically benefits the victimized and majority groups, whether some elements have negative effects on one group while serving the needs of the other, and whether including these elements achieves greater healing and reconciliation than is attained with the six interpersonal elements alone.

One predictor of whether governments chose to address identity and justice concerns is the number of years between the apology and the injustice. We examined the correlation between the number of years since the end of the injustice and the contents of the apologies across both subsets. Apologies for more recent injustices were more likely to include praise for the majority group ($r = -.51, p = .02$) and current system ($r = -.42, p = .06$), as well as statements that dissociate the present system from the injustice ($r = -.42, p = .06$). These three elements theoretically defend against threats to the system and the social identity of the majority. The data are only suggestive, because our sample is small and includes a potential confound between the type of injustice and the timing of the apology (the eight apologies for slavery followed the longest delays). As more governments apologize for historical injustices, we will be able to examine more thoroughly the hypothesis that apologies for recent injustices are more likely to include statements designed to bolster social identity and faith in the current system.

The Effects of Partial Redress

Although several of the apologies in Table 13.1 included most or all of the 12 elements, many of the apologies in our sample were less comprehensive. Some elements may be more critical than others. Minow (2002), a legal scholar, speculated that an apology for a historical injustice that omits an offer of repair (e.g., in the form of financial compensation) seems insincere and manipulative. If members of a previously victimized group regard an apology from the government as insincere and manipulative, they are unlikely to benefit psychologically or be moved to reconcile. The element of repair is of special interest, as scholarly debates about the content of apologies often concern whether to include financial compensation along with expressions of remorse (Brooks, 1999).

Blatz (2008) studied whether withholding offers of repair undermines the effectiveness of an apology for a historical injustice. He based his research on Ross and Ward's (1995) theory of reactive devaluation. According to Ross and Ward, if one side offers X during negotiations but withholds Y, the receiving side will devalue X and show an increased appreciation of Y. Using this framework, Blatz (2008) predicted that victimized groups would be satisfied with an apology that did not include financial compensation if they had demanded only an apology. In contrast, if victimized groups demanded both an apology and compensation but received only an apology, they would devalue the apology and increase their desire for compensation. Similarly, victimized groups would devalue compensation and increase their desire for an apology if compensation was offered but an apology was withheld.

Blatz (2008) found support for these predictions in several studies. In one experiment, a group of Chinese Canadian university students read a one-page summary of the head tax. The head tax was a significant and discriminatory tax that the Canadian government levied on Chinese immigrants between 1885 and 1923 to limit Chinese immigration (Dyzenhaus & Moran, 2006). Participants read that Chinese Canadian lobby groups had demanded that the Canadian government express remorse and offer financial compensation for the head tax. Participants then read that the government had offered (a) neither, (b) an expression of remorse but no compensation, (c) compensation but no expression of remorse, or (d) both compensation and remorse. In line with the predictions derived from reactive devaluation theory, participants felt less forgiving when an apology or compensation was offered alone compared to when neither or both were offered, although the compensation-only contrasts did not reach significance.

Of course, governments cannot always satisfy all of the demands of a previously victimized group. In some circumstances, partial redress may be all that is politically, financially, or legally feasible. The government is then in the position of choosing between partial and no redress. Governments will base their decisions mainly on political considerations, but from a psychological standpoint, the "right" decision is not self-evident. In the interpersonal domain, transgressors might almost always be wise to say "sorry." There is little evidence that an expression of remorse will hurt a relationship, and there is some evidence that it might help (e.g., Scher & Darley, 1997; Schlenker & Darby, 1981). With political apologies for severe historical injustices, however, Blatz's (2008) research indicates that an unsatisfactory apology may sometimes be more psychologically harmful than no apology, at least in the short term.

Who Benefits From Redress?

Many legal scholars and historians have argued that a collective response to a historical injustice, such as official government apologies, is necessary to heal the wounds caused by past harms (e.g., Barkan, 2000; Brooks, 1999; Minow, 2002). These scholars assume that in the absence of amends, the wounds from an injustice continue to fester, causing resentment and conflict. As evidence, legal scholars and historians have noted that Japan's unwillingness to apologize officially for war crimes it committed during WWII has prevented reconciliation with harmed groups,[2] whereas Germany's provision of apologies and compensation to some victim groups has facilitated favorable relations with former enemies and victimized groups (Barkan, 2000). The Japanese and German situations differ in many ways, however, and it is difficult to draw firm conclusions. There remains relatively little research on when, how, and for whom political apologies are beneficial. Given the increasing frequency of political apologies, the time is ripe for an exploration of the effects of these apologies.

Legal scholars, such as Minow (2002), have argued that government apologies can foster forgiveness. But do apologies for historical injustices actually increase forgiveness? The evidence for this hypothesis from social psychological research is underwhelming. Philpot and Hornsey (2008) asked Australian university students to read descriptions of five injustices committed against Australia. They then manipulated whether an apology was offered for these events. Across four studies, the results consistently demonstrated that even though participants were more satisfied with an apology than no apology, they did not report increased forgiveness.

Blatz (2008) examined the effects of government apologies on members of a previously victimized minority and members of the nonvictimized majority. In one study, he tested the impact of an extremely comprehensive apology offered by Prime Minister Steven Harper of Canada for the Chinese head tax. Blatz surveyed students of Chinese and non-Chinese heritage at a Canadian university 1 month before and 1 month after Harper offered his official apology and compensation of $20,000 to head tax payers or their surviving spouses. Both Chinese and non-Chinese participants were generally satisfied with the redress package. Compared to the Chinese participants, however, the non-Chinese participants evaluated the apology more favorably, regarded the apology as more effective, were less cynical about the government's motives for apologizing, and were less likely to note inadequacies in the compensation package.

Ironically, members of the nonvictimized majority seemed more impressed by the apology. This pattern is perhaps not surprising. Most political apologies are offered for gross human rights violations, such as racial discrimination, slavery, sexual abuse, and genocide. Because of the severity of these acts, any form of redress fails to completely restore justice for the victims and their group. Even the sincerest of apologies cannot turn back the clock and eliminate the harm, and even the most generous of compensation packages cannot repair the damage. This inadequacy of redress relative to the magnitude of the injustice is likely more obvious to the previously victimized group than to the nonvictimized group (Minow, 2002). Also, experimental research on negotiations indicates that people evaluate their own side's offers more favorably than equivalent offers by the opposition (Curhan, Neale, & Ross, 2004; Ross & Ward, 1995). Similarly, members of the majority group may judge redress that is offered on their behalf as more satisfactory than do members of the victimized group. If the impact of the apology on the majority group is (with hindsight) not entirely surprising, it is almost entirely overlooked in the scholarly literature. Most discussions of government apologies focus on the victimized group. Any discussion of members of the nonvictimized group usually centers on the circumstances in which they might be willing to support a government apology (Brooks, 1999).

It is important to note that none of the studies reviewed in this section included a sample of the direct victims of the injustice. Conceivably, victims of the original injustice would report greater appreciation of redress than would their descendents or other members of their ethnic group. Apologies for historical injustices, however, often occur long after the injustice, and few if any direct victims remain. The median passage of time was 59 years between the end of the injustice and the government apologies that we reviewed. If apologies for historical injustices are to promote forgiveness and reconciliation, they need to be effective for the victims' descendents and broader social group.

The existing research concerning the impact of government apologies suggests that it may be better to give than to receive redress. Contrary to the concerns of some scholars and politicians (Brooks, 1999), there was no evidence in these studies of a majority group backlash against the apology or the victimized group. In fact, Blatz (2008) found that members of the majority *increased* their support for the apology after it had been offered rather than responded with hostility to redress offered by their government. Public opinion polls conducted before and after government apologies show comparable effects. For example, 68% of Australians supported Kevin Rudd's apology for the Stolen Generations 10 days after it was offered compared to 55% of Australians who supported the apology 4 days before it was provided (Metherell, 2008). Similarly, polls conducted in Canada on public support for a government apology for abuse in Aboriginal residential schools showed that support jumped dramatically after an apology was offered (Akkad, 2008; "Reconciliation With Aboriginals Possible for Two-in-Five Canadians," 2008). Such findings are consistent with system justification theory (Jost & Banaji, 1994). Members of the nonvictimized majority justify a lack of government apology when none occurs and then justify an apology when it does occur. If additional research corroborates these findings, government leaders could perhaps be less concerned about the negative political repercussions of apologizing for historical harms.

Summary and Conclusions

We discussed when governments apologize, how they apologize, and the reactions to their apologies. The government apologies in our sample were generally quite extensive and much more comprehensive than the interpersonal apologies that appear in the literature (Meier, 1998). Indeed, some of these apologies could serve as textbook examples of what an apology should be according to various authors (Bavelas, 2004; Lazare, 2004; Tavuchis, 1991). One cannot judge the merits of an apology, however, by examining only its contents. To examine the effectiveness of an apology, one also has to assess the reactions of members of the previously victimized group and the nonvictimized majority. For example, although both Australian apologies for the Stolen Generations included all 12 elements that we assessed in political apologies, Kevin Rudd's apology has generally received more positive reactions from Aborigines and other Australians than John Howard's (Smith, 2008). One significant difference is that Howard had previously refused to apologize, arguing that Australia had no need to atone for past injustices; in contrast, Rudd had promised to apologize during his election campaign. This prior opposition from Howard may have caused recipients of the apology and other Australians to suppose that his Motion of Reconciliation was insincere. A second difference between the apologies is that Rudd explicitly used the words *apologize* and *sorry*, whereas Howard expressed *regret*. Although an expression of regret communicates empathy for the victims' suffering, an apology statement communicates a willingness to accept moral responsibility for the injustice (Thompson, 2002). Many people quoted in the Australian media noted this distinction between an apology and regret, calling for a "real" apology after Howard's Motion of Reconciliation in 1999 (e.g., "Apology Still Needed," 2007; "Rudd Promises Apology to Aborigines," 2007). Thus, although our coding of the elements within political apologies informs us about *how* governments apologize for historical harms (and perhaps their intentions), it cannot directly speak to how members of the previously victimized minority and nonvictimized majority regard these apologies.

We reported studies by Blatz (2008) that do not support Minow's (2002) contention that an apology for a historical injustice will be deemed inadequate if it fails to include offers of repair. What mattered, according to Blatz's data, was addressing the specific demands of those receiving the apology. He found that words alone could be effective but not if the victim group had demanded financial compensation as well. One unexplored question concerns whether apologies and compensation satisfy unique psychological needs. Scholars have theorized that apologies restore trust and faith in the social order, whereas compensation indicates that the government is sincerely sorry and recognizes the victims as important members of society (e.g., Bright-Fleming, 2008; de Grieff, 2008; Minow, 2002). It remains to be demonstrated whether compensation and the other 11 elements of an apology that we have identified serve their proposed psychological functions.

There are many other remaining questions concerning political apologies. For example, do offers of compensation need to be framed as symbolic rather than restorative to be effective? Members of previously victimized groups likely believe that no amount of money can right the original wrong. Consequently, they may regard compensation that is framed as restorative rather than symbolic as inadequate, even if the amount of compensation is actually greater. Also, we noted that many government apologies occur in response to political pressure. Are apologies less effective if they seem politically pressured rather than spontaneous? There is little research on the differential effectiveness of pressured versus spontaneous apologies in the interpersonal domain. One study of young siblings found that spontaneous apologies appeared to be more effective than apologies mandated by parents (Schleien, Ross, & Ross, 2008). A study of college students, however, found that apologies demanded by observers of a transgression were as effective as spontaneous apologies (Risen & Gilovich, 2007). Our guess is that victimized groups will not penalize a government for apologizing in response to their demands. They may simply conclude that the government finally understands their perspective. Members of the victimized group may be less enamored with the apology, however, if they believe that it is insincere, for example, if they think the government is apologizing to buy their votes rather than to address a wrong.

To this point, legal scholars and historians have conducted much of the relevant research on political apologies. These scholars tend to assume that apologies will be effective, especially if accompanied by compensation. The research presented in the current chapter suggests that political apologies will not necessarily have the transformative powers that are commonly attributed to them. From a social psychological perspective, the issue is not whether political apologies are effective but when, how, and why they are effective. We presented research that begins to address these questions.

Notes

1. A document presenting all 21 apologies is available at https://artsweb.uwaterloo.ca/~kschuman/political_apology_OS/
2. Some Japanese government officials, including Prime Minister Murayama, have apologized, but the Japanese parliament (Diet) has not officially endorsed these individual apologies.

References

Akkad, O. E. (2008). School abuse apology widely backed. Retrieved July 3, 2008, from http://www.theglobeandmail.com/servlet/story/RTGAM.20080614.apology14/EmailBNStory/National/

Apology for study done in Tuskegee. (1997). Retrieved April 17, 2007, from http://clinton4.nara.gov/textonly/New/Remarks/Fri/19970516–898.html

Apology still needed. (2007). Retrieved May 22, 2008, from http://www.brisbanetimes.com.au/news/national/apology-still-needed/2007/10/11/1191696080523.html

Apology to residential school students: Harper on Parliament Hill. (2008). Retrieved June 14, 2008, from http://network.nationalpost.com/np/blogs/posted/archive/2008/06/11/apology-to-residential-school-students-harper-on-parliament-hill.aspx

Armenian genocide dispute. (2006, October 12). *BBC News*. Retrieved April 28, 2007, from http://news.bbc.co.uk/2/hi/europe/6045182.stm

Assembly concurrent resolution no. 270. (2008). State of New Jersey. 212th legislature.

Backhouse, C. (1999). *Colour-coded: A legal history of racism in Canada, 1900–1950*. Toronto: University of Toronto Press.

Banfield, J. C., Blatz, C. W., & Ross, M. (2008). [Association between perceived suffering and endorsement of redress among minority and majority groups]. Unpublished raw data.

Barkan, E. (2000). *The guilt of nations: Restitution and negotiating historical injustices.* New York: Norton.

Bavelas, J. (2004). *An analysis of formal apologies by Canadian churches to First Nations.* Victoria, British Columbia, Canada: Centre for Studies in Religion and Society, University of Victoria.

Blatz, C. (2008). *How members of majority and victimized groups respond to government redress for historical harms.* Unpublished doctoral dissertation, University of Waterloo.

Blatz, C. W., Ross, M., & Starzyk, K. (2008). *Who opposes redress for historical harms and why?* Manuscript submitted for publication.

Branscombe, N., & Doosje, B. (2004). *Collective guilt: International perspectives.* New York: Cambridge University Press.

Bright-Fleming, E. (2008). When sorry is enough: The possibility of a national apology for slavery. In M. Gibney, R. E. Howard-Hasmann, J.-M. Coicaud, & N. Steiner (Eds.), *The age of apology: Facing up to the past* (pp. 95–108). Philadelphia: University of Pennsylvania Press.

Brooks, R. L. (Ed.). (1999). *When sorry isn't enough: The controversy over apologies and reparations for human injustice.* New York: New York University Press.

Controversial Turkish novelist wins Nobel. (2006). Retrieved May 30, 2008, from http://www.cbsnews.com/stories/2006/10/12/print/main2084113.shtml.

Curhan, J. R., Neale, M. A., & Ross, L. (2004). Dynamic valuation: Preference changes in the context of face-to-face negotiation. *Journal of Experimental Social Psychology, 40,* 142–151.

De Grieff, P. (2008). The role of apologies in national reconciliation processes: On making trustworthy institutions trusted. In M. Gibney, R. E. Howard-Hasmann, J.-M. Coicaud, & N. Steiner (Eds.), *The age of apology: Facing up to the past* (pp. 120–134). Philadelphia: University of Pennsylvania Press.

Doosje, B., Branscombe, N. R., Spears, R., & Manstead, A. S. R. (1998). Guilty by association: When one's group has a negative history. *Journal of Personality and Social Psychology, 75*(4), 872–886.

Expressing profound regret for Alabama's role in slavery. (2007). Retrieved June 20, 2007, from http://www.legislature.state.al.us/searchableinstruments/2007rs/resolutions/sjr54.htm

Florida senate and house express profound regret for slavery. (2008). Florida Legislature. Retrieved May 9, 2007, from www.myfloridahouse.gov/SECTIONS/Documents/loaddoc.aspx?DocumentType-PressRelease&FileName-101.

Harper's speech. (2006, June 22). *Globe and Mail.* Retrieved April 14, 2007, from http://www.theglobeandmail.com/servlet/story/RTGAM.20060622.wspeech0622/BNStory/Technology/

Hatamiya, L. T. (1999). Institutions and interest groups. In R. L. Brooks (Ed.), *When sorry isn't enough: The controversy over apologies and reparations for human injustice.* New York: New York University Press.

Japanese Internment National Redress. (1988). House of Commons Debates, 34th Parliament, 2nd Session, number 15, pp. 19499–19500.

Jost, J. T., & Banaji, M. R. (1994). The role of stereotyping in system-justification and the production of false consciousness [Special issue]. *British Journal of Social Psychology, 33*, 1–27.

Lazare, A. (2004). *On apology*. New York: Oxford University Press.

Lerner, M. J. (1980). *The belief in a just world: A fundamental delusion*. New York: Plenum.

Meier, A. J. (1998). Apologies: What do we know? *International Journal of Applied Linguistics, 8*, 215–231.

Metherell, M. (2008, February 18). PM said sorry—and so said more of us. *Sydney Morning Herald*. Retrieved March 5, 2008, from http://www.smh.com.au/news/national/pm-said-sorry--and-so-said-more-of-us/2008/02/17/1203190653987.html

Ministry of Foreign Affairs of Japan. (1995a). *Statement by Prime Minister Tomiichi Murayama on the occasion of the establishment of the "Asian Women's Fund."* Retrieved April 14, 2007, from http://www.mofa.go.jp/policy/women/fund/state9507.html

Ministry of Foreign Affairs of Japan. (1995b). *Statement by Prime Minister Tomiichi Murayama "On the occasion of the 50th anniversary of the war's end."* Retrieved April 14, 2007, from http://www.mofa.go.jp/announce/press/pm/murayama/9508.html

Minow, M. (2002). *Breaking the cycles of hatred*. Princeton, NJ: Princeton University Press.

Motion of Reconciliation. (1999). *Commonwealth of Australia Parliamentary Debates, 39*, 9207.

Negash, G. (2006). *Apologia Politica: States and their apologies by proxy*. Oxford, UK: Lexington Books.

Philpot, C. R., & Hornsey, M. L. (2008). What happens when groups say sorry: The effect of intergroup apologies on their recipients. *Personality and Social Psychology Bulletin, 34*(4), 474–487.

PM Rudd's "sorry" address. (2008). Retrieved March 20, 2008, from http://www.theage.com.au/articles/2008/02/12/1202760291188.html?page=fullpage#contentSwap2

PM's article for the New Nation newspaper. (2006, November 27). Retrieved April 17, 2007, from http://www.number-10.gov.uk/output/Page10487.asp

Polling report: Race and ethnicity. (2008). Retrieved March 24, 2008, from http://www.pollingreport.com/race.htm

Reconciliation with Aboriginals possible for two-in-five Canadians. (2008). Angus Reid poll. Retrieved June 28, 2008, from http://www.angusreidstrategies.com/index.cfm?fuseaction=news&newsid=221&page=2

Risen, J. L., & Gilovich, T. (2007). Target and observer differences in the acceptance of questionable apologies. *Journal of Personality and Social Psychology, 92*(3), 418–433.

Ross, L., & Ward, A. (1995). Psychological barriers to dispute resolution. In M. P. Zanna (Ed.), *Advances in experimental social psychology* (Vol. 27, pp. 255–310). San Diego, CA: Academic Press.

Rudd promises apology to Aborigines. (2007). Retrieved May 22, 2008, from http://english.aljazeera.net/NR/exeres/EE94C382-4E22-4915-A1A1-477DD91303C0.htm

Sahdra, B., & Ross, M. (2007). Group identification and historical memory. *Personality and Social Psychology Bulletin*, *33*, 384–395.

Scher, S. J., & Darley, J. M. (1997). How effective are the things people say to apologize? Effects of the realization of the apology speech act. *Journal of Psycholinguistic Research*, *26*, 127–140.

Schleien, S. M., Ross, H. S., & Ross, M. (2008). *Young children's apologies to their siblings*. Manuscript submitted for publication.

Schlenker, B. R., & Darby, B. W. (1981). The use of apologies in social predicaments. *Social Psychology Quarterly*, *44*, 271–278.

Schumann, K., & Ross, M. (2008). [Diary investigation of apologies in everyday life]. Unpublished raw data.

Smith, T. (2008). *The letter, the spirit, and the future: Rudd's apology to Australia's Indigenous people*. Retrieved May 9, 2008, from http://www.australianreview.net/digest/2008/03/smith.html

Starzyk, K., & Ross, M. (2008). A tarnished silver lining: Victim suffering and support for reparations. *Personality and Social Psychology Bulletin*.

Tajfel, H., & Turner, J. C. (1986). The social identity theory of intergroup conflict. In S. Worchel & W. G. Austin (Eds.), *Psychology of intergroup relations* (pp. 7–24). Chicago: Nelson Hall.

Tavuchis, N. (1991). *Mea culpa*. Stanford, CA: Stanford University Press.

Thompson, J. (2002). *Taking responsibility for the past: Reparation and historical justice*. Cambridge, UK: Polity.

Viles, P. (2002). *Suit seeks billions in reparations*. Retrieved August 24, 2006, from http://archives.cnn.com/2002/LAW/03/26/slavery.reparations

We won't pay: The homepage for those drawing the line. (2007). Retrieved June 4, 2008, from http://www.wewontpay.com/index.html

Author Index

E

F

G

H

I

J

K

L

M

N

O

P

Q

R

S

T

U

V

W

Y

Z

Subject Index

D

E

F

I

K

L

M

N

S

W

Y